Soil Chemistry
with Applied Mathematics

Cristian P. Schulthess

University of Connecticut
Storrs, CT 06269–4067, USA
c.schulthess@uconn.edu

Order this book online at www.trafford.com
or email orders@trafford.com

Most Trafford titles are also available at major online book retailers.

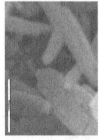

Cover photo: Goethite viewed under scanning electron microscope (SEM). Dark tones are holes and cavities, while the light tones are the mineral. Scale of 500 nm is as shown at bottom of photo. The original B&W photo, shown here, is courtesy of Dr. M.D. Raigón of the Universidad Politécnica de Valencia in Valencia, Spain, as part of a joint project with Dr. C.P. Schulthess involving goethite.

Print information available on the last page.

ISBN: 978-1-4120-3585-9 (sc)

Trafford rev. 07/27/2018

 www.trafford.com

North America & international
toll-free: 1 888 232 4444 (USA & Canada)
fax: 812 355 4082

for
Maria Elena

*C*ontents

3.   Soil Organic Matter   106

4.   Soil Mineralogy   158

Preface

I would like to express my gratitude to numerous people who encouraged and assisted me in the writing of this book. I was particularly motivated by my classroom students at the University of Connecticut, where I have taught soil chemistry since 1992 to soil scientists, geologists, natural resource scientists, environmental scientists, and environmental engineers. I wrote this book based on my experiences with them, and it is intended to be read by both undergraduate and graduate students, students well versed in mathematics as well as by those that are not, and students with a major degree in chemistry as well as those with just some understanding of general chemistry. To cover this range of backgrounds among the students, I wrote this book with a fair amount of detailed information to satisfy the needs of the more advanced students. Much effort was then made in various places to explain how complex equations or concepts were developed. The text in these places is practically "holding your hand" as it walks you through to the conclusion equations or concepts. This was done to satisfy the needs of the students who forgot how to apply their mathematical skills, which includes all of us from time to time. This book is intended to offer the student a rigorous explanation of modern soil chemistry principles. As a result, I hope that this book is one that will be kept by the student in his or her personal library long after the course is completed. I hope that this is not only a good textbook, but also a good reference book for the practicing professional in various environmental fields.

I owe many thanks to all of my graduate students for their stimulating discussions. I thank Pamm Kasper, R. Dean Rhue, and Baoshan Xing for their careful and thoughtful reviews of the manuscript. I am also eternally grateful for the encouragement by my wife, Margie Faber, and all my family.

Chapter 1:
Overview of Topics

[1.1]   The Big Picture

Soil is a natural body on the earth's surface that interacts with meteorologic, hydrologic, biologic and geologic activities over time. It is an open system that is constantly evolving and transforming. It is a complex system with boundaries and characteristics that are easy to identify in some soils and very difficult in others. For a given soil, various factors, such as the typical depth of rainwater infiltration over a period of time, slowly gives the soil a set of unique characteristics. Working backward based on our knowledge of soils, we can often anticipate some general soil characteristics to exist in a given landscape with reasonable accuracy. For more accurate and reliable information about a given soil, however, we must get closer to it.

As a soil transforms or evolves, soil horizons develop that record the history of the soil's past. Soil profiles are vertical exposures of the horizons present in soils. See Table 1-1 for a general description of horizon designations. To learn about them, it will sometimes be necessary to dig around and expose these profiles. Studying and classifying soils is a difficult but fascinating science that is commonly referred to as pedology, or the study of soil genesis and classification. Although each soil is a unique individual, there are five basic factors governing the development of soil profiles: parent material, climate, biota, topography and time. There were more than 19,000 individual soils (species) recognized in 1999 by the USDA Classification System. Table 1-2 gives an overview of the 12 orders used in soil classification. Table 1-3 and the example therein describes the 6 categories, or levels, used in the soil taxonomy system. The family name of a soil may change over time as improvements to the soil classification system are made. But the series name never changes once it is given to a soil that is determined to be uniquely different from other soils. When publishing results on research performed on a soil, one should specify both the series and family name.

Note that a soil name is given to a soil with numerous unique characteristics (e.g., horizon thickness, horizon arrangement, diagnostic features present, soil structure, and chemical properties). A portion of a soil (e.g., a B horizon subsample) is often removed and

Table 1-1: Master soil horizon designations used by USDA in soil taxonomy. These horizons can vary dramatically in depth from soil to soil, and not all horizons need be present in a given soil.

Horizon Designation	Properties
O	Surface organic layer; dominated by organic matter (OM).
L	Limnic material in Histosols; OM or minerals deposited in aquatic environments.
A	Surface mineral layer; mineral layer with most OM; clay depletion is common.
E	Elluvial horizon; loss of OM, clay, Fe or Al.
B	Horizon of maximum development below the O, A or E; accumulation by illuviation of clay and other materials is common.
C	Parent material; few or no pedogenic processes present.
R	Bedrock.
W	Water within or beneath the soil.

Table 1-2: Broad description of the 12 orders used in soil taxonomy. As an instructional aid, the formative element of the order's name is underlined. See Table 1-3 for an example of how the formative element is used in soil taxonomy.

Order	Description
Ge<u>li</u>sols	Very cold soils. Permafrost present.
Hi<u>sto</u>sols	Highly organic soils (>20% organic matter).
Sp<u>odo</u>sols	Soils with a horizon of illuvial amorphous material composed of organic matter and Al, with or without Fe.
An<u>di</u>sols	Soils from volcanic origins having a low organic content but large amounts of Al and Fe.
O<u>xi</u>sols	Highly weathered mineral soils, typically high in Fe and Al oxides.
Ve<u>rti</u>sols	Clayey soils with cracks that open and close periodically with the region's wet–dry seasonal cycles.
Ari<u>di</u>sols	Soils of arid regions, often having high Na and Mg content and low Ca content.
Ul<u>ti</u>sols	Highly weathered, clayey soils with a low affinity for cations.
Mo<u>lli</u>sols	Mineral soils high in organic matter content.
Al<u>fi</u>sols	Moderately weathered, loamy or clayey soils; clay accumulation in B horizon.
Ince<u>pti</u>sols	Young, weakly developed soils, excluding sandy soils.
E<u>nti</u>sols	Young, very weakly developed soils, including sandy soils.

brought to the lab. When this occurs, the sample is technically no longer a soil but rather a soil constituent or a disturbed soil subsample. A *pedon* is the smallest volume of soil needed to fully describe all of the soil's characteristics. Accordingly, to truly transport a soil sample to the lab, one would need to transport the entire undisturbed pedon, which is often mechanically not feasible to do. Good results can be obtained from "undisturbed" soil columns, which are cylindrical cores of soil inside a PVC pipe that has been pushed into the

Table 1-3: The six categories used in soil taxonomy, and an example of how a soil family name is constructed.

Taxonomic Category	Construction of the name:	Main category that name is based on:	Example & Comments
			Bold type in name used as an instructional aid only.
Order	Base word + i/o + sol	Principle soil forming processes	**Inceptisol** (Young soil with a few simple features of soil formation.)
Suborder	Prefix + formative element of the order to which it belongs.	Soil wetness & temperature regimes	**Aquept** (Evidence of water table present within 50 cm of the soil surface. Redoximorphic features present within 50 cm.)
Great Group	Prefix + suborder name	Various items: soil horizons & arrangement, temperature regimes, base saturation	**Humaquept** (Significantly higher organic matter content in surface layer relative to other aquepts.)
Subgroup	Separate adjective added to the great group name	Central concept or gradation or other concept about the soil	**Histic** Humaquept (Thick O horizon present; probably a very wet soil.)
Family	Modifiers added to the subgroup name	Soil use related properties: particle-size distribution, mineralogy, temperature regime	**Sandy, mixed, mesic** Histic Humaquept (Sandy = sand or loamy sand texture, <50% very fine sand; mixed = no single mineral predominates; mesic = average annual temperature 8–15°C, seasonal variation ≥5°C.)
Series	Unique name used. Family name is not part of the series name.	Series named for the locality where the particular soil was first described.	**Scarboro** (Of the many individual soils in this family, the Scarboro series was first described near Scarboro, ME, in 1915.)

ground, removed, and taken to the lab. Researchers often build small "labs" in the field and typically make use of buried lysimeters, which collect liquids that flow through at various depths and locations in the "undisturbed" natural soil environment.

From the discussion above, we notice that soil chemists need to be keenly aware of the contributions of soil physics. Through the process of preferential flow, the decayed root channels and various cracks and fissures in the soil profiles will greatly accelerate the movement of nutrients and contaminants to lower soil horizons. The water transporting various natural or man-made chemicals will eventually reach the groundwater table if the water volume is greater than what is evaporated back to the atmosphere or taken up by the plants (transpiration). The combined effect of evaporation and transpiration is called *evapotranspiration*.

There is also soil biological activity that will greatly accelerate chemical reactions as the soil constituents pass through the organism's digestive tract or processes. Accordingly, soil biology and soil microbiology also play an enormous role in the fate of chemicals in soils as well as on the remediation of contaminated soils. The long and the short of this is that soil science is a challenging field of study that involves various areas of expertise. Some of these include pedology, soil physics, soil biology, soil microbiology, soil mineralogy and soil chemistry.

Soil chemistry tries to describe the chemical reactions present in soils by analyzing either undisturbed whole soil samples, disturbed soil samples, or soil constituents. For example, we may choose to study the retention of organic compounds by an undisturbed whole soil sample. Such a research approach is useful, particularly for determining the amount of chemical reaction that may occur in a field with the same conditions as the test soil. It is an *applied science* approach. The down side of working with complex whole-soil samples is that it is often too difficult to discern *why* a particular reaction occurred in the first place. It also often fails to predict how a different soil sample will react.

Conversely, we may choose to study the retention of organic compounds by a very pure subcomponent of a soil (such as goethite). Your chances are greatly improved here for correctly finding *why* (or *how*) a particular reaction has occurred on the soil constituent. It is a *basic science* approach. In principle, the sum of the results of many single-component experiments, which serve as "building blocks" of information, will correctly explain the behavior of whole soils. The transferability of knowledge about the chemical reactivity of one test soil to predict the reactivity of another field soil relies heavily on our detailed knowledge of why the chemical reaction took place at all in the test soil. To be clear, soil chemistry is a mixture of both applied and basic research. Each research style may differ in its short-term objectives, but they both contribute significantly to our understanding of soils and soil chemistry.

Soil mineralogy focuses on the internal nature of each and every particle found in soils, while soil chemistry focuses on the reactivity of soil constituents, which obviously includes the soil particles. Accordingly, a soil chemist is generally much more concerned about the surface characteristics of each particle (namely, at the solid–liquid interface), than with the internal characteristics of each particle. Nevertheless, a thorough knowledge of soil mineralogy is essential to perform meaningful soil chemical research. Soil mineralogy reveals to us, for example, the presence of reactive interlayers in clays, pore channels in zeolites, the bond characteristics found at the particle's surface, and much more.

In soil chemistry, the solid components present in soils typically play the most important role in the fate of many nutrients and contaminants. This is because the chemical reactions that exist at the solid–liquid interface of soil particles strongly affect the mobility of all the ions and molecules present in the matrix. As a result of this, soil chemistry has become highly specialized in the characteristics and reactivity of the solid–liquid and solid–air interfaces of particles. The amount and type of exposed surface area in a soil sample will significantly control the amount and kind of chemical activity present.

The more exposed surface area the soil particles have per unit volume of soil, the more solid–liquid (or solid–air) interface area present also. It turns out that the smaller the particle sizes present in a soil sample, the larger the total surface area present per unit volume of soil. Not surprisingly, therefore, much attention is given to the reactivity of the smallest particles present in soils. Table 1-4 shows the size of soil separates (or particles) as defined by the USDA Classification System. A soil sample will contain a combination of soil separates. Figure 1-1 shows the name of the textural class based on the distribution percentage of soil separates present.

A soil with a loam textural class in its B horizon, for example, has clay and about equal amounts of sand and silt. Using this B horizon as an example, let us make a rough estimate of the surface area of a 100-gram sample using a hypothetical surface area distribution of each of its components. Assume the sample is 45 g sand, 45 g silt, and 10 g

Table 1-4: Size of the soil separates defined by the USDA. Fractions larger than 2.0 mm are rock fragments, smaller than 2 mm are fine-earth fractions. If the minerals forming the rocks (>2 mm) are not strongly cemented, then the prefix "para-" is added. Only the fine-earth fractions of a soil are used to determine the soil textural class.

Soil Separate		Diameter, mm		Length, mm
Non-flat Rock Fragments (>2 mm):			**Flat Rock Fragments (>2 mm):**	
	boulders	>600	boulders	>600
	stones	250–600	stones	380–600
	cobbles	75–250	flagstones	150–380
pebbles	coarse pebbles	20–75	channers	2–150
or gravel	medium pebbles	5–20		
	fine pebbles [a]	2–5		
Fine-earth Fraction (<2 mm):				
	very coarse sand	1.0–2.0		
	coarse sand	0.5–1.0		
sand	[medium] sand	0.25–0.5		
	fine sand	0.10–0.25		
	very fine sand	0.05–0.10		
	silt	0.002–0.05		
	clay	<0.002		

[a] Fine pebbles are also referred to as pea stones.

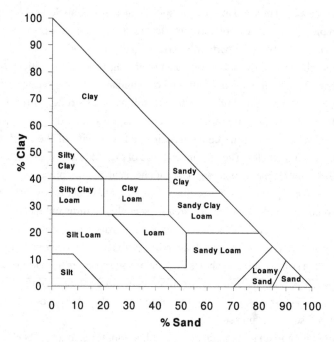

Figure 1-1: The 12 USDA soil textural classes, based on the sand, silt, and clay content of the fine-earth fractions of the soil sample (<2 mm). Note that the percent of all three fractions must add up to 100%. The percent silt is known if those of sand and clay are also known.

clay. As a rough estimate, assume that the soil sample has a total surface area of 4.5 m^2 due to the sand alone (45 g/sample $\times$ 0.1 m^2/g = 4.5 m^2/sample), 45 m^2 due to the silt fraction alone (45 g/sample $\times$ 1.0 m^2/g = 45 m^2/sample), and 500 m^2 due to the clay fraction alone (10 g/sample $\times$ 50 m^2/g = 500 m^2/sample). Note that the surface area numbers do vary widely in soils. Nevertheless, we can see from this example that while the clay fraction is only 10% of the total mass, it could easily account for 90% (or more) of the total inorganic solid surface that is interacting with the soil's liquid phase. These small particles have the largest areas (per unit volume or mass of sample) and, hence, the greatest influence on the chemical processes present in the soil matrix. Because of their high surface area, much attention is given to the identification, characterization, and reactivity of the clay fraction of soil samples.

[1.2]   Chapter Topics

Nearly all natural chemical reactions occur in water or somehow involve water. This is certainly true in soils, which also typically have a very high relative humidity due to their trapped water. To understand soil chemical reactions, we need to understand the basics of air, water, and solids, as well as the nature of the soil components and their chemical reactivity, particularly at the interfaces (air–water, air–solid, and water–solid). Accordingly, most of the chapters in this book are focused on the chemical characteristics of each of these components. Water is the most important reaction medium in the biosphere — its impact on both biological and geological processes cannot be underestimated. Chapter 2 (Soil Atmosphere & Soil Solution) is a review of basic atmospheric and aquatic chemistry.

Chapter 3 (Soil Organic Matter) addresses an important reactant to soil formation and evolution. Soil organic matter is the only source of food for many soil organisms as well as the by-product of most biological activity in soils. It plays an important role in soil fertility, retention of water and contaminants, soil structure and soil stability. Soil organic matter is a fascinating subject, but very difficult to study in the lab due to its complex and diverse nature. The major problem in the study of soil organic matter is its extraction from soils without destruction or alteration of its native physical and chemical characteristics.

This is followed by two chapters on the soil inorganic mineral phases. Chapter 4 (Soil Mineralogy) is a basic review of many diverse minerals commonly found in soils, with the emphasis being on the internal structure of minerals. Chapter 5 (Surface Characteristics & Analysis) is a continuation of Chapter 4, but the emphasis is on the portion of solid minerals that borders the interface with other media (namely, liquids or air).

With selected topics, detailed mathematical explanations will be found in the text. Understanding the basic science behind each topic is closely tied to our understanding of the mathematics and corresponding physical chemical explanations. No one can expect to improve or modify existing mathematical equations, or to create new ones, without an understanding of the derivation processes involved. However, it is also rare for a scientist to be well versed in all aspects of mathematics. Without the constant practice, we will typically not remember how to derive certain equations. Fortunately, most of the derivations of the mathematics involved are in fact rather easy to follow, but some are difficult. Accordingly, this book tries to walk you through each step in the derivation of many equations.

This book is focused primarily on the chemical structure of soils and soil chemical processes. Some of the chapters include discussions on the chemical nature of other environments, such as rivers, oceans, or even artificial environments (e.g., laboratories). These are included to emphasize the differences or similarities of the various natural environments, as well as the value or limitations of the samples collected from alternate sources. Some of the topics discussed will have little or no mathematical explanations, while others will have a great deal of mathematics. You may choose to skip the mathematical explanations and jump to the concluding equations or resulting discussions. This is perfectly reasonable and, in fact, recommended for most readers. If you ever wish to revisit a particular subject in more detail, the math will still be here for you to review at that future date as needed.

Many topics discussed in this book will only give a quick review of the basics. Soil chemistry is a difficult subject to study, but the reason for this may surprise you. The types of chemical reactions that can occur with each soil component are limited in scope due to the narrow range of temperature and pressure conditions present in most soils. Some of the complexity of soil chemistry is a result of the complex nature of the mixture consisting of many diverse components present in soils. The chemical reactions involved with each of the components of soils are sometimes quite simple, but this is nearly always masked by an array of other simple reactions that are occurring concurrently with other soil components present in the mixture. It is clearly easier to understand the chemical reactions of a clean, well characterized soil component than actual whole soil samples.

All topics about chemistry, regardless of their application, will eventually involve a discussion of the chemical bonds present. Chemical reactivity and stability are a function of the behavior of chemical bonds present in compounds under various natural or applied environments. The chemical bonds and electron sharing between the elements are basic to nearly all the topics discussed in this book. Accordingly, this introductory chapter closes with a review on the nature of chemical bonds and electronegativity. Other basic chemical concepts will be reviewed as needed in each of the subsequent chapters.

[1.3]   Covalent & Ionic Bonds

The elements that form all molecules and crystal structures are held together by numerous types of bonds. *Covalent bonds* form from an equal sharing of electrons between two elements. The bond distance of covalent bonds are smaller than the ionic radius of the two elements. An example of covalent bonding is orthorhombic native sulfur (S_8), shown in Figure 1-2. Another example is pyrite (FeS_2), where the FeS_6 octahedra share edges.

In an *ionic bond*, the two elements are held by the electrostatic attraction of a charged negative anion to a charged positive cation. Typically, an outer electron from one element (the one destined to become a cation) is fully sequestered by the other element (the one destined to become an anion) and the electron is not shared in the resultant ionic bond. The halite crystal structure (a NaCl crystal) offers an excellent example of ionic bonding. As Figure 1-2 illustrates, there are no unique NaCl molecules present. The NaCl is in octahedral coordination with one Na^+ cation surrounded by six Cl^- anions. Ionically bonded crystals generally have high melting points, moderate hardness and specific gravity, and are poor conductors of heat. Ionic crystals are also poor conductors of electricity because the ions, which neither gain nor lose electrons easily, are tightly packed with the nondirectional ionic bond evenly spread over the ions. The symmetry of the resultant crystal is generally high.

The elements in a crystal can contain a mixture of both ionic and covalent character, particularly when the difference in the electronegativity between the elements is small. Quartz (SiO_2) is an example of this. Quartz has a high melting point, as if no discrete molecules were present, which is an ionic character. However, it is not soluble in solvents of high dielectric constant (see Section 2.13.2) and does not conduct electricity when melted, which are both characteristic of covalent bonds.[2]

It should be noted that there is no sharp boundary between covalent and ionic bonding. This becomes more obvious through studies of the Molecular Orbital (MO) Theory, where a covalent component is introduced into the electrostatic viewpoint. Since the bond between two elements is rarely uniquely covalent or ionic, the resultant bond will often display both covalent and ionic characteristics. The range of the strength of these bonds forms a continuous spectrum of possibilities whose extremes have been labeled as either completely covalent or ionic. Based on bond energy calculations (see Section 1.5), the Si–O, Al–O, and Fe–O bonds are nearly equally covalent and ionic.

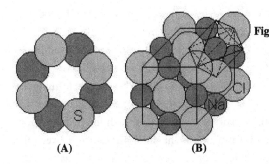

Figure 1-2: Illustration of the (A) S_8, and (B) NaCl crystal structures. The 6-fold coordination resulting in an octahedron sharing edges is also drawn. The crystal structure of pyrite (FeS$_2$) is similar to that shown here for NaCl, however replace each Na for Fe and each Cl for S$_2$.

(A)   (B)

[1.4]   Other Types of Bonds

The *van der Waals bond* is similar to the ionic bond but it involves neutral molecules rather than charged ions. Neutral molecules can attract each other electrostatically because they form slightly positive and negative regions within each molecule. Each of the slightly charged regions in one molecule will attract the slightly oppositely charged region in another molecule. The weak net attraction is about $^1/_{100}$ of an ionic or covalent bond.

According to the modified covalent theory of the metallic bond, a *metallic bond* is similar to the covalent bond but each atom shares its electrons with all its neighbors. The shared electrons form a "sea" of electrons due to their ability to move around. The mineral may exhibit a metallic luster, which is a typical property of metallic bonds. Metals are good conductors of electricity, good conductors of heat, and good reflectors of radiant energy. They are also ductile and malleable. An example of a metallic bond is pyrite (FeS$_2$), where some metallic bonding is attributed to the sharing of electrons between the metal atoms. Note, however, that the Fe–S and S–S bonds are covalent in character.

In the band theory of the metallic bond, the valence electrons are delocalized (which means that the electrons do not belong to any particular atoms). The atomic orbitals combine to form molecular orbitals. The close packing of the crystal causes overlapping of electronic energy levels to occur, which results in the formation of bands of electronic energy levels. In metal crystals or liquids (not gases), the number of molecular orbitals increases as the number of atoms present are increased, which in turn causes a decrease in the differences between the successive energy levels. For example, a small crystal may contain about 10^{18} atoms, and the resultant successive energy levels are essentially continuous (known as the electron band). The filling of the electron bands (such as the filling of the s, p, and d orbitals) and the energy spacing of the bands determines whether the substance is a conductor (metal), nonconductor (insulator, or nonmetal), or semiconductor (metalloid). Insulators have large energy gaps (known as forbidden zones) with bands that are completely filled or completely empty. Conductors have partially filled bands. With many electronic energy levels in a band that is partially filled, any electron within the unfilled band can absorb energy and be promoted within the band. Semiconductors have either defects in the lattice that introduce additional energy levels into which the electrons can move, or narrow forbidden gaps that allow electrons to jump from the top of the valance band into the conduction band.

Semiconductors will act as insulators at absolute zero presumably because the valence electrons lack the thermal energy needed to complete these jumps.

A *coordinate covalent bond* occurs when one of the two atoms involved in the bonding furnishes all of the electrons of the electron-pair bond. Two examples are $H_3N: +$ H^+ forming $H_3N:H^+$ (the ammonia supplies both electrons to the hydrogen ion, forming the ammonium ion), and $H_2O + H^+$ forming H_3O^+ (the oxygen in the water molecule has two unshared electron pairs, where one of these pairs is used to bond with the hydrogen ion, forming the hydronium ion). The mode of formation is the chief difference between the coordinate covalent bond and normal covalent bonds; they are indistinguishable once established.

Hybrid bonds are covalent bonds in which the bonding electrons are in two or more different orbitals and the bonding electrons have a mixture of the properties of the original orbitals. The resulting radii may be different from the radii of their normal covalent radii, but this effect does not seem to be important for most minerals when determining their coordination characteristics. Hybrid bonds are formed by a number of elements; numerous examples are shown in Table 1-5.

Crystal structures will typically contain a mixture of ionic and covalent bonds. An example is $CaCO_3$. The planar CO_3 bonds are covalent, with 6 CO_3 forming an octahedron around each Ca. However, the Ca–CO_3 bonds are ionic and easily broken in water:

$$CaCO_3(s) + H_2O \rightarrow Ca^{2+} + CO_3^{2-} \qquad [1\text{-}1].$$

Another type of bond is the *hydrogen bond*, which is an electrostatic interaction between electronegative atoms (F, N, or O) in one molecule and H atoms bound to electronegative atoms in another molecule. Rather than share electrons to form the bond between the electronegative atoms, H^+ is simultaneously attracted to the two atoms of high electronegativity. Examples are FHF, OHF, NHF, OHO, NHO, NHN, and OHCl, and can also occur between atoms of the same molecule (such as the intramolecular H bonding of proteins and polymers). There are also BHB bridges in borane (BH_3) compounds, which are informally known as "banana bonds" because of the shape of the bridges formed. These are not covalent bonds because the H atom only has one electron in a $1s$ orbital and is not able to form more than one covalent bond. It is instead a strong dipole–dipole attraction. Note that the H proton is not shared equally by the two electronegative atoms in the H bond. Using water as an example, the O atom attracts the electrons in the O–H bonds. This results in an electrostatic attraction of the positively charged H proton with a lone pair of electrons on an oxygen atom in a neighboring molecule. The strength of H bonds is about 5 to 10% of that of ordinary covalent bonds.

[1.5]   Electronegativity

Electronegativity (x) is an index that attempts to represent the electron-attracting tendency of an atom, where high values indicate a strong attraction for electrons. How evenly the electrons are shared, or not shared, between two atoms plays an important role in the characteristics of the bond as being either ionic, covalent, or a mixture of the two. There are

Table 1-5: Examples of hybrid bonds and their corresponding shapes and coordination number.

Electron Geometry [a]	Molecular Geometry [a]	Diagram [b]	Coordination Number [c]	Hybridization [d]	Examples
Linear	Linear	 180°	2	sp	$BeCl_2$, $Ag(NH_3)_2^+$, CO_2 (O=C=O)
Trigonal planar	Trigonal planar	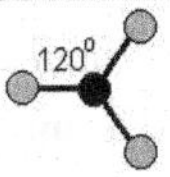 120°	3	sp^2	BF_3, NO_3^-, CO_3^{2-}
	Bent or Angular	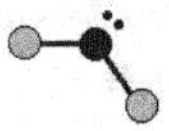	2		SO_2 (O–S=O, 120°)
Tetrahedral	Tetrahedral	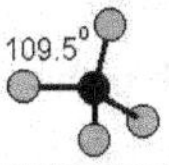 109.5°	4	sp^3	CH_4, $AlCl_4^-$, NH_4^+, $Zn(CN)_4^{2-}$
	Trigonal pyramidal	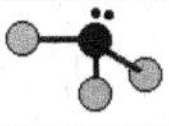	3		NH_3 (107.3°), AsH_3
	Bent or Angular		2		H_2O (104.5°), OF_2
Trigonal bipyramidal	Trigonal bipyramidal	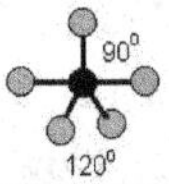 90° 120°	5	sp^3d	PCl_5
	Sawhorse or Seesaw	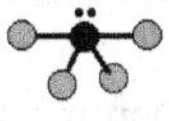	4		SF_4
	T-shaped	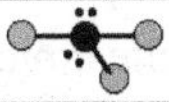	3		ClF_3
	Linear		2		XeF_2
Octahedral	Octahedral	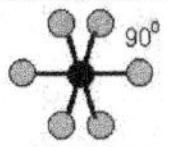 90°	6	sp^3d^2	SF_6, $Co(NH_3)_6^{3+}$
	Square pyramidal	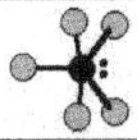	5		BrF_5
	Square planar	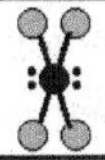	4		XeF_4, $Ni(CN)_4^{2-}$, $Pt(NH_3)_4^{2+}$

[a] Electron pairs that are not involved in a chemical bond are called lone electron pairs. As a result of the possibility of lone pairs, the molecular geometry may differ from the electron geometry.

[b] Lone electron pairs form large orbitals that push the other orbitals away. Accordingly, the angles formed decreases as the number of lone electron pairs increases. Bond angles are listed for some molecules.

[c] The coordination numbers shown here refer to the number of elements surrounding the central atom when the molecule is in the gas or liquid state, but it may be different in crystals. For example, $BeCl_2$ is shown here to be linear, but in solid form it has a tetrahedral structure with shared edges.

[d] The number of σ bonds is easily deduced from the hybridization, but note that π bonds may also be present. For example, O=C=O with sp hybridization has 2σ + 2π bonds.

several methods for determining the electronegativity of an atom or ion. Three methods will be discussed here to highlight the types of parameters typically involved in quantifying the electronegativity values.

The concept of electronegativity was originally introduced in 1932 by Linus Pauling[3], who received the Nobel Prize in Chemistry in 1954 "for his research into the nature of the chemical bond and its application to the elucidation of the structure of complex substances." Pauling defined electronegativity, which only has meaning in a bond, as the "power of an atom in a molecule to attract electrons to itself."[4] The Pauling method assumes that the difference in calculated bond energy from the experimental bond energy can be taken as a measure of the ionic character of a bond. Using HCl as an example, the covalent bond energies, $D_{covalent}$, are calculated as

D_{Cl-Cl} = 243 kJ mol^{-1}, from experimental data,

D_{H-H} = 436 kJ mol^{-1}, from experimental data,

D_{H-Cl} = 339.5 kJ mol^{-1}, from the arithmetic mean = (243 + 436)/2 ,

D_{H-Cl} = 325.5 kJ mol^{-1}, from the geometric mean = $\sqrt{(243)(436)}$.

From experimental data, the actual bond energy for H–Cl (D_{H-Cl}) is 432 kJ mol^{-1}, where the energy difference is attributed to the ionic character of the bond:

$$D_{ionic} = D_{measured} - D_{covalent} \qquad [1\text{-}2].$$

From the arithmetic mean, D_{ionic} = 432 – 339.5 = 92.5 kJ mol^{-1} = 0.96 eV, where 1 electron volt = 1eV = 96.48 kJ mol^{-1}. From these values, the H–Cl bond is roughly estimated to be 21% ionic (= 92.5/432) and 79% covalent (= 339.5/432). Now, the D_{ionic} value is assumed to be a measure of the electronegativity difference of the two atoms of the bond. Pauling found that if one uses the square root of this value, then a self-consistent set of electronegativities is obtained. The use of the square root is quite arbitrary, as is also the step that follows. With an arbitrary reference point for the hydrogen atom of x_H = 2.05, the electronegativity values will fall in the convenient range of 1.0 to 4.0. Obtaining a convenient working range is sufficient justification for the insertion of arbitrary steps. Accordingly, the electronegativity for the Cl atom is calculated by

$$x_{Cl} = 2.05 + \sqrt{0.96} = 3.0 \qquad [1\text{-}3].$$

Pauling (1960)[4] used the arithmetic mean for normal covalent bonds and the geometric mean for alkali metals, but noted that the geometric mean generally gives similar results if we insert a correction factor of 0.88. Accordingly, from the HCl geometric mean, D_{ionic} = 432 – 325.5 = 106.5 kJ mol^{-1} = 1.10 eV, and x_{Cl} = 2.05 + 0.88 $\sqrt{1.10}$ = 3.0.

The Mulliken[5] method assumes that the attraction of an ion on a pair of bonding electrons is a function of the attraction of the free ion for an electron (i.e., the ionization energy, I) and the attraction of the neutral atom for an electron (i.e., the electron affinity, A):

$$x_M = \frac{I + A}{2} \qquad [1\text{-}4],$$

where the Mulliken electronegativity (x_M) is a simple average of I and A (in electron volts). The ionization potential is the energy required to remove an outer electron from the neutral atom. The energy given off when an electron is added to a neutral gaseous atom is referred to as the electron affinity. Note that the atom in question need not be neutral. An outstanding feature of this electronegativity definition is its allowance for atom hybridization and atom charge. Mulliken's definition allows for direct determination of ion electronegativities because ionization potential and electron affinity are generally available for ions. Somewhat surprisingly, considering the different definitions used by the two researchers, the Mulliken electronegativities (x_M) often correlate well with the Pauling values (x_P):

$$x_P = 0.34x_M - 0.2$$ [1-5].

The Allred–Rochow[6] electronegativity scale is based on the electrostatic force of attraction (F) between the nucleus and the valence electrons:

$$F = \frac{e^2 Z_{eff}}{r^2}$$ [1-6],

where r = covalent radius = distance between the electron and the nucleus, e = electron charge, and Z_{eff} = effective charge of the nucleus on the electron. If Z = atomic number of the nucleus = number of protons, then $Z_{eff} = Z - S$, where S = shielding parameter of the nuclear charge on the electron caused by all the other electrons. The maximum value of S is $Z - 1$, which does not occur because it would mean that all the remaining electrons exert a full shielding effect of the nucleus on the outermost electron. In other words, the value of Z_{eff} lies between +1 and +Z, and its value can be determined experimentally or estimated to within 10%, such as with Slater's rules.[7,8] The eZ_{eff} value is the effective charge at the electron due to the nucleus and its surrounding electrons. This Allred–Rochow electronegativity (x_{AR}) coincides well with Pauling electronegativities when expressed as:

$$x_{AR} = 0.744 + 0.359\frac{Z_{eff}}{r^2}$$ [1-7].

The electronegativity is a function of size of the atom, which includes the space occupied by the electrons (resulting in an electron screening effect), and the nuclear charge of the atom (affecting the degree of attraction for the electrons). The general tendency is for an atom to seek eight electrons on the outer shell, which is referred to as the inert gas structure and follows what is known as the octet rule. With respect to the electron screening effect, the more s orbital character present in a hybrid bond, the higher the value of the orbital electronegativity. A strong s orbital presence in a hybrid bond (such as an sp configuration) results in a decreased shielding of the p electrons due to a reduction in the number of s electrons. The s electron has a deeper penetration of the electron core than the p electrons. In general, there is a large difference in the s and p orbital electronegativities by about a factor of 2.

Table 1-6 shows the periodic table with the electronegativity of the elements. These values are to be used only as a rough estimate of each atom's ability to attract electrons. In general, as one goes horizontally across the table the electronegativity increases, and as one goes vertically down the table the electronegativity decreases. Some exceptions exist due to the intervention of the d electrons and other effects associated with very large atoms. If two ions are held by an ionic bond, the negative ion is the more electronegative element. Even if the bond is strongly covalent, the bond between two different atoms will be a polar bond, with the negative end occupied by the atom with the higher electronegativity.

A large difference in the electronegativity of two ions suggests a strong likelihood that an ionic bond is holding them together. Conversely, when $\Delta x = 0$, such as when they are the same element, then the bond is covalent. For most bonds, an electronegativity difference

Table 1-6: Pauling's electronegativity values of the elements. Note that the electronegativity value varies with the oxidation number of the element, increasing as the oxidation state increases. [9] For example, $x = 1.65$ for Fe^{2+}, but $x = 1.8$ for Fe^{3+}. Only the common oxidation states are listed.

1	2	3	4	5	6	7	8	9	10	11	12	13	14	15	16	17	18
₁H 2.1																	₂He –
₃Li 1.0	₄Be 1.5											₅B 2.0	₆C 2.5	₇N 3.0	₈O 3.5	₉F 4.0	₁₀Ne –
₁₁Na 0.9	₁₂Mg 1.2											₁₃Al 1.5	₁₄Si 1.8	₁₅P 2.1	₁₆S 2.5	₁₇Cl 3.0	₁₈Ar –
₁₉K 0.8	₂₀Ca 1.0	₂₁Sc 1.3	₂₂Ti 1.5	₂₃V 1.6	₂₄Cr 1.6	₂₅Mn 1.5	₂₆Fe 1.8	₂₇Co 1.8	₂₈Ni 1.8	₂₉Cu 1.9	₃₀Zn 1.6	₃₁Ga 1.6	₃₂Ge 1.8	₃₃As 2.0	₃₄Se 2.4	₃₅Br 2.8	₃₆Kr –
₃₇Rb 0.8	₃₈Sr 1.0	₃₉Y 1.3	₄₀Zr 1.4	₄₁Nb 1.6	₄₂Mo 1.8	₄₃Tc 1.9	₄₄Ru 2.2	₄₅Rh 2.2	₄₆Pd 2.2	₄₇Ag 1.9	₄₈Cd 1.7	₄₉In 1.7	₅₀Sn 1.8	₅₁Sb 1.9	₅₂Te 2.1	₅₃I 2.5	₅₄Xe –
₅₅Cs 0.7	₅₆Ba 0.9	₅₇La * 1.1	₇₂Hf 1.3	₇₃Ta 1.5	₇₄W 1.7	₇₅Re 1.9	₇₆Os 2.2	₇₇Ir 2.2	₇₈Pt 2.2	₇₉Au 2.4	₈₀Hg 1.9	₈₁Tl 1.8	₈₂Pb 1.8	₈₃Bi 1.9	₈₄Po 2.0	₈₅At 2.2	₈₆Rn –
₈₇Fr 0.7	₈₈Ra 0.9	₈₉Ac ** 1.1															

*	₅₈Ce 1.1	₅₉Pr 1.1	₆₀Nd 1.2	₆₁Pm –	₆₂Sm 1.2	₆₃Eu –	₆₄Gd 1.1	₆₅Tb 1.2	₆₆Dy –	₆₇Ho 1.2	₆₈Er 1.2	₆₉Tm 1.2	₇₀Yb 1.1	₇₁Lu 1.2
**	₉₀Th 1.3	₉₁Pa 1.5	₉₂U 1.7	₉₃Np 1.3	₉₄Pu 1.3	₉₅Am 1.3	₉₆Cm –	₉₇Bk –	₉₈Cf –	₉₉Es –	₁₀₀Fm –	₁₀₁Md –	₁₀₂No –	₁₀₃Lr –

of 1.7 or higher results in a greater ionic character percentage and a weaker covalent character in the bond. Always remember that there is no sharp boundary between covalent and ionic bonding.

The ionic character percentage of a bond can be calculated based on the observed dipole moment (μ_{obs}) of the molecule and from the Debye equation:

$$\wp = \frac{N}{3}\left(\alpha + \frac{\mu^2}{\epsilon_o 3kT}\right) = \frac{(\epsilon/\epsilon_o - 1)}{(\epsilon/\epsilon_o + 2)}\frac{M}{\rho} \qquad [1\text{-}8],$$

where ϵ/ϵ_o = dielectric constant, ϵ_o = permittivity constant, ρ = density, M = molar concentration, k = Boltzmann constant, T = absolute temperature, N = Avogadro's number, $\wp$ = molar polarization, and μ = molecular dipole moment. Note that the average dipole moment (μ_{av}) in the direction of the field is equal to $\mu^2 E/(3kT)$, where E = electric field strength. Plotting $\wp$ against $1/T$ gives the average dipole moment as its slope, and the molecular dipole moment is easily determined. If the bond is completely covalent, then $\mu_{obs} = 0$. If the bond is completely ionic, then the theoretical ionic dipole moment (μ_{ionic}) would be

$$\mu_{ionic} = (\text{charge of an electron}) \times (\text{bond length}) \qquad [1\text{-}9].$$

The ionic character percentage is calculated based on a linear extrapolation of μ_{obs} between these two extremes (zero and μ_{ionic}):

$$\text{ionic character \%} = \frac{\mu_{obs}}{\mu_{ionic}} \times 100\% \qquad [1\text{-}10].$$

For HCl, $\mu_{ionic} = (0.1602 \times 10^{-18}\ \text{C})(1.275 \times 10^{-10}\ \text{m}) = 20.42 \times 10^{-30}\ \text{m C} = 6.12$ debyes, $\mu_{obs} = 1.07$ debyes, and ionic character % = $(1.07/6.12)100\% = 17\%$. Based on dipole moment calculations, Figure 1-3 illustrates the general relationship between the electronegativity difference between two elements of a molecule and the ionic character of the bond between them.

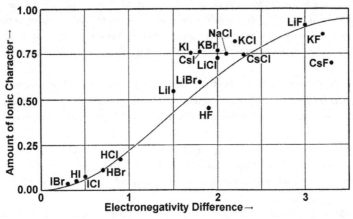

Figure 1-3: Electronegativity difference and ionic character fraction. Experimental points are based on observed values of the electrical dipole moment of diatomic molecules. Curve relates the ionic character of a bond to the electronegativity difference of the two atoms, where $y = 1 - \exp[-(x_A - x_B)^2/4]$. *Reprinted from Pauling (1960)[4], copyright © 1960 by Cornell University, with permission from Cornell University Press.*

References Cited

[1] Hillel, D. 2001. Ideas for the role of the soil in the environment and in human welfare. Crop Science, Soil Science and Agronomy News (CSA News) 9:8–9.

[2] Brownlow, A.H. 1979. Geochemistry. Prentice-Hall, Englewood Cliffs, NJ.

[3] Pauling, L. 1932. The nature of the chemical bond. IV. The energy of single bonds and the relative electronegativity of atoms. J. Am. Chem. Soc. 54:3570–3582.

[4] Pauling, L. 1960. The nature of the chemical bond. 3rd ed. Cornell Univ. Press, Ithaca, NY.

[5] Mulliken, R.S. 1934. A new electroaffinity scale; Together with data on valence states and on valence ionization potentials and electron affinities. J. Chem. Phys. 2:782–793.

[6] Allred, A.L., and E.G. Rochow. 1958. A scale of electronegativity based on electrostatic force. J. Inorg. Nucl. Chem. 5:264–268.

[7] Slater, J.C. 1930. Atomic shielding constants. Phys. Rev. 36:57–64.

[8] Clementi, E., and D.L. Raimondi. 1963. Atomic screening constants from SCF functions. J. Chem. Phys. 38:2686–2689.

[9] Gordy, W., and W.J.O. Thomas. 1956. Electronegativities of the elements. J. Chem. Phys. 24: 439–444.

Questions:

1.   How much interdisciplinary interest and cooperation do you feel exists between the various fields that address environmental problems? Speak with instructors and other field professionals, if possible, about how their specialty cooperates with, and/or solicits the help of, professionals in other environmental areas of expertise. Are there any clear boundaries between geology, soil science, studies of natural resources, environmental science, environmental engineering, and environmental policies and laws?

2   How do you view yourself in soil chemistry? For example, you may be an environmentalist (such as a soil scientist) that needs to know the basic chemical explanations of various natural soil processes. You may be a chemist that seeks to identify chemical processes in nature. You may even be both as an environmental chemist (geochemist or soil chemist). Your answer to this question may evolve or change significantly over the years.

3.   Nearly all human activity affects the environment, which is basically sectioned into air, water, and land (or atmospheric, aquatic, and terrestrial) components. Discuss how the ecological health of one of these components can affect the health of the others.

4.   When the air warms too much (global warming) or has too much smog or smells bad, we tend to notice. When the water is murky or tastes bad, we tend to notice. Why is soil quality important and how poor does it need to be before most people take notice?

5.   In the library there are several journals that address numerous soil related problems. Is there a range of awareness in these journals about how diverse soils really are? Compare the detail of the soils' classification and descriptions presented in the articles. Now, more importantly, how does this information, or lack thereof, affect the transferability of information (e.g., observed retention of chemical components, or rate of chemical reactions) from their test field to another one in your neighborhood? Note that the amount of details present will

change based on the publication years that are being surveyed. (See also "My friend, the soil – A conversation with Hans Jenny", J. Soil Water Conserv., 1984, 39(3):158–161).

Chapter 2:
Soil Atmosphere & Soil Solution

From ocean water to vadose water,
some basic principles never change.

Water is the most important molecule on the planet. It is a critical component of all known life forms. Water was involved in the development of life since its beginnings in primitive Earth. By extension of water's impact on Earth, NASA scientists argue that life probably existed on Mars if water was present. Water is also involved in the weathering of minerals, the transport of nutrients, the mobility of contaminants, and the reactivity of most organic and inorganic compounds. Electric fields are affected by the presence of water, particularly when dissolved salts are present. Water is a complex medium that can hold dissolved salts, inorganic compounds, organic compounds, and gaseous molecules. Water is involved in soil chemistry, geochemistry, soil biochemistry, and soil atmosphere chemistry. Clearly, water is a very important component of soils and, as such, needs to be well understood if soils are to be well understood. Accordingly, the physical characteristics and chemical reactivity of water are presented in this chapter.

Air is another important component to the biogeochemical activity of soils. The chemical nature of the soil atmosphere is also presented in this chapter, particularly in the first two sections. Although oxygen (O_2) and carbon dioxide (CO_2) are important gases in the biosphere, nearly all of their biochemical reactivity (such as electron transfer and respiration) and geochemical reactivity (such as weathering of minerals) occur when these gases are dissolved in water. This is also true for all other gases. Accordingly, the emphasis that this chapter places on the soil atmosphere in the subsequent sections is on the reactivity of the atmospheric gases in water, and particular emphasis is placed on the reactivity of $CO_2(aq)$.

[2.1]   Soil Atmospheres

Burrowing animals (such as moles, worms, and ants) leave channels in the soil that are filled with air. Plant roots will leave empty root channels in the soil after they die and decay. In addition to the impact of biological activity on soil, soils typically form structures known as soil aggregates, or peds, that are densely packed clumps of small particles. There are ample void spaces between the peds. At a smaller range of pore diameters, there are also void spaces to be found between the individual particles that form the ped. The smallest range of soil pore diameters are found inside each individual mineral or particle in the soil. All of these pore sizes are present in nearly all soils. All of the soil's void space is filled by either water or air, depending on the climatic conditions of the area. The components in the soil water and soil air greatly influence the reactivity and fate of compounds present in the soil.

The chemical composition of the soil air is controlled predominantly by the chemical composition of air above the soil. The air moves in and out of the soil daily with expansion in the warm daylight and contraction in the cool nighttime hours. There is also gas diffusion controlled by the high to low concentration gradient of the particular molecules present in the atmosphere and the soil air. Fluctuations in soil water content further induces bulk air movement in soils, such as by upward bouyancy of the air bubble, by a forced piston flow ahead of the moving water, or as entrapped air in the moving water[1]. The most significant mechanism for air transfer, at least near the surface, is the low-pressure-induced mass flow or suction effect that occurs when wind gusts blow across the soil surface. Note that, as is demonstrated in an atomizer, moving air has a lower pressure than stationary air.

The major gaseous components in the air above the soil are N_2, O_2, Ar, and CO_2 (Table 2-1). These values are relatively constant everywhere in the planet's outdoors. The percent of the gaseous components remains essentially unchanged with altitude within the troposphere (which is 8 to 20 km thick) and the stratosphere (which is 20 to 50 km above ground). The atmosphere is deeper and has better defined layers at the equator than at the poles, but depth varies with season and latitude (the stratosphere, for example, is hard to detect at the poles). The formation of ozone (O_3) near the ground often follows a lightning storm. However, the well known "ozone layer" is at about 20 to 25 km in the stratosphere, while ozone concentrations in the troposphere are much lower but vary considerably at lower altitudes with geographical location. Water vapor pressures reduce with height throughout the depth of the atmosphere. Variations present in the troposphere, such as steam, dust and smoke, will reduce the percentage of the other gases slightly. Other variables include swamp gases (such as H_2S formation) and distance to fossil fuel burning or spills (namely, release

Table 2-1: Composition of the atmosphere. This is the 1976 US Standard Atmosphere, where the air is assumed to be dry, homogeneously mixed, and at low altitude. *Data from Lide (2002).*[2]

Gaseous Component	Volume % in Atmosphere	Gaseous Component	Volume % in Atmosphere
N_2	78.084	He	0.000524
O_2	20.9476	CH_4	0.0002
Ar	0.934	Kr	0.000114
CO_2	0.0314 [a]	H_2	0.00005
Ne	0.001818	Xe	0.0000087

[a] The average values for atmospheric CO_2 vary monthly, and annual averages near sea level are rising, contributing to a global warming effect. Average annual CO_2 values at Mauna Lou Observatory in Hawaii were 0.031691% in 1960, 0.032568% in 1970, 0.033869% in 1980, 0.035419% in 1990, and 0.036940% in 2000.[2]

of organic vapors). The actual concentration of the gases present will depend on the pressure and temperature of the area. For low altitudes, changes in atmospheric pressure with altitude above sea level follows the barometric formula:

$$P_h = P_o e^{-\frac{mgh}{kT}}$$
[2-1],

where P_h = pressure at altitude h in meters, P_o = pressure at sea level, m = mass of one molecule ($m_{avg} \approx 29$ amu), g = force of gravity (≈ 9.8 m s^{-2}), k = Boltzmann's constant (1.3806×10^{-23} J K^{-1}), and T = absolute temperature. Save for areas with temperature inversions near the ground, the air temperature will generally decrease by 9 °C/km (or 5 °F/1000 feet). For reference, the highest peak on Earth is Mt. Everest at 8,850 m.

Indoors, such as in a lab where people are breathing and instruments are running, the amount of gaseous O_2 and CO_2 present are greatly affected. This is also true in soil environments, where the biological members of the soil community are breathing in O_2 and respiring CO_2 in the poorly vented pockets of soil air. In addition to the respiration of CO_2 by the plant roots and the animals present in the soil, soils are home to an enormous population of microorganisms that affect the quality of the soil air. Denitrifying bacteria will increase the percentage of gaseous nitrogen compounds (NO_x) in the soil atmosphere. Methane and sulfide gases can be concentrated by microbial activity in some soils that are under reducing conditions (such as in water-logged soils). Geological processes can also affect the concentration of gases in soils, such as the release of radon (Rn) through the Earth's crust. And finally, the physical entrapment of liquids in soils results in a higher partial pressure of various volatile solvents, particularly water. Entrapped water drives the relative humidity within many soils to above 98%.

The CO_2 gas will react quickly with water to form $CO_2(aq)$ and its various species, namely H_2CO_3, HCO_3^-, and CO_3^{2-}. All of these molecules are referred to collectively as inorganic C. This gas, consequently, strongly affects the pH of the soil water, and its aqueous components are highly reactive with other soil constituents. The effect would be stronger if soils were not well buffered environments. A thorough understanding of the behavior of CO_2 in soils is an important milestone toward our understanding of soil chemistry. Table 2-2 describes the general distribution of CO_2 in air and water. In soils and deep subsurface environments, the inorganic C concentrations vary spatially with mineral composition (carbonate dissolution) and depth (gaseous partial pressure), and seasonally with biological activity (respiration). Some measurements of soil CO_2 concentrations are nearly three orders of magnitude greater than that in the atmosphere.

As Table 2-2 shows, high partial pressures of CO_2 are not necessarily found very deep in the soil, but rather are often found very close to the surface. In noncalcareous soils, the primary cause of high $CO_2(g)$ concentrations in the soil pore spaces is microbial activity. Buyanovsky and Wagner (1983)[3] measured the CO_2 content under different crops in Missouri for over a year and noted a strong correlation to soil temperature (Figure 2-1). Presumably, the higher soil temperatures encourage a higher microbial activity. Similar findings were observed by Castelle and Galloway (1990)[4] in an acidic Virginia forest soil.

The physical entrapment and slow release of $CO_2(g)$ to the atmosphere plays an

Table 2-2: Typical range of inorganic carbon concentrations in the environment. TIC = total (aqueous) inorganic carbon. P_{CO_2} = partial pressure of $CO_2(g)$. P_{CO_2} values listed for soils are maximum concentrations reported, although they varied seasonally and with depth.

Site Description	Concentration	Reference
Atmosphere	$P_{CO_2} \approx 10^{-3.45}$ atm	[2]
Groundwater	P_{CO_2} up to $10^{-0.7}$ atm	[5]
Groundwater	TIC up to $10^{-2.2}$ M	[5]
River water	TIC $\approx 10^{-3}$ M	[5]
Surface marine water	TIC $\approx 10^{-2.71}$ M	[5]
Deep marine water	TIC up to $10^{-2.63}$ M	[5]
Missouri corn, wheat & soybean field, at 25 cm	$P_{CO_2} \approx 10^{-1.15}$ atm	[3]
Virginia deciduous forest, at 30 cm	$P_{CO_2} \approx 10^{-1.85}$ atm	[4]
New England spruce–fir forest, in B horizon	$P_{CO_2} \approx 10^{-1.90}$ atm	[6]
Gasoline-contaminated site, at 61 cm	$P_{CO_2} \approx 10^{-0.86}$ atm	[7]

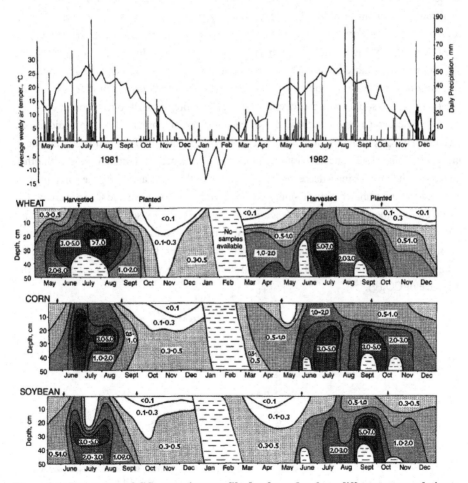

Figure 2-1: Percentage of CO_2 at various profile depths under three different crops relative to time along with a plot of mean air temperature and daily precipitation data. *From Buyanovsky and Wagner (1983)[3], with permission.*

important role in maintaining high inorganic C levels in soil air. The tortuosity of the soil pore structure generally ensures long retention times of gases in soil. It is interesting to note that soil moisture, which tends to lower the soil's porosity and CO_2 diffusivity, often does not correlate well with the soil gas concentration levels. Apparently, small decreases in gas diffusivity or soil moisture (i.e., rate of gas escape) are not as critical as the impact of temperature (i.e., rate of gas production) in soil environments. A nice example of the importance of soil aeration was given by Fernandez and Kosian (1987)[6], who described the fluctuation of CO_2 concentrations in Maine soils according to the soil horizon (Figure 2-2). The lower variability in the O horizon, which is high in nutrients and biotic activity, was attributed to the rapid exchange of gases with the atmosphere through the large pores of the O horizon.

Unusually high $CO_2(g)$ concentrations in some soil areas indicates a high microbial activity for that area. Testing for this gas is relatively easy and inexpensive, and detection of high values in soil air samples may suggest that organic contaminants are also present. The organic contaminant serves as a rich food source for microorganisms, which in turn respire $CO_2(g)$. An interesting application of this relationship was made by Robbins et al. (1990)[7], who observed an increase in CO_2 content near gasoline-contaminated areas. Thus, the respiration of the high microbial activity on the contaminant can aid in determining the location of the gasoline, which is deeper and slightly more difficult to monitor.

The CO_2 gas is in equilibrium with various forms of inorganic C in the liquid phase. Accordingly, CO_2 in the soil atmosphere also affects the fate of other chemical compounds present in soils. High concentrations of inorganic C will behave competitively with other anions in solution, causing a decrease in the retention of the other anions by the soil minerals. Conversely, and somewhat surprisingly, low concentrations of inorganic C can increase the retention of some aqueous anions (such as SO_4^{2-} and SeO_4^{2-}) by soil minerals.[8,9] Another effect of this gas on solid surfaces, but probably not critical to our understanding of most soils, is its ability to modify the behavior of very dry surfaces. In very dry environments, for example, the initiation of water vapor adsorption on a dry silica gel is prevented in the presence of $CO_2(g)$.[10]

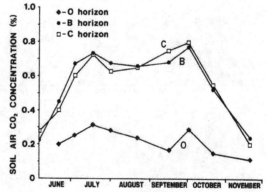

Figure 2-2: Forest soil air CO_2 concentrations at the Mud Pond watershed in eastern Maine. Values are at various depths during the 1985 growing season. *From Fernandez and Kosian (1987)[6], with permission.*

[2.2]   Sampling Soil Atmospheres

One method for the measurement of CO_2 for long-term experiments is the use of a sealed chamber on the soil surface with a NaOH solution in a glass jar. The alkaline solution traps the CO_2(g) that diffuses out of the soil and forms a more neutral product. The amount of CO_2 trapped by the alkaline solution is quantified by titrating the solution with acid.

A direct measurement of CO_2 and other gases (namely, O_2) can be made using gas chromatography. The gas sample is collected with a syringe and a long steel tube that is pushed into the soil. The steel tube can be left in place for later sampling, with only the sealed syringe transported to the lab.

A variation on this method involves replacing the syringe and gas chromatography with a gas-sensitive tube. Using a hand pump to draw air up the steel tubing, the air is forced through another tube that is filled with a solid base (Si gel) coated with a chemical reagent (an amine) that changes color when reacted with CO_2 (from white to violet). Draeger tubes (by National Draeger, Pittsburg, PA) are commonly used this way. As the air moves through the column, a color change occurs progressively in the inside of the tube. The color change can be seen easily because the tube walls are transparent. After a fixed quantity of air is passed through the tube, the location of the color front in the tube correlates with the amount of CO_2(g) in the soil air sampled. These tubes are much cheaper than the up-front costs for a gas chromatography instrument. Tubes are also available to detect numerous other gaseous compounds.

A fourth method worth mentioning is a portable infrared (IR) detector. Using a steel tube and a pump to withdraw soil air samples, partial pressures of soil CO_2(g) are very easily and quickly obtained. The CO_2 content correlates well with its absorption of infrared light.

None of the methods discussed are satisfactory for all types of soils, soil conditions, and research objectives. It is particularly difficult to study gaseous transformations in soils under field conditions. The most common shortcoming of soil air field-sampling techniques is the cross contamination with or infiltration of atmospheric air. Accordingly, most research on these transformations has been performed by analyzing atmospheres off-site in sealed flasks or bottles specially fitted for gas sample removal and analysis.

[2.3]   Nature of Water

The H_2O molecule has a tetrahedral structure (see Table 1-5) with electron pairs on two of the tetrahedral points and protons on the remaining two tetrahedral points. The bond angle between the O–H bonds is 104.51°. The dipole moment of polyatomic molecules is a result of the vector sum of the individual dipole moments of the bonds of the molecule (refer to Equation [1-8]). The observed dipole moment of water is $\mu_{obs} = 1.85$ debyes; hence, the dipole moment of each O–H bond is $\mu_{OH} = 1.85/(2 \cos 52.25°) = 1.51$ debyes. The O–H bond length is estimated to be 0.09575 nm.[2] Note that the radius of an oxygen atom is slightly larger (0.13 to 0.14 nm), and that the center of the hydrogen cation is literally inside oxygen's outer electron shell. The radius of an OH⁻ anion is 0.137 nm.[2]

Although the average chemical composition of water is H_2O, it is constantly dissociating and recombining with other water molecules to form the hydronium cation (H_3O^+). Typically, equations that describe the hydronium cation will use just H^+ for short. The reaction

$$H_2O \rightleftharpoons H^+ + OH^-$$ [2-2],

has an equilibrium constant $pK_w = -\log K_w = 14$ at 25°C, where

$$K_w = (H^+)(OH^-) = 10^{-14}$$ [2-3].

Note that the "w" subscript serves to remind us that the activity of water, which varies only slightly in most natural reactions, is factored into the K_w value. That is, $K_w = K_{[2-2]}(H_2O)$; read also the related discussions in Sections 2.13.3 and 2.16.2. The exact value of K_w will vary slightly with the temperature and ionic strength of the solution. The activity of H^+ ions affects the reactivity of a vast number of compounds in nature. It is often necessary, therefore, to discuss the concentration of this aqueous ion when describing natural processes. The range of hydronium values allowed in water is from about 1 to 10^{-14} mol L^{-1}, which is too large and cumbersome to work with. In 1909, S.P. Sørensen introduced a pH scale, known as the "potential of hydrogen activity," where

$$pH = -\log (H^+)$$ [2-4].

A pH value of 7 indicates a neutral solution, a value below 7 indicates an acidic solution, and a value above 7 indicates an alkaline solution. Placing 7 in the middle, the pH range of water is customarily said to be from 0 to 14 (see Question #2 at end of this chapter). In general practice, pH is measured with a hydrogen electrode (e.g., a glass electrode with a fixed H^+ ion concentration inside) as one half of the cell, plus a reference electrode (e.g., a calomel electrode) as the other half cell. Both electrodes are in the sample solution. The pH meter determines the H^+ activity by measuring the change in electromotive force (emf) or potential (E, in V units) between the two electrodes based on the Nernst equation:

$$pH = (E - E_R) \frac{F}{2.303 \, RT}$$ [2-5],

where E_R = potential across the glass electrode calibrated with a standard buffer solution, F = Faraday's constant (96,489 C mol^{-1}), R = gas constant (8.3143 J K^{-1} mol^{-1}), and T = absolute temperature. Since 1 C = 1 J V^{-1}, Equation [2-5] becomes $pH = (E - E_R)/0.05915$ at 25°C.

The water's proton is a very mobile element that is constantly jumping from one water molecule to another, and thus forming the hydronium cation (H_3O^+). This mechanism, which is illustrated in Figure 2-3, was first suggested by Grotthuss in 1805[11] to explain the high conduction and mobility of ions. Note, however, that Grotthuss's mechanism is only applicable to the H_3O^+ and OH^- ions. Dissolved ions move slowly through a tortuous path in water even under an applied electric field. Conversely, the H_3O^+ and OH^- ions move 5 to 10 times faster due to the proton's additional ability to jump from molecule to molecule. The mobility values (in velocity ÷ voltage units, or m/s ÷ V/m) are: 36.3×10^{-8} m^2 V^{-1} s^{-1} for H^+, and 20.5×10^{-8} m^2 V^{-1} s^{-1} for OH^-.[12] The mean lifetime of H_3O^+ (τ_{H^+}) prior to jumping back onto an OH^- anion to reform an H_2O molecule is much shorter than the mean lifetime of H_2O

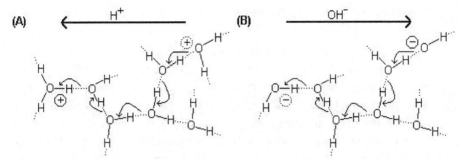

Figure 2-3: Movement of (A) protons in an acidic medium and (B) hydroxyls in an alkaline medium by the Grotthuss mechanism. Arrows show the proton jumps from molecule to molecule. The old (prior to the proton jumps) and new (after the proton jumps) locations of positive and negative charges are also shown.

or mean interval between jumps (τ_{H_2O}):

$$\tau_{H^+} \approx 10^{-12} \text{ seconds} \qquad \text{[2-6]},$$

$$\tau_{H_2O} \approx 5 \times 10^{-4} \text{ seconds} \qquad \text{[2-7]}.$$

There is much H bonding between the molecules of water, which is the primary reason for water's high boiling point (100°C) and melting point (0°C) temperatures. In the absence of H bonding, the boiling point of water would be about −80°C. The structure of liquid water consists of H-bonded framework regions with interstitial monomers occupying some of the cavities enclosed by the framework. Its structure is random in any given location due to the molecular motions, but it tends toward a specific structure as it cools. The highest density of water is reached at 4°C.

Water that is near a hydrophilic mineral surface will be attracted to that surface. The first layer of water will form H-bonds with the oxygen atoms (and hydroxyl molecules) that are part of the mineral surface. The altered electron distribution of the water molecules makes it easier for them to form H-bonds with other water molecules. As this proceeds, a more ordered structure than that of ordinary water is built out from the mineral surface. If a preferred orientation of the water molecules relative to the surface is established, then the thin film of water near the surface is said to have an ice-like structure. Although this thin film near the surface is water and not ice, its properties may be slightly different from the properties of H_2O molecules in the bulk solution. In this layer, the specific volume is lower, the viscosity is greater, and the resistance to ionic diffusion is greater than in ordinary water. Anderson and Low (1958)[13] estimated that this layer may reach out to distances of at least 6 nm from a bentonite surface. Ice-like structures are also encountered around hydrated compounds, including macromolecules. In solid H_2O-cages known as clathrates or gas hydrates, the ice forms a H-bonded host lattice that encages a great variety of small guest atoms or molecules. Water is sometimes also described as a partially broken down structure of ice. Based on the ice-like nature of water near surfaces, it is desirable for us to also understand the general nature of ice.

[2.4]   Structure & Properties of Ice

The triple point of water (where ice, water, and vapor can coexist) is at 0.01°C (273.17 K) and 6.1 mbar. Ice is abundant on the earth's surface, and if it all melted, sea level would rise about 70 m. There are many structures that ice can have at various temperatures and pressures, and the formation of the 14 known structures (I to XIV) can be easily described using phase diagrams (temperature–pressure diagrams). However, the only natural ice on earth consisting of ordered crystals of pure H_2O(s) is hexagonal ice (ice Ih) (Figure 2-4). (See Section 2.12 for structure of clathrates.) At ambient pressures, a cubic modification (ice Ic) can be synthesized if water vapor is condensed on a very cold surface (between -80 to -130°C), while an amorphous solid forms if the condensation occurs below -130°C.

Ice Ih has a highly symmetrical local structure when viewed in very small time- and space-averaged atomic densities, but the high translational symmetry is not retained well at the crystallographic unit cell level due to the atomic motions present. Consequently, ice Ih is a highly disordered crystalline material. Nevertheless, even the lower symmetry of the local configurations resulting from the static or dynamic disorder will obey Pauling's four ice rules[14]: (1) O–H distances and angles of ice resembles H_2O in the gas phase, (2) two H atoms form H bonds with two of the four O atoms that surround the H_2O molecule tetrahedrally, (3) only one H atom lies along each O...O axis, and (4) the interaction of non-adjacent molecules will not stabilize any one of the many alternate H_2O configurations that would also satisfy the preceding conditions with reference to the others (that is, all satisfying configurations are equally possible).

From Pauling's ice rule number 4, there are six equally possible orientations for the H bonds around each O atom in ice Ih. That is, given four H bonds per O, where two H atoms are at 0.101 nm (2H form the H_2O molecule) and two are at 0.174 nm (2H belong to two

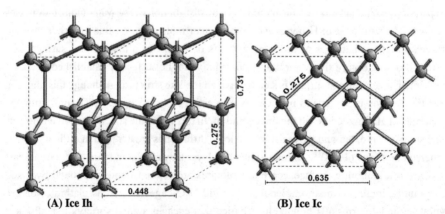

(A) Ice Ih   (B) Ice Ic

Figure 2-4: Ideal structure of (A) ice Ih, and (B) ice Ic at 100 K. The protons (not shown) lie in asymmetric position, and the H bonds formed are shown as rods. Cell dimensions are: 0.275 nm for O...O; $a = 0.448$ nm, $c = 0.731$ nm for ice Ih; and $a = 0.635$ nm for ice Ic. Average bond angle is 109.5°. Each H lies 0.101 nm from one O atom and 0.174 nm from the other. Note holes in structure, which result in a lower density for ice Ih (0.917 g cm^{-3}) than water (0.9998426 g cm^{-3}) at 0°C. After Kamb (1968).[15]

neighboring H_2O molecules), there are six possible ways of arranging the H atoms around each O atom. The H_2O orientation of each tetrahedron in the ice Ih structure is completely random. This is referred to as *proton disorder*, or *water-molecule orientation disorder*. Only ice II, VIII, and IX have ordered proton orientations. It is for this reason that the H atoms are not shown in Figure 2-4. The proton disorder of ice Ih does not become ordered with further cooling of the mineral.

There are various common defects that merit mention here. Near the surface of ice Ih, many dangling bonds are present that result in liquid-like properties. Inside the ice structure, point defects include H_3O^+ and OH^- ions, as well as substitutional or interstitial impurity atoms or ions. There are also crystal dislocations (line defects) and subgrain or grain boundaries (surface defects), the latter in polycrystalline ice. Important defects are the D- and L-bond defective sites, where the H bonds are broken and violate Pauling's third ice rule. When the H_2O molecule rotates to another orientation without compensation from the neighboring molecules to which it is H bonding, two defects will occur: some of the O...O contacts will have no protons between them (L defect), while others have two associated protons (D defect). These defects strongly influence the electrical and mechanical properties of ice.

While water is a conductor of electricity, ice is a strong insulator (electrical conductivity = 1.1×10^{-8} S m^{-1} at -10 °C). The very weak conduction that can be measured in ice is carried by the protons. The proton jumps, similar to those shown in Figure 2-3, occur collectively along a favorably oriented chain of H bonds, resulting in a very high mobility in the ice structure. The mobility of orientational L and D defects facilitate the reorientation of the molecular dipoles in ice. Similar to the high mobility of protons in water compared to aqueous ions, the proton mobility in ice Ih (8×10^{-5} m^2 V^{-1} s^{-1}) is about 10 times greater than the mobility of other ions in most solid materials.

[2.5] Freezing-Point Depression

Soils generally have a high content of dissolved salts or neutral compounds that lower the freezing point of water. In dilute solutions, the freezing-point depression can be estimated by the van't Hoff equation, expressed as:

$$\frac{dT}{dm_B} = \frac{-RT^2}{55.5\,\Delta H}$$

[2-8],

where m_B = molality of solute B (mol kg^{-1}), 55.5 = molality of water, T = temperature (273.16 K for water), R = gas constant (8.3143 J K^{-1} mol^{-1}), and ΔH = enthalpy of fusion of ice (6,010 J mol^{-1}). Since the change in m_B relative to pure water is m_B, Equation [2-8] simplifies to:

$$\Delta T_{fp} = -1.860\,(vm_B)$$

[2-9],

where ΔT_{fp} = freezing-point temperature depression (°C). A new factor is inserted to correct for all the solutes actually present when dissolution of salts occurs. That is, v = equivalent number of aqueous species dissolved (e.g., 1 for sucrose, 1 for non-electrolytes, 2 for NaCl, 3 for $CaCl_2$; but use 5/2 for NaCl + $CaCl_2$, where 5 = 1Na+1Ca+3Cl, and 2 = 1NaCl+1$CaCl_2$, because ν is the ratio of all species "after" divided by all species "before" dissolution in the water).

Deviations from Equation [2-9] are apparent at concentrations $>0.2\ \nu m_B$, particularly for electrolytes with high valence values. Figure 2-5 illustrates the freezing-point depression of various salt solutions, where the deviations from Equation [2-9] are $MgSO_4 > Na_2SO_4 > NaCl$, which follows their respective valences. The deviations from the limiting slope of -1.86 are partially attributed to ion pairing formed by the oppositely charged ions. Ion pairing decreases the true concentration of dissolved chemical compounds and ions in solution, and its strength is proportional to the product of the ionic charges.

A phase diagram at sub-zero temperatures is illustrated in Figure 2-6 for the NaCl–water system. At high NaCl concentrations ($>5.17\ m$), the initial crystalline phase to

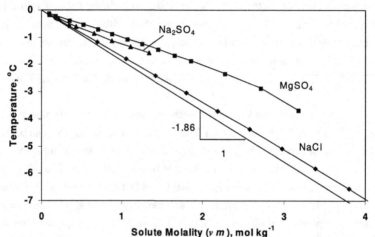

Figure 2-5: Freezing-point depression of various salt solutions. The limiting slope of -1.86 from Equation [2-9] is also shown. *Data from Lide (2002).*[2]

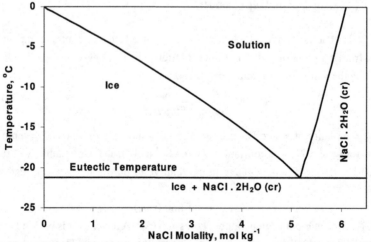

Figure 2-6: Stability diagram for the NaCl–H$_2$O system. The eutectic temperature is the lowest freezing point for any possible mixture of the components present in the solid solution. *Data from Lide (2002)*[2]*, and Marion (1995).*[16]

form as the temperature drops is $NaCl \cdot 2H_2O(cr)$. As it forms, the salt concentration decreases. At $-21.2°C$, the concentration of NaCl has decreased to 5.17 m and a mixture of ice and $NaCl \cdot 2H_2O(cr)$ begins to precipitate. Conversely, at low NaCl concentrations (<5.17 m), the initial phase to form as the temperature drops is ice. As the ice forms, the salt concentration remaining in solution increases. The ice continues to form as the temperature lowers, and the salt concentration continues to increase in the liquid phase. Finally, at $-21.2°C$ and at a NaCl concentration of 5.17 m, a mixture of ice and $NaCl \cdot 2H_2O(cr)$ begins to precipitate. Other eutectic temperatures for pure salt solutions are: $-0.34°C$ for $CaCO_3$, and -50.4 °C for $CaCl_2$.

With this illustration, it is easy to see that a salt solution will not freeze completely unless the temperature drops below the eutectic point of the mixture. This is also true with soil solutions, but in soils there is the added effect of the capillary interaction of the soil solid surfaces with the liquid solution. This effect tends to lower the freezing point of the soil solution below the eutectic point of a simple salt mixture. Note that adsorbed ions on the solid phase can also affect the nature of water near the solid–liquid interface, which, in turn, further affects the freezing point of the soil solution. In spite of the fact that at high temperatures (that is, above freezing) the H_2O molecules in the thin-film region are more ordered then the liquid bulk H_2O molecules, each soil particle will have a very thin film of water around it that resists freezing into an ice Ih structure at low temperatures.

As ice forms, anions are generally preferentially absorbed into the ice phase, resulting in a charge separation or freezing potential. This phenomenon is known as the Workman–Reynolds Effect.[16,17] Note that isomorphic substitution of ions in crystals is relatively easy when they are of similar size. This is the case here with anions because, compared to the cations, they are closer in size to the OH^- ions of ice.

[2.6] Surface Tension & Contact Angles

The surface tension (γ), sometimes referred to as surface energy, of a liquid tries to minimize the amount of exposed surface area of the liquid. It is the reason that droplets form, and supplies the force that allows insects to walk on water. All phase boundaries have this property and behave this way, but it is most apparent for deformable liquid surfaces.

With water, the interior of the liquid forms up to four hydrogen bonds with neighboring molecules. Each molecule in the interior of the liquid is attracted to other molecules equally from all sides. At the water–air interface, the molecules are only attracted inward and the tendency of the surface to move inward results in a tendency to minimize the exposed surface area of the water. The surface tension (γ) is force (F) per unit length (ℓ) or energy per unit area, with units $N\ m^{-1}$ or $J\ m^{-2}$. Figures 2-7 and 2-8 illustrate contact angle (θ) and a few ways to measure surface tension. Table 2-3 lists the surface tension of various liquids across a liquid–vapor (or liquid–gas) interface (γ_{LV}).

We have defined γ_{LV} and we can measure it. We can similarly define γ_{SV} as the surface tension across a solid–vapor (solid–gas) interface and γ_{SL} as the surface tension across a solid–liquid interface. However, we cannot measure these latter two properties

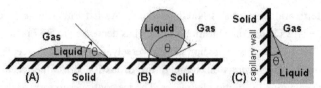

Figure 2-7: Illustrations of contact angles (θ): (A) small contact angle when liquid "wets" the solid, (B) large contact angle when liquid is "repelled" by the solid, and (C) contact angle formed at the crown of the meniscus resulting from capillary rise of a liquid that wets the solid wall.

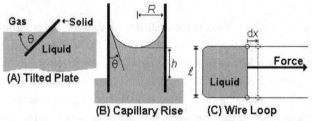

Figure 2-8: (A) Measuring contact angle (θ) by the tilted plate method. (B) Measuring capillary rise (h) in a capillary tube of radius (R); if θ is known, then the surface tension (γ_{LV}) can be calculated using Equation [2-12]. (C) Measuring γ_{LV} using a wire loop, where γ_{LV} = maximum force that the liquid can support without breaking. In the wire loop method, $\gamma_{LV} = F/(2\ell)$, where F = applied force, ℓ = wire size, and 2 = two sides of film.

Table 2-3: Surface tension (γ_{LV}) of various liquids. Note that the values decrease with increase in temperature. *Data from Lide (2002).*[2]

Liquid		γ_{LV}, mN m^{-1} (or g s^{-2})					
		0 °C	10 °C	20 °C	25 °C	50 °C	100 °C
n-hexane	C_6H_{14}		19.42		17.89	15.33	
ethanol	$H_3CCH_2\text{-OH}$		23.22		21.97	19.89	
toluene	$C_6H_5\text{-}CH_3$		29.71		27.93	24.96	19.01
2-ethoxyethanol [a]	$C_2H_5OC_2H_4\text{-OH}$				28.35	26.11	21.62
ethylene glycol	$HO\text{-}C_2H_4\text{-OH}$				47.99	45.76	41.31
water	H_2O	75.64	74.23	72.75	71.99	67.94	58.91
mercury	Hg		488.55		485.48	480.36	470.11

[a] Also known as ethylene glycol monoethyl ether (EGME).

because the solid will support a shear stress and will not deform the way a liquid will. Young's (1805) equation describes the balance of the forces at equilibrium in the horizontal direction as follows:

$$\gamma_{SV} - \gamma_{SL} = \gamma_{LV}\cos(\theta) \qquad [2\text{-}10],$$

where θ = contact angle. This equation assumes that the surface is smooth, which may be corrected by multiplying the right-hand side with a roughness factor ($\beta > 1$). It also assumes that the surface is chemically homogeneous. A heterogenous surface may be corrected by splitting the left-hand side into component fractions ($f_1, f_2, \dots, f_n$). More importantly, Equation [2-10] also assumes that the solid surface is in a state of equilibrium, which it isn't since it cannot deform. This lack of equilibrium might not really be problematic because the parameters on the left-hand side of the equation seek the difference in surface tension of two interfaces and the same solid is involved in both interfaces. Regardless, since γ_{SV} and γ_{SL} cannot be measured, Equation [2-10] still eludes definitive empirical verification and remains highly controversial.

Capillary rise and capillary flow results when a liquid is able to wet the solid phase (Figure 2-7A). Capillary rise is a vertical, typically controlled, version of capillary flow (Figure 2-8B). At equilibrium in capillary rise, the vertical component of the surface tension equals the weight of the liquid column:

$$2\pi R\, \gamma_{LV} \cos(\theta) = \pi R^2\, hg\, \Delta\rho \qquad [2\text{-}11],$$

where R = radius of the capillary tube (m), h = height of capillary rise (m), g = force of gravity (9.8 m s^{-2}), $\Delta\rho$ = difference in the density of the liquid and its surroundings (g m^{-3}), and γ_{LV} is in mN m^{-1} (or g s^{-2}). The force of attraction up the tube is equal to the size of the contact edge ($2\pi R$) times the surface tension adjusted by the contact angle, while the volume of the liquid equals $\pi R^2 h$. Equation [2-11] is an approximate relationship because the weight of the liquid in the crown of the meniscus is ignored. If the h and θ values are known, then the surface tension can be calculated using

$$\gamma_{LV} = \frac{Rhg\, \Delta\rho}{2 \cos(\theta)} \qquad [2\text{-}12].$$

Capillary rise is an important mechanism in soils for maintaining a high moisture content well above the water table levels. Assuming $\cos(\theta) \approx 1$, rearranging Equation [2-12] and solving for water, we can estimate the height of capillary rise in a soil as a function of the size of the pore spaces (using h and R in mm units):

$$h \approx 15/R \qquad [2\text{-}13].$$

For example, assume that particles of a fine sandy soil are spheres, are all nearly the same size (0.10 mm diameter) and are closely packed in the soil. Depending on how the particles are packed, the radius of the pore space can be calculated (see Table 4-3 for identical calculations, but using instead atoms in crystals). As a rough estimate, the diameter of pore spaces in soils is usually taken to be approximately equal to 0.4 × diameter of particles. Applying these numbers to Equation [2-13], we obtain an estimated capillary rise of 75 cm above the water table. The capillary rise will be much higher if the soil components are variable in size. That is, if the spaces formed between the average-sized particles are filled by a few of the smaller particles, then the final interparticle pore diameter of the soil will also be smaller.

It is important to note that these capillary forces increase as the pore diameter decreases. Intraparticle pore spaces (pore spaces formed in the cracks and cavities of each particle) are orders of magnitude smaller than the diameters of interparticle pore spaces (pore spaces formed by the spaces between particles). A great deal of water is retained in soils in these very small intraparticle pore regions, even when the surrounding interparticle pore spaces have been depleted of their water.

Contact angle is a property of liquids that is closely related to surface tension. Surface tension is a two-phase phenomenon (liquid–gas), while contact angle is a three-phase phenomenon (solid–liquid–gas). All three phases should be specified when describing the contact angle of a liquid. The contact angle (θ) is the angle in the liquid that is formed at the junction of the three phases. When the angle is small, then the liquid is said to "wet" the solid (Figure 2-7A) and the solid surface is hydrophilic. The liquid here is attracted to the solid phase to a greater extent than its cohesive attraction to other liquid molecules. When the contact angle is large, then the liquid is said to "repel" the solid (Figure 2-7B) and the solid surface is hydrophobic. Here, the liquid's cohesive force is stronger than its attractive force to the solid.

[2.7]   Ionic Strength & Activity Calculations

A very difficult concept to grasp for many students of aquatic chemistry is the need to express the amount of dissolved ions in activity terms rather than concentration terms. When one studies the behavior of ions in solution, it often becomes apparent that the reactivity of a given experimental matrix shifts as the number of the ions (expressed in concentration terms) increases. Conversely, this apparent shift in reactivity is minimized when the number of ions present is expressed in activity terms. The activity of an ion is a measure of the number of ions present in the matrix that can participate in chemical reactions, while the concentration of an ion is a measure of the number of ions believed to be truly present in the matrix.

The activity of an aqueous ion (or compound) is the "effective chemical concentration" of the ion in solution. As the amount of solute dissolved in water increases, the clustering properties of water change significantly. This, in turn, affects the solute's mobility and effective reactivity in water. Also note that the chemical properties of a dissolved solute are masked (or screened) by the water molecules and the other ions in solution. This screening effect increases as the concentration increases.

The activity (a) of an ionic species is defined as the active mass of the species in molar units (mol L^{-1} or simply M). It is equal to the concentration (c) of the species in solution (also in molar units) times a mean activity coefficient (f):

$$a = fc \qquad \text{[2-14]}.$$

In chemical equations where the amount of an ion present is represented by symbols, it is customary to use brackets [ ] when the symbols are expressed in concentration terms, or parentheses ( ) when the symbols are expressed in activity terms; for example, use [HCO_3^-] for aqueous concentration of bicarbonate anions and (HCO_3^-) for activity of bicarbonate anions. Note that both of these refer to dissolved aqueous species. For adsorbed ions or compounds, curly braces { } are customarily used, and the surface activity units are typically $\mu mol/m^2$.

The activity of a species can be determined by several methods, such as:
- from diffusion coefficients,
- from electromotive force (EMF) cells (for $c < 0.1\ M$),
- from freezing points,
- from solubilities,
- from vapor pressures (the isopiestic method for $c > 0.1\ M$).

Not all instruments measure the activity of an ion in solution. A pH electrode and ion selective electrodes measure the activity of the ions based on calibrations with known activities of standard solutions or buffers. However, an inorganic carbon analyzer, which is based on infrared measurements, measures the concentration of inorganic C in the sample solution. Similarly, an atomic absorption spectrometer measures the concentration, not the activity, of dissolved metal ions.

Since most instruments do not measure or estimate activity, it is often not experimentally practical or possible to directly measure the activity of an ion in solution. Fortunately, the concentration is easily measured, or it may be known based on the actual mass of the specific solutes added to the solvent. The activity of the ions in the solution are then calculated based on one of various equations used to estimate the mean activity coefficient (f).

The derivations of the various equations to estimate the mean activity coefficient are difficult to follow. They are, however, of vast importance to mathematical predictions on the fate of dissolved chemical compounds, predictions on the quality of our global environment or local soils, and numerous other reactivity predictions that make extensive use of activity or concentration terms. Accordingly, this section will briefly introduce the basic components of the mathematical derivations of the mean activity coefficient (f).

The most common approach to predict activity coefficients is based on the ion-pairing or *ion-association model*. This model assumes that long-range electrostatic forces between ions influence the activity of the ions. Depending on the ionic strength (I) of the solution, one of four equations are commonly used (the A, B, and a parameters will be defined as each equation is derived in the subsections that follow):

Debye–Hückel
(for $I < 10^{-2.3}$ M):
$$\log f = -Az^2\sqrt{I}$$
[2-15],

Extended Debye–Hückel
(for $I < 0.1$ M):
$$\log f = \frac{-Az^2\sqrt{I}}{1 + Ba\sqrt{I}}$$
[2-16],

Güntelberg
(for $I < 0.1$ M and several electrolytes):
$$\log f = \frac{-Az^2\sqrt{I}}{1 + \sqrt{I}}$$
[2-17],

Davis
(for $I < 0.5$ M):
$$\log f = \frac{-Az^2\sqrt{I}}{1 + \sqrt{I}} + 0.2\,I$$
[2-18].

The ionic strength (I) in the equations above is defined as:

$$I = \frac{1}{2}\sum_{1}^{s} c_i z_i^2$$
[2-19],

where c_i = molar concentration of species i (mol L^{-1}), and z_i = valence of species i.

In the derivations of Equations [2-15] to [2-18] that follow, the "ional" concentration (Γ) will be used instead of ionic strength (I) because this keeps the derivations shown consistent with the terminology found in most thermodynamic texts, where:

$$\Gamma = \sum_{1}^{s} c_i z_i^2 \qquad [2\text{-}20].$$

The derivations shown also define the ionic strength in molality units:

$$I = \frac{1}{2} \sum_{1}^{s} m_i z_i^2 \qquad [2\text{-}21]$$

or,

$$\Gamma = 2\rho I \qquad [2\text{-}22],$$

where m_i = stoichiometric molality (mol kg^{-1}) of species i, and ρ = density of the solution.

[2.7.1] Deriving the Debye–Hückel Equation: $\log f = -Az^2 \sqrt{I}$

A.   Assuming that the Maxwell–Boltzmann distribution is maintained, the electrical potential (Ψ) of an electrolyte solution undisturbed by external forces is given through the capacitance model as:

$$\Psi = \frac{-ze\kappa}{\epsilon} \qquad [2\text{-}23],$$

where $1/\kappa$ = mean radius of an ionic atmosphere, ϵ = dielectric constant, z = valence of the ion, and e = electronic charge. The electrical potential (Ψ) here is due to point charges that are surrounded by ions of opposite sign. Also,

$$\kappa = \sqrt{\frac{4\pi e^2 N_A \Gamma}{1000 \epsilon kT}} \qquad [2\text{-}24],$$

where N_A = Avogadro's number, and kT = Boltzmann's constant × absolute temperature.

B.   The electrostatic contribution to chemical potential (u_i) of an ion (from Güntelberg) is:

$$\Delta u_i = \int_{0}^{z_i e} \Psi(e_i)\, de_i \qquad [2\text{-}25],$$

where e_i = instantaneous charge. Assuming a linear superposition of ionic atmospheres, where

$$(\text{constant})\, \Psi = e_i = z_i e \qquad [2\text{-}26],$$

Equations [2-23], [2-25], and [2-26] are combined and solved:

$$\Delta u_i = \int_{0}^{\Psi} (\text{constant})\Psi\, d\Psi = \frac{(\text{constant})\Psi^2}{2} = \frac{z_i e \Psi}{2} = \frac{-(ze)^2 \kappa}{2\epsilon} \qquad [2\text{-}27].$$

C.   The partial molal free energy ($\overline{G}$) of an electrolyte is given by:

$$\overline{G} = \overline{G}_N^o + vRT \ln(f N_i) \qquad [2\text{-}28],$$

where $\overline{G}_N^o = \overline{G}$ in the standard state, v = number of cations + anions, N_i = mean ionic mole fraction, and $f = a/N_i$ = "rational" activity coefficient or the stoichiometric mean ionic mole fractional activity coefficient. Note that the mean activity coefficient (f) can be measured in any of several ways, but is impossible to experimentally resolve this term into the individual activity coefficients. That is

$$f = (f_A^m f_B^n)^{1/(m+n)}$$ [2-29]

for the electrolyte $A_m B_n$. If $A_m B_n$ is a precipitate, then

$$K_{sp} = [A]^m[B]^n f_A^m f_B^n = [A]^m[B]^n f^{(m+n)}$$ [2-30].

We may split Equation [2-28] into two fractions:

$$\overline{G} = \overline{G}_N + \overline{G}_f$$ [2-31],

where

$$\overline{G}_N = \overline{G}_N^o + \nu RT \times \ln N_i$$ [2-32]

and

$$\overline{G}_f = \nu RT \times \ln f$$ [2-33].

$\overline{G}_N$ is the limiting law for dilute solutions of unionized solutes, and $\overline{G}_f$ is for ionized solutes. $\overline{G}_N$ would also apply to electrolytes if they were not charged. Thus, for completely ionized electrolytes in dilute solutions, the deviation of electrolyte solutions from ideality can be attributed entirely to electrostatic forces between the ions:

$$\overline{G}_f = N_A \Delta u_i$$ [2-34].

Combining Equations [2-27], [2-33] and [2-34] gives:

$$\Delta u_i = \frac{\overline{G}_f}{N_A} = \frac{\nu RT}{N_A} \ln f = kT \ln f_i = \frac{-(ze)^2 \kappa}{2\epsilon}$$ [2-35],

where

$$\ln f = \frac{1}{\nu} \sum_1^s \nu_i \times \ln f_i$$ [2-36].

Solving for $\ln f_i$ gives the limiting expression for the mean activity coefficient of an ion (of any kind) in a solution containing s kinds of ions at concentrations $n_1, n_2, n_3, ..., n_s$ per cm^3. Combining Equation [2-35] with [2-24]:

$$\ln f_i = -z_i^2 \sqrt{\frac{\pi e^6 N_A \Gamma}{1000(\epsilon kT)^3}}$$ [2-37].

Substituting Equation [2-37] into [2-36], the mean activity of an electrolyte dissociating into p kinds of ions is:

$$\ln f = \frac{-1}{\nu} \sum_1^p \nu_i z_i^2 \sqrt{\frac{\pi e^6 N_A \Gamma}{1000(\epsilon kT)^3}}$$ [2-38].

This is one form of the Debye–Hückel "limiting law" for mean activity coefficients.[18] Equation [2-38] may also be stated as:

$$\ln f = -S \sqrt{\Gamma}$$ [2-39],

where

$$S = \frac{1}{\nu} \sum_1^p \nu_i z_i^2 \sqrt{\frac{\pi e^6 N_A}{1000(\epsilon kT)^3}}$$ [2-40].

Note that the summation in S uses unambiguous terms and is confined to the ions $1, ..., p$ resulting from the dissociation of the electrolyte to which f refers. Note also that Γ is a function of all the ions 1, ..., s in the solution regardless of the source (see Equation [2-20]).

D.   Using the following definitions:

γ = activity coefficient when the concentration (m) is expressed in molal units; γ is also called the "practical" activity coefficient, and

M_w = molecular weight of the solvent,

then

$$a = \gamma m \qquad [2\text{-}41].$$

Thus, for dilute solutions:

$$\ln f = \ln \gamma + \ln\left(1 + \frac{mvM_w}{1000}\right) \qquad [2\text{-}42].$$

The right-hand term disappears in Equation [2-42] for dilute solutions. Solving for log γ, for $c \ll 1$ and using Equation [2-22]:

$$\log \gamma = \log f = \frac{-S\sqrt{\Gamma}}{2.303} = \frac{-S\sqrt{2\rho I}}{2.303} \qquad [2\text{-}43],$$

or

$$\log \gamma = -A z^2 \sqrt{I} \qquad [2\text{-}44],$$

where, for a given ion, $v = 1$ and $p = 1$ in Equation [2-40], and

$$A = \frac{S\sqrt{2\rho}}{2.303\, z^2} = \frac{1}{2.303}\sqrt{\frac{2\pi e^6 N_A \rho}{1000(\epsilon k T)^3}} \qquad [2\text{-}45].$$

Equations [2-44] and [2-15] are essentially the same. For dilute solutions, $\rho = 1$, $m_i = c_i$, and the numerical value of I is the same using either concentration units (Equation [2-19]) or molal units (Equation [2-21]). Note that f is sometimes called the "activity coefficient" rather than the "mean activity coefficient". This is because in dilute solutions γ and f are numerically the same.

For pure water:

$$A = 1.824 \times 10^6 (\epsilon T)^{-3/2} \cong 0.5 \text{ at } 25°\text{C} \qquad [2\text{-}46]$$

when the constants in Equation [2-45] are defined as: $\rho = 1$ kg L^{-1}, $e = 4.80223 \times 10^{-10}$ esu, $k = 1.38054 \times 10^{-16}$ erg °C^{-1} molecule^{-1}, $N_A = 6.02252 \times 10^{23}$ mol^{-1}, $\epsilon = 80.37$ at 20°C, and 78.54 at 25°C. The electrostatic unit (esu) refers to the c.g.s. system, where 1 esu of electric charge = 1 statcoulomb = 3.335641×10^{-10} C. The coulomb (C) is an SI unit, and $e = 1.6022 \times 10^{-19}$ C. All electrostatic units have the prefix "stat-" attached to their names.

[2.7.2] Deriving the Extended Debye–Hückel Equation: $\log f = \dfrac{-Az^2\sqrt{I}}{1 + Ba\sqrt{I}}$

A.   A more exact equation for the electrical potential (Ψ) of an electrolyte solution is obtained as follows (compare with Equation [2-23]):

$$\nabla \cdot \nabla \Psi(r) = \kappa^2 \Psi(r) \qquad [2\text{-}47],$$

and solving yields

$$\Psi(r) = \frac{A'e^{-\kappa r}}{r} - \frac{A''e^{\kappa r}}{r} \qquad [2\text{-}48],$$

where r = distance of ionic atmosphere from the ion. In solving, note that at $r = \infty$, $\Psi(\infty) = 0$; therefore, $A'' = 0$ and Equation [2-48] becomes:

$$\Psi(r) = \frac{A'e^{-\kappa r}}{r} = \frac{z_i e}{\epsilon r} + \Psi^*(r) \qquad [2\text{-}49],$$

where $(z_i e/\epsilon r)$ = potential at r due to point charge $z_i e$, and $\Psi^*(r)$ = potential of the ionic atmosphere at r. Rearranging, one obtains:

$$\Psi^*(r) = \frac{A'e^{-\kappa r}}{r} - \frac{z_i e}{\epsilon r} \qquad [2\text{-}50].$$

Now let a = ionic parameter = minimum average distance to which ions (both positive and negative) can approach each other. In order that the field of the ion and its atmosphere $[\partial\Psi(r)/\partial r]$ be continuous when $r = a$, then $[\partial\Psi(r)/\partial r] = [-(z_i e)/(\epsilon r^2)]$ = the field of the ion alone. Thus,

$$\frac{\partial\Psi^*(r)}{\partial r} = 0 = \frac{-A'e^{-\kappa r}(1 + \kappa a)}{a^2} + \frac{z_i e}{\epsilon\, a^2} \qquad [2\text{-}51],$$

solving for A' :

$$A' = \frac{z_i e}{\epsilon}\, \frac{e^{\kappa a}}{1 + \kappa a} \qquad [2\text{-}52],$$

and combining Equations [2-50] and [2-52] gives:

$$\Psi^*(a) = \frac{-z_i e}{\epsilon}\, \frac{\kappa}{1 + \kappa a} \qquad [2\text{-}53].$$

B.   Again assume the potential is proportional to the charge and employ the Güntelberg charging process (from Equations [2-25] and [2-27]):

$$\Delta u_i = \int_0^{z_i e} \Psi(e_i)\, de_i \qquad [2\text{-}54],$$

$$\Delta u_i = \int_0^{\Psi^*} (\text{constant})\Psi^*\, d\Psi^* = \frac{(\text{constant})\Psi^{*2}}{2} = \frac{z_i e\Psi^*}{2} \qquad [2\text{-}55].$$

Combining Equation [2-55] with Equations [2-53] and [2-35] yields:

$$\Delta u_i = \frac{-(z_i e)^2}{2\epsilon}\, \frac{\kappa}{1 + \kappa a} = kT \times \ln f_i \qquad [2\text{-}56].$$

Solving for f_i and f (see Equation [2-36]) for p kinds of ions:

$$\ln f = \frac{1}{\nu}\sum_1^p \nu_i \ln f_i = \frac{-1}{\nu}\sum_1^p \frac{\nu_i z_i^2 e^2}{2\epsilon kT}\, \frac{\kappa}{1 + \kappa a} \qquad [2\text{-}57],$$

substituting for κ (Equation [2-24]), Γ for $2\rho I$ (Equation [2-22]) and log γ (Equation [2-42]) for dilute solutions:

$$\log \gamma = \log f = \frac{-1}{2.303\ v} \sum_{1}^{P} v_i z_i^2 \ \frac{\sqrt{\dfrac{2\pi e^6 N_A \rho I}{1000(\epsilon kT)^3}}}{1 + a\sqrt{\dfrac{8\pi e^2 N_A \rho I}{1000\ \epsilon kT}}} \qquad [2\text{-}58],$$

or

$$\log \gamma = \frac{-Az^2\sqrt{I}}{1 + Ba\sqrt{I}} \qquad [2\text{-}59],$$

where A = same as Equation [2-45], and

$$B = \sqrt{\frac{8\pi e^2 N_A \rho}{1000\ \epsilon kT}} \qquad [2\text{-}60];$$

$B = 50.28(\epsilon T)^{-\frac{1}{2}}$ for a given in angstrom units; $B \cong 0.329$ for pure water at 25°C. For similar reasons as those given following Equation [2-44], Equations [2-59] and [2-16] are essentially the same for dilute solutions. Values for the ionic parameter a for 130 ions were determined by Jacob Kielland (1937)[19] and his results are shown in Table 2-4. Note that most of the a values are larger than the hydrated radii of the ions, although some are nearly the same. Hence, the a parameter, which was defined as the "minimum average distance of approach", does not equal the radii or diameters of the ions, but is closely associated with them.

[2.7.3] The Güntelberg Approximation: $\log f = \dfrac{-Az^2\sqrt{I}}{1 + \sqrt{I}}$

For estimating unknown activity coefficients, Güntelberg proposed to let $a = 3.0$ in Equation [2-59]. Thus,

$$Ba = 0.986 \approx 1.0 \qquad [2\text{-}61]$$

and substituting $Ba = 1.0$ into Equation [2-59] for dilute solutions (where $\log \gamma = \log f$):

$$\log \gamma = \frac{-Az^2\sqrt{I}}{1 + \sqrt{I}} \qquad [2\text{-}62].$$

The values obtained for γ using Equation [2-62] are usually too small. It would be better to choose from Table 2-4 a value of a for an ion of similar charge and chemical nature.

[2.7.4] Empirical Expressions: $\log f = \dfrac{-Az^2\sqrt{I}}{1 + \sqrt{I}} + 0.2\,I$

Guggenhein suggested that the parameter a be fixed at 3.0 and that the data be fitted by adding a linear term. For dilute solutions, where $\log \gamma = \log f$, we obtain a new one-parameter equation:

$$\log \gamma = \frac{-Az^2\sqrt{I}}{1 + \sqrt{I}} + \dot{B}I \qquad [2\text{-}63].$$

Equation [2-63] is sometimes referred to as the "B-dot" equation, and the extended

Table 2-4: Values of the parameter *a* in the Extended Debye–Hückel equation for various ions.
Adapted from Kielland (1937)[19], copyright © 1937, with permission from the American Chemical Society.

a, Å	Ions
	$\mid z \mid = 1$
9	H^+
8	$(C_6H_5)_2CHCOO^-$, $(C_3H_7)_4N^+$
7	$[OC_6H_2(NO_3)_3]^-$, $(C_3H_7)_3NH^+$, $CH_3OC_6H_4COO^-$
6	Li^+, $C_6H_5COO^-$, $C_6H_4OHCOO^-$, $C_6H_4ClCOO^-$, $C_6H_5CH_2COO^-$, $CH_2CHCH_2COO^-$, $(CH_3)_2CCHCOO^-$, $(C_2H_5)_4N^+$, $(C_3H_7)_2NH_2^+$
5	$CHCl_2COO^-$, CCl_3COO^-, $(C_2H_5)_3NH^+$, $(C_3H_7)NH_3^+$
4.5	CH_3COO^-, CH_2ClCOO^-, $(CH_3)_4N^+$, $(C_2H_5)_2NH_2^+$, $NH_2CH_2COO^-$
4–4.5	Na^+, $CdCl^+$, ClO_2^-, IO_3^-, HCO_3^-, $H_2PO_4^-$, HSO_3^-, $H_2AsO_4^-$, $[Co(NH_3)_4(NO_2)_2]^+$
4	$(COOH)CH_2NH_3^+$, $(CH_3)_3NH^+$, $C_2H_5NH_3^+$
3.5	OH^-, F^-, NCS^-, NCO^-, HS^-, ClO_3^-, ClO_4^-, BrO_3^-, IO_4^-, MnO_4^-, $HCOO^-$, $H_2(citrate)^-$, $CH_3NH_3^+$, $(CH_3)_2NH_2^+$
3	K^+, Cl^-, Br^-, I^-, CN^-, NO_2^-, NO_3^-
2.5	Rb^+, Cs^+, NH_4^+, Tl^+, Ag^+
	$\mid z \mid = 2$
8	Mg^{2+}, Be^{2+}
7	$[OOC(CH_2)_5COO]^{2-}$, $[OOC(CH_2)_6COO]^{2-}$, (Congo Red anion)$^{2-}$
6	Ca^{2+}, Cu^{2+}, Zn^{2+}, Sn^{2+}, Mn^{2+}, Fe^{2+}, Ni^{2+}, Co^{2+}, $C_6H_4(COO)_2^{2-}$, $CH_2(CH_2COO)_2^{2-}$, $(CH_2CH_2COO)_2^{2-}$
5	Sr^{2+}, Ba^{2+}, Ra^{2+}, Cd^{2+}, Hg^{2+}, S^{2-}, $S_2O_4^{2-}$, WO_4^{2-}, $CH_2(COO)_2^{2-}$, $(CH_2COO)_2^{2-}$, $(CHOHCOO)_2^{2-}$
4.5	Pb^{2+}, CO_3^{2-}, SO_3^{2-}, MoO_4^{2-}, $[Co(NH_3)_5Cl]^{2+}$, $[Fe(CN)_5NO]^{2-}$, $(COO)_2^{2-}$, $H(citrate)^{2-}$
4	Hg_2^{2+}, SO_4^{2-}, $S_2O_3^{2-}$, $S_2O_6^{2-}$, $S_2O_8^{2-}$, SeO_4^{2-}, CrO_4^{2-}, HPO_4^{2-}
	$\mid z \mid = 3$
9	Al^{3+}, Fe^{3+}, Cr^{3+}, Sc^{3+}, Y^{3+}, La^{3+}, In^{3+}, Ce^{3+}, Pr^{3+}, Nd^{3+}, Sm^{3+}
6	$[Co(ethylenediamine)_3]^{3+}$
5	citrate^{3-}
4	PO_4^{3-}, $[Fe(CN)_6]^{3-}$, $[Cr(NH_3)_6]^{3+}$, $[Co(NH_3)_6]^{3+}$, $[Co(NH_3)_5H_2O]^{3+}$
	$\mid z \mid = 4$
11	Th^{4+}, Zn^{4+}, Ce^{4+}, Sn^{4+}
6	$[Co(S_2O_3)(CN)_5]^{4-}$
5	$[Fe(CN)_6]^{4-}$
	$\mid z \mid = 5$
9	$[Co(SO_3)_2(CN)_4]^{5-}$

Debye–Hückel equation (Equation [2-59]) often replaces the Güntelberg approximation term shown above. Data can obviously be more accurately represented when more adjustable parameters are used, such as:

$$\log \gamma \; = \; \frac{-Az^2\sqrt{I}}{1 + \sqrt{I}} \; + \; b_1 I \; + \; b_2 I^2 \; + \; b_3 I^3 \; + \; \cdots \qquad [2\text{-}64].$$

However, Equation [2-64] and others of this type ignore any attempt to correlate physical interpretations with observed data. That is to say, *we may gain precision on a value, but we also lose understanding of that value.*

Davis proposed that $\dot{B}$ in Equation [2-63] be set equal to 0.2, such that

$$\log \gamma \; = \; \frac{-Az^2\sqrt{I}}{1 + \sqrt{I}} \; + \; 0.2\,I \qquad [2\text{-}65].$$

Figure 2-9 shows other values for $\dot{B}$ in Equation [2-63]. Note that $\dot{B}$ consistently shows a large variation at low molality yet becomes nearly constant at high molality. Keep in mind that $\dot{B}$ has not yet been given any physical significance. However, we know that even the extended Debye–Hückel equation (which has a physical interpretation for each parameter) does not correctly describe the activity of a species in solution at high ionic strength either. Therefore, equations such as that proposed by Davis (Equation [2-65]) are used often.

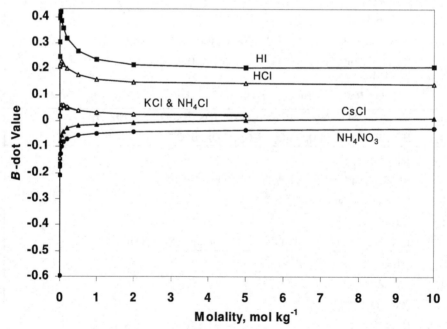

Figure 2-9: The apparent ion interaction coefficient $\dot{B}$ for Equation [2-63] for various 1–1 monovalent electrolytes, based on the mean activity coefficient as a function of molality data. *Data from Lide (2002).*[2]

[2.7.5] Specific Ion-Interaction Model: The Pitzer Equation

The Debye–Hückel equation and its subsequent modifications (Equations [2-15] to [2-17]) are based on several assumptions. These ion-association models assume that the ion is a hard charged sphere, but many ions are not really spherical and not all electrolytes are completely dissociated into charged cations and anions (that is, not all electrolytes are strong electrolytes). They also assume that the ion feels the electrical field from all other ions, and that the solvent's dielectric constant is constant. However, the electric field should not be limited to only coulombic forces (such as Equation [2-23] or [2-47]), but should also consider short-range potential forces induced by each of the neighboring ions. Local variations in the dielectric constant of the solvent (such as water) are also likely as a result of the clustering of the solvent molecules. Because of these limitations in the theory, the ion-association models do not predict activity coefficients in solutions with high ionic strength.

The most widely used model for high ionic strength conditions is the ion-interaction model developed by Pitzer and co-workers in the 1970s.[20] This model is generally restricted to maximum concentrations of about 6 M [21], but applications of up to 20 M across a temperature range of 0 to 250°C have been reported.[22]

Based on a virial expansion of Gibb's excess free energy, Pitzer developed a virial expansion of the Debye–Hückel equation:

$$\ln \gamma_i = \ln \gamma_{DH_i} + \sum_j B_{ij} X_j + \sum_j \sum_k C_{ijk} X_j X_k + \cdots \qquad [2\text{-}66],$$

where γ_i = activity coefficient of ion i; γ_{DH_i} = extended Debye–Hückel activity coefficent of ion i (Equation [2-16]); $X_{j,k}$ = molar concentration of ionic species j and k; B_{ij} = coefficient for specific interaction among ion pairs as a function of ionic strength; and C_{ijk} = coefficient for specific interaction among three ions (value is independent of ionic strength). More virial coefficients can be added for the specific interaction among larger groups of ions as needed.

Note that the virial coefficients, or interaction coefficients, and their dependence on ionic strength and charge can probably be derived from theory because these coefficients are related to potentials of mean force in the solvent. Nevertheless, they are ultimately evaluated empirically.[20] The Pitzer interaction coefficients are fitted to activity measurements on various solutions with each containing only the ions i and j (or i, j, and k). The multiple interactions by the various ions (j and k) with a specific ion (i) are assumed to be additive. Due to the importance of inorganic C to soil environments, much attention has been given to these ions in the literature (such as He and Morse, 1993[23]). Articles such as these should be reviewed for a clearer understanding of the application of the Pitzer equation. The details of the Pitzer equation are not included in this section due to length constraints. An excellent review on this topic, however, is presented in a book edited by Pitzer (1991).[24]

One problem with the Pitzer equation is that there are only a few ion-interaction values available in the literature. If the necessary ion-interaction parameters are lacking, then you may use the corresponding values of a chemically analogous species. For example, you may use the interaction parameters of Nd^{3+} for Pu^{3+} since the values for Pu are not available due to the extreme difficulty in maintaining Pu in the Pu(III) oxidation state.[22]

[2.7.6] Activity Coefficients of Neutral Species

The activity coefficient (γ) of a neutral species is obtained empirically. Beginning with either Equation [2-62] or [2-63], since $z = 0$ for neutral species, they reduce to

$$\log \gamma = \dot{B}I \qquad \qquad [2\text{-}67],$$

or

$$\log \gamma = b_1 I + b_2 I^2 + b_3 I^3 + b_4 I^4 \qquad \qquad [2\text{-}68],$$

where I = ionic strength (Equation [2-19]), and the $\dot{B}$ and b_i coefficients are empirical adjustable parameters. A more exact and theoretically sound determination can be obtained with the Pitzer equation (Equation [2-66]) with $\ln \gamma_{DH} = 0$ since $z = 0$ for neutral species, or with virial expressions of Equation [2-67] based on neutral species-interaction models.[25]

Equation [2-67] is commonly mentioned in theoretical discussions. The $\dot{B}$ fitting parameter is called the salting coefficient, where a positive value indicates a *salting out* effect (solubility decreases as ionic strength increases) and a negative value indicates a *salting in* effect (solubility increases as ionic strength increases, which is a rare occurrence).

In practice sometimes, the activity of a neutral species is instead set equal to the activity coefficient of CO_2(aq) as measured in pure NaCl solutions of the same ionic strength. The assumption is that the activity coefficient of the neutral species will be similar to that of neutral CO_2(aq) at the same concentrations. For this approach, Equation [2-68] can be used to fit the data and to estimate the activity coefficient of the neutral species based on the corresponding b_i values for CO_2(aq). The Pitzer equation is also recommended for this approach.

[2.8]   Range of Ionic Strength (I) Values

Very tiny liquid stratospheric aerosols (diameter < 1 μm), which hover 20 to 50 km above the earth, are mostly water droplets concentrated in sulfuric acid.[21] Other gases, such as HCl, HBr, HOCl and HNO_3, will also dissolve in these aerosols. The resultant ionic strength of these acidic droplets is very high (7 to 15 M).[21]

Closer to ground level, the ionic strengths of droplets are generally very low, but much variation has been observed (Table 2-5). Typical values for rainwater I range from almost 0 to 2 mM. The electrolyte concentration in an average downpour may be much more dilute than in the raindrops of a very light sprinkling, where the droplets are nearly completely evaporated back to the atmosphere before reaching the ground. It may also decrease over time for a given rainstorm event, particularly in areas with much haze or smog. Nevertheless, the ionic strength of rainwater is often low enough for the activity coefficient of its dissolved electrolytes to be well described by the ion-association models.

Some variations in ionic strength in raindrops (diameter 0.1 to 3 mm), or droplets in clouds or fog (diameter 1 to 100 μm), also occur as a result of air pollutants. Note that when a fog first forms, the concentrations of its chemical components are high. As the fog develops, the droplets grow in size and the concentration of its chemical components drop due to dilution. Finally, as the air is heated, evaporation of water decreases the size of the

Table 2-5: Properties of atmospheric aqueous aerosols. *From Martin (1984)[26], with permission.*

Particle or Drop	Typical Radius, μm	Volume Fraction, cm^3 water/m^3 air	pH	Ionic Strength, molality units, mol kg^{-1}
Haze	0.03 – 0.3	$10^{-5} – 10^{-4}$	1 – 8	≈ 1
Clouds	10	0.1 – 3	3 – 6	$10^{-3} – 10^{-2}$
Fog	10	0.02 – 0.2	2 – 6	$10^{-3} – 10^{-2}$
Rain	200 – 2000	0.1 – 1	4 – 5	10^{-4}

droplets and the concentrations of its chemical components return to high values.[27] The haze over urban environments generally have a high ionic strength (typically $\approx 1\ M$, see Table 2-5), but they can potentially also reach much higher ionic strength values (up to 23.7 M if droplets are saturated with NH_4NO_3; note, $K_{so}^{20C} = 23.7\ M$ for NH_4NO_3).[21] Accordingly, we should not assume that the ionic strength of all rainfall (or dew) over a particular soil to be nearly the same.

Ocean spray on coastal soils, however, will have a higher ionic strength than most other sources of precipitation (up to 6 M, or saturation of NaCl).[21] The ionic strength of sea water is 0.7 M. The higher I values of ocean spray presumably are due to the partial evaporation of the droplets and the resultant enhanced concentration of the electrolytes dissolved in them.

Conversely, the ionic strength of soil solutions varies dramatically. In moist climates, the ionic strength of the soil solution is generally very low, and for these environments the ion-association models are more than adequate for estimating ion activity coefficients. In saline soils, in drier soils, or in partially frozen soils, the ionic strength of the soil water can be very high. For these reasons, equations that estimate activity coefficients under higher ionic strength conditions are often needed.

[2.9]   Electrical Conductivity (EC)

Our present understanding of electrolyte dissociation, such as the dissociation of NaCl in water into Na^+ and Cl^- ions, is attributed to S.A. Arrhenius, who received the Nobel Prize in Chemistry in 1903 for his theory of electrolytic dissociation. In 1887 he postulated that the dissolution of an electrolyte in an aqueous solution could lead to electrolytic dissociation and the conversion of an appreciable fraction of the electrolyte to free ions. He observed the variation of conductance with concentration, and attributed it to the partial dissociation of the electrolyte, which can be described for dilute solutions as a linear correlation:

$$EC = c\ EC_{equiv} \qquad [2\text{-}69],$$

where EC = electrical conductivity of the solution (S m^{-1}), c = concentration (equivalents L^{-1}, or equivalents m^{-3}), and EC_{equiv} = equivalent conductivity (sometimes expressed as Λ).

Equivalent conductance (EC_{equiv}) is the area (m^2) needed to attain one equivalent concentration of electrolyte between electrodes of a unit distance apart (1.0 m). Expressed differently, the EC_{equiv} value is a measure of the current-carrying ability of an equivalent of solute. Its units are $m^2\ ohm^{-1}\ equiv^{-1}$ or $S\ m^2\ equiv^{-1}$ (the siemens unit, S, replaces the mho unit). Often, resistance (R) is measured instead of conductivity, where

$$EC\ =\ \kappa/R \qquad [2\text{-}70].$$

The cell constant (κ) equals the electrode separation distance divided by the electrode area ($\kappa = d/A$), which numerically adjusts for the fact that most electrodes do not have a 1.0 m^2 surface area that is 1 m apart. Typical cells used for dilute solutions have κ values of 0.1 cm^{-1}, but cells with κ values of around 0.4 to 1 cm^{-1} are common for environmental water analysis. Resistivity can be measured instead of conductivity. The unit for electrical resistance (R) is ohm, but the units are ohm m for electrical resistivity (ρ), where $\rho = R/\kappa$ and $EC = 1/\rho$.

Strong acids and bases will dissociate completely in water. Weak acids and bases will not dissociate completely. The degree of dissociation (α) is equal to the equivalent conductance (measured EC over known concentration) divided by the limiting equivalent conductivity (EC_o):

$$\alpha\ =\ EC_{equiv}\ /\ EC_o \qquad [2\text{-}71].$$

Since the degree of dissociation decreases with concentration due to ionic strength effects, we obtain the limiting equivalent conductivity at infinite dilution, that is $EC_{equiv} \rightarrow EC_o$ as $c \rightarrow 0$. The limiting equivalent conductivity (EC_o) is specific to each electrolyte. From Kohlrausch's law of independent migration of ions, the conductance of an electrolyte at infinite dilution depends on the independent migration of the cations and anions:

$$EC_{equiv}\ =\ \lambda_+ + \lambda_- \qquad [2\text{-}72],$$

where $\lambda_\pm$ are the equivalent ionic conductances of the ions (Table 2-6).

For pure water, $EC_{pure\ H_2O} = 0$ if no electrolytes are present and no dissociation occurs. However, we know from Equation [2-2] that H^+ and OH^- ions are always present in water. Accordingly, the true electrical conductivity of pure water at pH 7 is calculated as follows:

$$EC_{H_2O,\ pH\ 7}\ =\ c\ EC_{equiv}\ =\ (c_{H^+})\lambda_{H^+}\ +\ (c_{OH^-})\lambda_{OH^-}$$

$$=\ 10^{-7}\ \frac{mol}{L}\ \times \left[(349.65 + 198) \times 10^{-4}\ \frac{S\ m^2}{mol}\right] \times 10^3\ \frac{L}{m^3}\ \times 10^3\ \frac{mS}{S}$$

$$=\ 0.0054\ mS\ m^{-1},\ or\ 0.054\ \mu S\ cm^{-1},\ or\ 5.4 \times 10^{-5}\ dS\ m^{-1} \qquad [2\text{-}73].$$

Taking the inverse, the resistivity of ultra pure water at pH 7 is 18.5 MΩ cm. In practice, distilled water will have an EC range of 0.5 to 3 $\mu S\ cm^{-1}$ due to the presence of some entrained salts and absorption of atmospheric CO_2 (which increases as the hot distilled water cools, or as it ages in storage). Fresh distilled, deionized water is very close to ultra pure water.

The electrical conductivity (EC in dS m^{-1}) of soil solutions will generally correlate strongly ($r^2 > 0.98$) with their ionic strength (I in mol L^{-1})[28]:

$$I\ \cong\ m(EC) + b \qquad [2\text{-}74],$$

Table 2-6: Equivalent ionic conductivities of selective ions at infinite dilution and 25°C. *Data from Lide (2002).* [2]

Ion	λ $10^{-4}\ m^2\ S\ mol^{-1}$	Ion	λ $10^{-4}\ m^2\ S\ mol^{-1}$
H^+	349.65	OH^-	198.0
Li^+	38.66	Cl^-	76.31
Na^+	50.08	ClO_4^-	67.3
K^+	73.48	NO_3^-	71.42
NH_4^+	73.5	HCO_3^-	44.5
$1/2\ Mg^{2+}$	53.0	$1/2\ CO_3^{2-}$	69.3
$1/2\ Ca^{2+}$	59.47	$1/2\ SO_4^{2-}$	80.0
$1/2\ Fe^{2+}$	54.0	$1/3\ PO_4^{3-}$	69.0
$1/3\ Fe^{3+}$	68.0	Acetate$^-$	40.9

where the coefficient m ranges from 0.011 to 0.016, and b ranges from 0.0 to -0.002. From Equation [2-19] we see that this correlation estimates the *total dissolved salt* (TDS) concentration in $mol_c\ L^{-1}$ and, due to the dissociation of the dissolved salts, it also estimates the total concentration of cations or anions in $mol_c\ L^{-1}$. Arbitrary values of $m = 10$ and $b = 0$ are commonly used to determine the total concentration of cations or anions in the soil solution based on the measured EC value (in $dS\ m^{-1}$):

$$\text{(concentration of cation or anion in mmol}_c\ L^{-1}) \approx 10 \times EC \qquad [2\text{-}75].$$

Another common practice is to assume an average molecular weight of 64 g mol^{-1} for the salts dissolved in the soil solution, which yields from Equation [2-75]:

$$\text{TDS (in mg L}^{-1}) \approx 640 \times EC \text{ (in dS m}^{-1}) \qquad [2\text{-}76].$$

A more precise method for determining TDS is to carefully weigh the salt residue after evaporating an aliquot of the soil solution. This approach, however, is not as practical as a simple EC measurement in the field. This is because the evaporation technique for determining TDS may still have minor errors due to some salts or compounds being carried off in the vapors (such as inorganic C in the form of $CO_2(g)$ or volatile low-molecular-weight organic compounds), plus the errors from inadvertently weighing some hygroscopic water in the sample (such as on hydroscopic salts).

The variance in the above equations ([2-74] to [2-76]) arises from the diversity of the ions commonly found in the soil solution. As was noted in Table 2-6, different ions have different equivalent ionic conductivities; accordingly, differences in the ion mixtures present in soil solutions will display different EC values even if their I values were the same. The ions present in soil solutions come from various sources, such as dissolution of soil minerals, atmospheric deposition, rain, irrigation, reactions of aqueous metals with seasonally varying bicarbonate ions (HCO_3^-), cation exchange capacity of the soil minerals, buffering processes of the soil mixture, and plant root exudates to name but a few.

Caution needs to be taken when using a soil solution extract, or where water is added to push the soil water through the extraction column. This method causes a dilution effect by the added water on the ion concentrations and EC measurements, particularly if the contact time is short or if the cation exchange capacity (CEC) is low. Also note that the EC values are temperature dependent and need to be measured at 25°C whenever possible. For less accurate needs, the measured electrical conductivity can be roughly corrected to 25°C using

$$EC_{25\,C} \cong EC_T \times [1 + 0.019\,(25 - T)] \qquad [2\text{-}77],$$

where $EC_{25\,C}$ = EC at 25°C, EC_T = EC measured at temperature T in °C.[29] That is, EC increases by about 1.9% per degree centigrade.

There are four commonly measured EC values:

1. EC_w = EC of the undiluted/untreated natural soil solution or water sample (Collect the soil solution with suction if necessary.)

2. EC_e = EC of the soil solution extracted from a saturated soil paste (The soil paste has only enough distilled water added to saturate the soil. After equilibrating the paste for several hours, extract the liquid portion with suction through filter paper. Liquid extracted this way is also sometimes called "saturation extract".)

3. EC_p = EC of the saturated soil paste itself

4. EC_b = EC of the bulk field soil

Most of the EC of bulk soil samples is dependent on texture rather than soil moisture. The conductivity travels through the liquid or solids or coupled liquid–solids electrical pathways. In general, EC of clays > EC of silts > EC of sands. Efforts to correlate EC_b or EC_p values with soil texture, cation exchange capacity, and other soil physical or chemical properties are regularly attempted.

Clearly, since the presence of solid components affects the measured EC values, it is important that solids be absent from the liquid extracts used for EC_w or EC_e determinations. High electrical conductivity in the soil solution or irrigation water is indicative of a high dissolved salt content. High EC values may suggest toxic or hazardous levels of some soluble ions present.

Saline soils are soils that have sufficiently high levels of salts present to adversely reduce plant growth. Salt tolerance varies significantly among plants. For example, the threshold salt tolerance of various grasses, forage crops, and some grain crops (e.g., corn, peanut, rice) is low, with EC_e values around 1.5 to 3.9 dS m^{-1}.[30] The threshold salt tolerance is higher for some grain and special crops (e.g., barley, cotton, sorghum, soybean, wheat), with EC_e values around 5.0 to 8.0 dS m^{-1}, and much higher for rye with tolerance to EC_e values of about 11.4 dS m^{-1}.[30] In general, a soil is considered saline if EC_e > 4 dS m^{-1} at 25°C[31], moderately saline if EC_e > 8 dS m^{-1}, and strongly saline if EC_e > 16 dS m^{-1}.

Saline soils have high EC_e values: up to 35 dS m^{-1}, and sometimes almost 50 dS m^{-1}.[28] For comparison, EC of irrigation supply water is typically <0.750 dS m^{-1}, potable water in the U.S. 0.3 to 1.5 dS m^{-1}, melted snow 0.02 to 0.042 dS m^{-1}, rain water 0.02 to 0.05 dS m^{-1}, and sea water 50 to 60 dS m^{-1} (sea water salinity = 33 to 37 g L^{-1}).

A high EC_e value may be due to a high sodium (Na$^+$) content. A Na$^+$ hazard occurs when there is a high Na$^+$ content (sodicity) in the soil. This causes (1) the soil's clay fraction to swell and disperse easily, particularly if 2:1 clays are present, (2) the soil's hydraulic conductivity or water permeability to decrease due to plugging of the soil pore channels, and (3) the soil's surface to crust badly when dry. High sodium and chloride concentrations can also be toxic to some plants (such as fruit and berry crops, and to other woody plants). Since sodic soils are problem soils specifically related to the Na^+ content, sodic soils are not defined based on the EC_e values (which could be high), but rather on the *sodium adsorption ratio* (SAR), which is defined as

$$SAR = \frac{[Na^+]}{\sqrt{[Ca^{2+}] + [Mg^{2+}]}}$$ [2-78].

A sodic soil has SAR > 13 for the saturation extract (that is, for the liquid portion of the soil paste sample).[31]

High EC_e measurements resulting from high bicarbonate concentrations may result in high pH conditions as well. If this occurs, then there may be precipitation and subsequent deficiency of various essential nutrients in the soil environment (such as iron or other micronutrients).

High EC_w measurements (including EC measurements of the irrigation water or other water sources) may be due to the presence of hard water. *Hard water* is water with high concentrations of minerals, particularly Ca^{2+} and Mg^{2+}. Hard water will have greater than 150 mg L^{-1} (or ppm) of Ca^{2+} and Mg^{2+}, and very hard water over 300 mg L^{-1}. Conversely, *soft water* has low concentrations of these cations (<75 mg L^{-1}), and the concentration of Na^+ ions is often also low. Hard water is a problem in washing because it reduces the performance of soap or detergent compared to soft water, and it forms a scum when it is combined with soaps (which in turn requires more soap to generate a lather and more water to wash away the scum). Hard water also causes the formation of $CaCO_3(s)$ scale in pipes and heating systems.

[2.10] Nature of Hydrated Ions

When an ion (or neutral molecule) is in true solution, there are various interactions between the ion and the water molecules. The electrostatic fields of the ions cause the surrounding water molecules to polarize and bind with the ion. This field may also influence more distant water molecules. The result is that the dissolved ions are hydrated, forming a sphere of water molecules (typically six or more molecules) with the ion at its center:

$$M^{z+}(g \text{ or } s) + nH_2O \rightarrow M(H_2O)_n^{z+}(aq)$$ [2-79].

The effective radius of the dissolved ion is known as its hydrated radius, and is always much larger than the ionic radius of the ion as it would exist in the interior of crystal structures. Kavanau (1964)[32] described the water regions around an ion as three concentric water regions: (1) an innermost region of structured, polarized, electrostructural water molecules; (2) an intermediate region where the water is less structured than ordinary water; and (3) an outer region of ordinary water. Note that some ions, especially large monovalent ions, do not completely form Region 1. According to Samoilov (1957)[33], the close water molecules, or nearest neighbor molecules, influence the ion's kinetic properties, such as viscosity and diffusion. The distant hydration, or distant water molecules involved in the hydration of the ion, influence the heat of hydration of the ion. The solvent water molecules in the first and second solvation shells are oriented by the ion's electric field, and these are not compatible with the normal mutual orientation of the solvent molecules in the bulk solution.

It is quite difficult to precisely quantify the total number of water molecules for complete hydration of a dissolved ion because the transition from ion-hydrating water molecules to bulk-water molecules is not necessarily a sharp boundary. Table 2-7 lists the hydration radius of numerous ions in dilute solutions. In concentrated salt solutions, some of the water molecules may be shared by the cations and anions present. It is presumed that the hydration numbers will increase with dilution.

In general, ions of small size and high charge will hydrate most strongly, resulting in larger hydration radii. This is because ions with high charge density bind more strongly with water molecules, resulting in larger water clusters. That is, the degree of hydration is large when the ionic potential (z^2/r_{ionic}) is large. Common trends in the periodic table on the hydration radius of ions are opposite to those of the ionic radius trends. Ionic radii generally increase right to left and top to bottom in the periodic table, while hydration radii of ions tend to increase in the opposite direction.

Most anions are larger than most cations, but if the ionic sizes and absolute value of the charges are similar, then the anions will be more strongly hydrated than the cations. This is because water's H^+ atoms can approach the anions quite closely (about 0.08 nm), while the oxygen atoms are not able to do this with the cations. Furthermore, the H bonds between water and some anions are not broken easily because of the lack of a nearby H-bonding partner in the ion hydration structure.

When chemical reactions occur involving ions, then the number of water molecules surrounding the ions will usually change. The ease and degree of hydration of ions is closely related to the solubility of the corresponding elements of a solid phase. Also keep in mind that most ions in solution can then react with the water molecules in such a way that they sequester at least one proton (H^+) or hydroxyl (OH^-). The result of these reactions is the presence of various aqueous species (for example, $H_3PO_4(aq)$, $H_2PO_4^-$, HPO_4^{2-}, and PO_4^{3-}), where each of the species formed will have a different hydration radius. The surrounding water molecules also pull on the oxygen atoms of these ions, slightly increasing their bond distance from the central element. Protonation of an oxygen atom will also cause it to increase its distance from the central element bonding to it. For example, in dilute solutions of CrO_4^{2-}, the Cr–O distance is 1.660 ± 0.001 Å, but in dilute solutions of $HCrO_4^-$ the Cr–O distance is 1.622 ± 0.003 Å and the Cr–OH distance is 1.800 ± 0.021 Å.[34]

Small or monovalent ions are considered to be water structure-making (e.g., Li^+, Na^+, Ca^{2+}, and Mg^{2+}), while large monovalent ions are considered to be water structure-breaking (e.g., K^+, NH_4^+, Rb^+, and Cs^+). As a result of this, the smaller the ionic radius of an ion of the same charge, the larger will be the hydrated radius of the ion. Small ions exhibit a strong bond with the hydrating water molecules of the first shell (Region 1) surrounding them, which results in a lower mobility of the local water molecules, some additional local order, and a higher apparent density for the solution water. Similarly, a higher apparent

Table 2-7: Hydration numbers, volumes, and radii of ions in solutions and crystals. Hydrated volumes (V_{ih}^∞) and hydration numbers ($h_{i,s}^\infty$) are based on Stokes radii corrected for a packing factor of 0.888 for dilute solutions. The hydrated radii (r_{ih}^∞) of the ions are calculated from the hydrated volumes (V_{ih}^∞) and the general formula for a sphere: $V_{ih}^\infty = Nk(4\pi/3)(r_{ih}^\infty)^3$, where N = Avogadro's number (6.0221367×10^{23} molecules/mole) and k = packing factor = 0.888. The hydration volumes, hydration radii, and the hydration numbers based on the Stokes radii describes the amount of water that accompanies the ion when it migrates under the influence of an applied electric field. Scale conversion: 1 pm = 10^{-12} m. *From data compiled by Marcus (1985)*[35] *and Brownlow (1979).*[36]

Cation	r_{ionic}, pm	V_{ih}^∞, cm^3 mol^{-1}	$h_{i,s}^\infty$	r_{ih}^∞, pm	Anion	r_{ionic}, pm	V_{ih}^∞, cm^3 mol^{-1}	$h_{i,s}^\infty$	r_{ih}^∞, pm
Li$^+$	82	125.9	7.4	383	F$^-$	123	103.7	5.5	359
Tl$^+$	97	93.6	5.0	347	Cl$^-$	172	93.6	3.9	347
Ag$^+$	123	99.4	5.9	354	Br$^-$		92.8	3.4	346
Na$^+$	124	109.0	6.5	365	I$^-$		92.8	2.8	346
K$^+$	181	94.4	5.1	348	NO$_3^-$		95.3	3.3	349
Rb$^+$	181	92.8	4.7	346	ClO$_4^-$		96.9	2.6	351
Cs$^+$	196	92.8	4.3	346	ClO$_3^-$		97.7	3.0	352
NH$_4^+$		94.4	4.6	348	MnO$_4^-$		100.3	2.9	355
					ReO$_4^-$		103.7	2.7	359
Be^{2+}	35	218.1	13.5	460	BrO$_3^-$		103.7	3.4	359
Cu^{2+}	54	147.8	10.3	404	IO$_3^-$		120.1	4.9	377
Mn^{2+}	73	189.6	12.2	439					
Ni^{2+}	77	147.8	10.3	404	CO$_3^{2-}$		137.1	7.1	394
Mg^{2+}	80	176.9	11.7	429	SO$_4^{2-}$		123.0	5.3	380
Zn^{2+}	83	178.2	11.3	430	SeO$_4^{2-}$		127.9	5.2	385
Co^{2+}	83	169.6	11.5	423	CrO$_4^{2-}$		120.1	4.9	377
Fe^{2+}	86	174.5	11.8	427	MoO$_4^{2-}$		128.9	4.8	386
Cd^{2+}	103	173.2	11.4	426	WO$_4^{2-}$		136.0	5.4	393
Ca^{2+}	120	156.7	10.4	412					
Sr^{2+}	133	156.7	10.4	412	Fe(CN)$_6^{4-}$		164.4	3.8	419
Pb^{2+}	137	143.4	9.5	400					
Ba^{2+}	150	146.7	9.6	403					
Al^{3+}	61	241.7	16.8	476					
Cr^{3+}	70	219.5	15.5	461					
Fe^{3+}	73	215.3	15.5	458					
Tm^{3+}	96	228.2	16.2	467					
Ce^{3+}	109	208.3	14.9	453					
La^{3+}	113	208.3	14.8	453					

density of water will also be exhibited by higher charged ions. Water structure-makers, or *kosmotropes*, exhibit a weak ion–water interaction (weaker than water–water interactions). Water structure-breakers, or *chaotropes*, exhibit a strong ion–water interaction (stronger than water–water interactions). Strongly hydrated ions (chaotropes) breakdown the water's tetrahedral network by interfering with the natural H bonding processes of the linked water molecules.

Kosmotropes, or weakly hydrated ions with large ionic size and low charge, form low-density water (an expanded water structure), but prefer to accumulate in normal density water (a collapsed water structure). Further addition of weakly hydrated ions fit into the expanded water network without affecting its H bonding and viscosity. Chaotropes, or strongly hydrated ions, form high-density water or normal-density water, but prefer to accumulate in low-density water. Further addition of these ions into the high-density water impacts the H bonding and viscosity of the structure. Expressed mathematically, the viscosity of the liquid as a function of the concentration of dissolved salts present is[37]

$$\eta / \eta_0 = 1 + Ac^{0.5} + Bc \qquad [2\text{-}80],$$

where η = viscosity of an aqueous salt solution, η_0 = viscosity of pure water at the same temperature, c = salt concentration ($c < 0.1\ M$), A = constant (always positive), and B = constant. The constant B is positive and the constant A is larger for strongly hydrated ions (chaotropes). Ions that are weakly hydrated (kosmotropes) have negative B coefficients and have a smaller effect on viscosity with concentration.

Ion pairs may form in water. If the ionic attraction of two ions in solution overcomes the hydration shells, then an ion pair will form. This will occur with two small ions each with large charge densities (such as $CaF_2(aq)$) because the strong ionic attraction overpowers the hydration shells that were trying to separate them. An ion pair will also form if the two ions are large and each has a small charge density (such as $AgI(aq)$) because these ions do not have strong hydration shells that can effectively keep them apart.

Ion pairs will not form easily if we mix this formula for ion pair formation. This is because a large ion with a low charge density is not able to break through the hydration shell surrounding a small ion with a high charge density. Accordingly, the ion pairs CaI_2 or AgF will not form easily. Strong ion pairs form with two kosmotropic ions or two chaotropic ions, but not with a kosmotropic ion and a chaotropic counter ion. Ion pair formation will also affect the salt's solubility, where high solubility of a salt will result when the ions that form the salt exhibit weak ion pair formation in water. The salt solubility data of the ion pairs mentioned above (units are in g of salt/kg of water)[38] are as follows: 2090 for CaI_2, 1820 for AgF, 1.6×10^{-2} for CaF_2, and 1.6×10^{-6} for AgI.

The hydration and stability of large organic compounds in water and salt solutions remains ill understood. It has long been known that dissolved salts can be ranked based on their ability to precipitate (or to destabilize) a mixture of hen egg white proteins, known as the Hofmeister series[39]. This series, also called the lyotropic series or chaotropic series, is now used to rank the ability of ions to destabilize the structure of proteins in water. Note that the relative positions of the ions in the series will vary with protein, pH, temperature, and with the counter ion present. Some anions and cations in the Hofmeister series are:

Most Destabilizing . . . . *Most Stabilizing*
(weakly hydrated anions)   (strongly hydrated anions)
Anions: $SCN^- > I^- > NO_3^- > Br^- > Cl^- > F^- > citrate^{3-} > acetate > HPO_4^{2-} > SO_4^{2-}$
Cations: $Al^{3+} > Ca^{2+} > Mg^{2+} > Na^+ = K^+ > NH_4^+ > [N(CH_3)_4]^+$
(strongly hydrated cations)   (weakly hydrated cations)   [2-81].

At low concentrations, the ions that stabilize the proteins bind weakly with them and enhance their hydration. Conversely, ions that destabilize the proteins bind strongly with them and decrease their hydration. That is, ionic binding with the proteins will affect the H-bonding ability of the proteins and increase their hydrophobic nature.

At high concentrations ($> 0.1\ M$), excess ions that are not bound to the proteins will be held instead by the water molecules, effectively competing with the proteins for the solvent. This results in a decrease in solvation of the proteins, which then aggregate and precipitate. Precipitating compounds following the addition of salts is called "salting out". Note that some salts will increase solubility and protein stability, and this effect is called "salting in".

Another explanation for the Hofmeister series involves the relative hydration behavior of the electrolytes present. Hydrophilic polymers and large organic compounds (such as proteins) that dissolve in water are surrounded by low-density water, which can be as large as the radius of an icosahedral water cluster (a water structure with 20 sides of about 1.5 nm-radius). Ions in solution will partition between these resultant low-density and normal-density waters, where chaotropes prefer low-density water and kosmotropes prefer normal-density water. Salts whose dissolved ions are distributed evenly in the normal-density and low-density water will stabilize the organic solubility environment and its hydrophobic junction zones (e.g., $(NH_4)_2SO_4$). Salts with both ions preferentially concentrating in either the low-density water (e.g., NH_4I or NH_4NO_3) or the normal-density water (e.g., $MgSO_4$) will destabilize the organic solubility environment.

[2.11] Structure of Water & Hydrated Ions

One of the most important properties of water molecules is their ability to H bond. For water, the average O...O distance is 0.276 nm, and the O–H···O bond angle is 180° (that is, they lie on a straight line).[35] The structure of water is generally described in terms of the average number of H bonds that occur per water molecule, which is between 0 and 4 for water.[35] It is these H bonds that allows the water molecules to aggregate into large clusters.

A useful structure for illustrating water clusters is a tetrahedral structure formed by a 14-molecule unit. This is shown in Figure 2-10, where four of the oxygen molecules are located in the corners of the tetrahedra. Other structural models for water based on x-ray diffraction data show similar water cluster models, albeit with some variations, such as: a 10-molecule cluster with the four corner water molecules missing[40], or only an 8-molecule

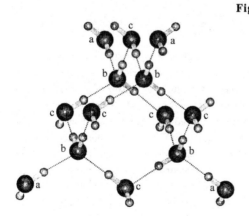

Figure 2-10: Arrangement of 14 water molecules in a slightly flattened tetrahedral unit, where three edges are 5% shorter than the other three. The four corners of the tetrahedra are labeled "a". There are four internal molecules labeled "b", and six "c" molecules that are positioned midway along the six edges of the tetrahedra. *Reprinted from Chaplin (1999)[41], "A proposal for the structure of water", Biophysical Chem. 83:211–221, copyright © 1999, with permission from Elsevier.*

cluster around an organic compound with the four corner and two of the oppositely positioned water molecules missing.[42]

There are various different kinds of clusters that probably exist in water[43], and many of these are suspected to co-exist in larger super clusters. An example of one proposed super cluster is shown in Figure 2-11, which is an icosahedral (20-sided) water cluster. This icosahedra may be formed from the merging of twenty of the smaller tetrahedral structures shown in Figure 2-10. Each water molecule in this super cluster can be described as belonging to one of several octameric (8-membered) structures believed to be the most probable clusters for water.[43]

The lack of precisely 72° angles in the tetrahedra shown in Figure 2-10 for a perfect fit in an icosahedral cluster is compensated by the bonding flexibility. The icosahedral water cluster shown in Figure 2-11 is 3 nm in diameter[41], which equals the limit of order in supercooled heavy water.[44] The expanded structure (ES) density ($\rho = 0.94$ g cm^{-3}) is smaller than the collapsed structure (CS) density ($\rho = 1.00$ g cm^{-3}).[41] The CS density is similar to water at 0°C ($\rho = 1.00$ g cm^{-3}). The ES density is similar to that of low-density water around macromolecules ($\rho = 0.96$ g cm^{-3})[45], supercooled water at −45°C ($\rho = 0.94$ g cm^{-3})[2], and low-density amorphous ice ($\rho = 0.94$ g cm^{-3}).[46,47] The CS water is dominant at high temperatures, but expands with further increases in temperature. The ES water is dominant at low temperatures. The transition from ES to CS produces positive entropy due to the less ordered structure, and a positive enthalpy as a result of the greater H bond bending.[41] The

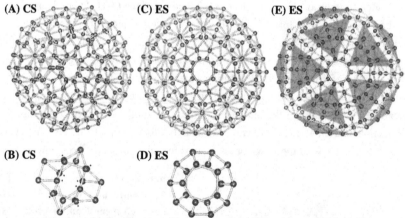

Figure 2-11: A 280-molecule icosahedral water cluster built with twenty tetrahedra illustrated in Figure 2-10. Only the network oxygen are shown for clarity. Each molecule in the cluster has four H bonds. No H bonds are broken when the cluster fluctuates from the ES to CS structure. The small figures on the bottom illustrate the dodecahedron present at the center of each icosahera: (B) for the puckered center in the high density CS, and (D) for the convex center of the low density ES. Drawing (E) illustrates how the tetrahedra in Figure 2-10 are combined to form this icosahedra. Drawing (E) is identical to (C), but the ten sides of the top half of the icosahedra are shaded, where each shaded triangle corresponds to the exposed side of each tetrahedra. *Reprinted from Chaplin (1999)[41], "A proposal for the structure of water", Biophysical Chem. 83:211–221, copyright © 1999, with permission from Elsevier; Drawing (E) was modified.*

density of a CS cluster if it were fully collapsed at higher pressure and with weaker H bonding ($\rho = 1.18$ g cm^{-3}) would be similar to that of high-density amorphous ice ($\rho = 1.17$ g cm^{-3}).[41,47]

Water super clusters have large interstitial cavities that can be occupied by solutes. The interstitial sites in the CS cluster can be occupied by weakly H-bonded molecules.[41] The cavities in the ES cluster can be occupied by structure-forming ions, or kosmotropes, where the dodecahedron in the ES center has an estimated radius of 3.94 Å.[41] However, regardless of the presence or absence of water super clusters, small clusters do form around solutes. These solutes strengthen the H bonding of the water structures and are well documented to exist in dodecahedral cavities in aqueous solutions.[41]

In concentrated solutions of various cations (<10 H$_2$O per ion, such as with Li$^+$, Ca^{2+}, Ni^{2+}, or Cu^{2+}), the dipolar axis of the water molecule lies at an angle of $50 \pm 10°$ from the cation–O line (Figure 2-12A).[35] In dilute solutions, the effects of the neighboring counter ions on the potential energy of the water in the hydration shell decrease, which in turn causes the angle shown in Figure 2-12A to decrease. The geometrical arrangement of the water molecules around Cl$^-$ anions forms nearly a straight line, with less then 10° deviation from the 180° angle of the Cl–D–O atoms (Figure 2-12B).

The presence of small, multivalent ions exert a strong central electric field that produce an orienting influence on the first and second hydration shells (and beyond).[35] The first hydration shell, in turn, exerts its own orienting effect on the outer water molecules. A third effect that needs to be considered is the tendency of water molecules to orient themselves in a tetrahedral structure, which will extend as near to the central ion as possible depending on the competition from the other two effects. Nonpolar regions of the central molecule or ion will lack the electric field effect and, hence, will enhance the normal tetrahedral structure of water around these nonpolar regions.

Both the polar and nonpolar regions of the central molecule will protect the nearby water molecules from the thermal motion of "free" water molecules and, thus, protect their structure from being disrupted by these natural collisions.[35] The net effect of these polar and nonpolar regions is that the oriented structure of the surrounding water molecules will be enhanced. In the nonpolar regions, the water molecules will be more tetrahedral and ice-like. In the polar regions, the water molecules will be oriented by the charge. An in-between region will exist between the polar and nonpolar regions consisting of disoriented water molecules, sometimes referred to as a "thawed" region (Figure 2-13).

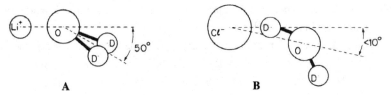

A   **B**

Figure 2-12: The geometrical configuration of water molecules around Li$^+$ and Cl$^-$ ions based on differential neutron diffraction of isotopically labeled atoms. *Reproduced from Marcus (1985)[35], "Ion solvation", copyright © 1985, with permission from John Wiley & Sons Limited.*

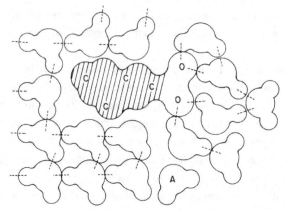

Figure 2-13: Schematic representation of a butanate anion in water. The hydrophobic portion of the anion is shaded. The arrangement of water molecules around this nonpolar region has an ice-like tetrahedral arrangement. The two oxygen atoms on the carboxylate group (the polar region of the anion) orient the water molecules by the charge and are H bonded to them. The water molecule marked "A" illustrates a water molecule that is in an unstructured region in between the two kinds of structured regions. *Reproduced from Marcus (1985)*[35], *"Ion solvation", copyright © 1985, with permission from John Wiley & Sons Limited.*

[2.12] Clathrates

Although discovered in 1810 by Sir Humphry Davy[48], gas molecules trapped inside cages of ice crystals are often overlooked in environmental studies. However, these gas hydrates, also known as clathrates, are very abundant in nature. They are considered low-pressure forms of ice because they are stable at ambient pressures in cold environments, or even at room temperature at pressures easily attainable within the earth's crust. They are only stable, however, when gas molecules are present in their cages.[49] For solid hydrates to form and remain stable, the concentration of the gas in the liquid phase (or seawater within the sediment) must exceed its solubility limit.[50]

Trapped gases and compounds in snow are not clathrate structures, and they would not form unless they melt and refreeze under pressure. Accordingly, clathrates are not found in our warm climate soils. We should still try, however, to be well familiar with them if we wish to broaden our understanding on the general nature of the hydration of hydrophobic molecules in water. This is because the aqueous water structure around hydrophobic groups are believed to be clathrate-like.[51]

Clathrates are found in numerous natural environments, most notably in the following places[50]: in the cold sediments under any of the world's oceans (typically under a minimum of 500 m of ocean water for CH_4 hydrates if temperatures are around 5°C, or 3,000 m if temperatures are around 22°C), and close to the surface in terrestrial permafrost regions, where the temperatures are always very low. There are over 100 large deposits now known around the world, including areas on or near the equator. Methane (CH_4) clathrate deposits are not presently mined in spite of the fact that they are huge deposits of natural gas — in some places holding more natural gas than conventional reserves. Note that melting 1 m³ of methane gas hydrate will yield about 0.79 m³ of water and 172 m³ of CH_4(g) at 1 atm and 15.6°C (60°F).[50] But mining this fuel is difficult because the density of CH_4 in the sediment mixture is often too low, resulting in the gas well being labeled a "dry hole." Even if these gas deposits are not mined and burned (producing $CO_2 + H_2O$), they may eventually release (or "outgas") into the atmosphere as a result of global warming. This would further accelerate global warming since CH_4(g) traps heat 25 times more effectively than CO_2(g).[50,52] Clathrate deposits in sediments are not very thick (often only a few hundred meters) because they destabilize as the earth's temperature increases with depth. However, natural gas often collects below the gas hydrate layer, and these areas are generally more cost effective to tap into.[50] Other places believed to be rich in clathrates include the ice cap of Mars (CO_2 hydrates), comets (CH_4 and NH_3 hydrates), and the plants Uranus and Neptune (CH_4 and NH_3 hydrates).[49]

Several clathrate structures are known, but only two will be discussed here. The most common unit cell is the body-centered packing of Structure I. Of all the naturally occurring clathrates, over 99% are methane clathrates[52], which are Structure I hydrates. The Structure I unit cell has 46 H_2O molecules in 2 pentagonal dodecahedra (12-sided polyhedra) and 6 tetrakaidecahedra (14-sided polyhedra), part of which is illustrated in Figure 2-14A. The lattice of Structure I is 12 Å across, where molecules 5.1 Å or less in diameter can fit inside the pentagonal dodecahedra (such as CO_2, but not CH_3Br), and molecules 5.8 Å or less in diameter can fit inside the tetrakaidecahedra (such as CH_3CH_3, but not $CH_3CH_2CH_3$).[49] Although all eight cages can be occupied by gas molecules, there is never 100% occupancy, albeit they are often close to it. Some gases that form Structure I hydrates are: N_2, O_2, Ar, Kr, Xe, CF_4, CH_4, C_2H_6, CO_2, H_2S, CH_3Cl, Cl_2, and cyclopropane (C_3H_6). These trapped gases do not interact with the clathrate water molecules. They also do not affect or modify the clathrate structure.[49] As a result of this, mixed gases in the hydrate are in solid solution, rather than in two or more separate crystal forms.

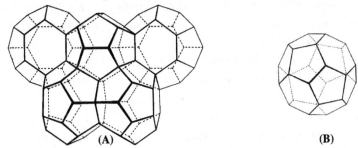

(A)   (B)

Figure 2-14: Illustration of clathrate structures. The oxygen atoms are at vertices, while the hydrogen atoms are on the lines between each of them. (A) The lattice of the Structure I hydrate. The pentagonal dodecahedron is formed by 20 water molecules and is shown here in the upper center of the structure. The remaining four cages shown are tetrakaidecahedra, each formed by 24 water molecules. (B) The hexakaidecahedron formed by 28 water molecules, which is a component of the Structure II hydrate. *From Miller (1973)[49], in Whalley et al. (ed.) "Physics and chemistry of ice", copyright © 1973, with permission from the Royal Society of Canada.*

The Structure II hydrate unit cell is 17 Å across, with 136 H_2O molecules forming 16 pentagonal dodecahedra and 8 hexakaidecahedra (16-sided polyhedra, illustrated in Figure 2-14B).[49] The hexakaidecahedra can hold molecules with diameter of 6.7 Å or less (such as propane, $CH_3CH_2CH_3$, but not butane, $CH_3CH_2CH_2CH_3$).[49] Some gases that form Structure II hydrates are: $CHCl_3$, C_2H_5Cl, $CBrClF_2$, propane, and cyclopropane.

[2.13] Factors Affecting Solubility

When a mineral is in contact with water, some of the elements that comprise the mineral will often detach from the mineral and become dissolved in the water. Typically, the first step for the solubility of a mineral $A_xB_y(s)$ is the formation of aqueous neutral molecules according to the reaction:

$$A_xB_y(s) \rightleftharpoons A_xB_y(aq) \tag{2-82},$$

where (s) denotes the compound in a solid phase, and (aq) denotes the compound dissolved in the liquid phase. Nearly all solids will dissolve in water, albeit to varying degrees. Some dissolve quickly and easily (such as NaCl), while others do not seem to dissolve at all (such as quartz). There are three primary factors that affect the solubility of a compound in water or other liquid phases: properties of the solute, nature of the solvent, and solution chemistry. Each of these factors are discussed below.

[2.13.1] Properties of the Solute

The two properties of the solute that influence the solubility of a compound are the lattice energy ($E_{lattice}$) and the solvation (or hydration) energy ($E_{hydration}$). These properties are also dependent on charges, sizes, and polarizability of the ions composing the solute. The *lattice energy* is the energy liberated when the free gaseous ions, initially at infinite distance apart, unite to form the ionic crystal lattice. The *hydration energy* is the energy liberated when an ion and a certain number of water molecules, initially at a large distance apart, come together to form a hydrated ion.

If the lattice energy < hydration energy, then the substance is soluble; otherwise it is relatively insoluble. The enthalpy (heat) of solution ($\Delta H_{solution}$) is defined as:

$$\Delta H_{solution} = E_{lattice} - E_{hydration} \tag{2-83}.$$

Note that it is the sign of the Gibbs free energy of the system (ΔG_{sys}) that determines if the reaction actually occurs, whether it be toward solubility or toward precipitation.

[2.13.2] Nature of the Solvent

There are three components of the solvent that affect solubility. These are the dielectric constant, the temperature, and the pressure of the solvent.

A. *The Dielectric Constant:* The relative dielectric constant (ϵ_r, now also referred to as the relative permittivity) is a measure of the ability of a substance to decrease the force of attraction between two oppositely charged particles relative to their attraction under vacuum. A solvent with a high dielectric constant results in a high solubility of ionic crystals. For two charged particles separated a distance r across a medium with permittivity ϵ, the electrostatic potential energy $= (q_1 q_2)/(4\pi\epsilon r)$ and force = energy/distance. Accordingly, the French engineer and physicist Coulomb (1736–1806) described the attractive force between two charged objects in 1785, resulting in Coulomb's Law:

$$\text{Force} = \frac{q_1 q_2}{4\pi\epsilon\, r^2} \qquad [2\text{-}84],$$

where $q_{1,2}$ = charge on each particle 1 and 2, r = distance between them, and ϵ = absolute permittivity of the medium. The units of ϵ in cgs are dyne cm^2 statcoulomb^{-2}, but in mks-SI the units are newton m^2 coulomb^{-2}. A common method to measure ϵ is by measuring the capacitance of the material (C), which is the ability of the material to store electricity. It is defined as

$$C = Q/V \qquad [2\text{-}85],$$

where Q = stored charge on one conductor, and V = difference in potential between the conductors, or difference between one conductor and earth (ground). Furthermore,

$$\epsilon_r = \epsilon/\epsilon_o = C/C_o \qquad [2\text{-}86],$$

where C_o = the capacitance of the condenser in vacuum, and ϵ_o = the absolute permittivity of free space (now known as the electric constant) = 8.854×10^{-12} F m^{-1} (or C^2 J^{-1} m^{-1}), or $1/(4\pi\epsilon_o) = 8.987\times10^9$ N m^2 C^{-2} (note that the C units here are coulombs). The dielectric material has, therefore, increased the capacitance of the condenser. The relative dielectric constant of various materials is listed in Table 2-8. Note that ϵ_r of vacuum = 1.0.

   B. *Temperature:* The effect of temperature on solubility is illustrated in Figure 2-15. Temperature increases can cause large increases in solubility of some substances, and little

Table 2-8: Relative dielectric constant (ϵ_r) of various solvents and solubility of NaCl in selected solvents. *Data from Lide (2002)[2] and Robbins (1967).[53]*

Medium		Temperature, °C	Dielectric Constant, ϵ_r	NaCl solubility at 25°C, g/100 g solvent
ice Ih [a]	H_2O	−30	99	
		0	96.5	
water	H_2O	0	87.7	
		10	83.8	
		20	80.1	
		25	78.3	36.12
		30	76.6	
		40	73.2	
methanol	CH_3OH	25	32.2	1.3
2-ethoxyethanol [b]	$C_2H_5OC_2H_4\text{-OH}$	25	13.38	
carbon tetrachloride	CCl_4	25	2.23	0.00
vacuum			1.0	

[a] At 0 kbar

[b] Also known as ethylene glycol monoethyl ether (EGME).

or no change for others. Although the solubility generally increases with temperature, it sometimes decreases with temperature. The solubility of a gas in water will always decrease with temperature. The slope (Δsolubility/ΔT) is related to the enthalpy change (ΔH) of solution (or of dissolution) of the particular solute, where a positive slope indicates a positive enthalpy change ($+\Delta H_{reaction}$) and a negative slope indicates a negative enthalpy change ($-\Delta H_{reaction}$). The rate of dissolution was first shown explicitly to be a function of temperature by S.A. Arrhenius around 1889:

$$\ln k = \frac{-E}{RT} + \text{constant} \qquad [2\text{-}87],$$

where k = reaction-rate constant, and E = activation energy. From this equation, J.H. van't Hoff, who received the first Nobel Prize in Chemistry in 1901, described the relationship of solubility and heat of solution as follows. From Equation [2-87]:

$$K_{eq} = \frac{e^{-E_f/RT}}{e^{-E_b/RT}} = e^{-(E_f - E_b)/RT} = e^{-\Delta H/RT} \qquad [2\text{-}88],$$

where K_{eq} = thermodynamic equilibrium of solution = $k_{forward}/k_{backward}$ (or k_f/k_b), and the constants carried over from Equation [2-87] cancel out. Taking the natural log and the derivative of Equation [2-88], we obtain what is now known as van't Hoff's isochore:

$$\frac{d \ln K_{eq}}{dT} = \frac{\Delta H_{reaction}}{RT^2} \qquad [2\text{-}89],$$

After integrating, Equation [2-89] can be expressed as:

$$\log K_{eq} = \log K_{eq,ref} - \frac{\Delta H_{reaction}}{2.303\ R} \left(\frac{1}{T} - \frac{1}{T_{ref}} \right) \qquad [2\text{-}90],$$

where $K_{eq,ref}$ = known equilibrium constant at a specific reference temperature (T_{ref}), and K_{eq} = equilibrium constant at a specific temperature (T). If sufficient information is available, the temperature dependence of the equilibrium reaction can be calculated empirically using

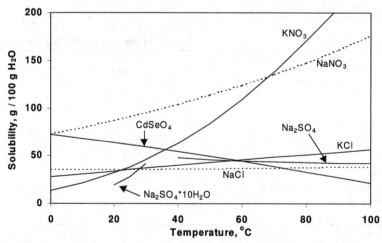

Figure 2-15: Influence of temperature on solubility. There is a solid phase change in the solubility data of sodium sulfate. *Data from Lide (2002).*[2]

a power function of the form:

$$\log K_{eq} = A + BT + \frac{C}{T} + D \log T \qquad [2-91],$$

where A, B, C and D are empirical adjustment parameters fitted to the known data set.

C. *Pressure:* Pressure has very little influence on the solubility of solids. This is because the net volume change is not very large. However, the solubility of gases in liquids is very dependent on pressure, as is discussed later in Section 2.14 with Henry's Law.

[2.13.3] Solution Chemistry

When a solid compound dissolves in water (Equation [2-82]), it generally further dissociates into aqueous ions or electrolytes. The net dissolution reaction can be expressed as

$$A_xB_y(s) \rightleftharpoons xA^{z+} + yB^{z-} \qquad [2-92].$$

This reaction, or any other similar process, will seek a specific equilibrium condition that satisfies the equilibrium constant for that particular system. At "infinite time", an equilibrium condition is established in the solid–liquid mixture such that the products over reactants equals a number specific for that mixture, known as the equilibrium constant. The equilibrium constant for Equation [2-92] can be expressed as

$$K = \frac{(A^{z+})^x(B^{z-})^y}{(A_xB_y(s))} \qquad [2-93],$$

or for Equation [2-82] as

$$K = \frac{(A_xB_y(aq))}{(A_xB_y(s))} \qquad [2-94].$$

But these expressions are not practical because the activity of a specific solid phase is constant. That is, the proximity of one A_xB_y molecule to another A_xB_y molecule on the solid surface does not change, and it will not change as long as the type of solid phase present remains the same. Although the solids activity in the denominator is a constant, there is no uniform convention on how to determine its value. The solids activity has been equated to the mole fraction (c_s) times its activity coefficient (γ_s). For pure solids (such as an Al oxide), $c_s = 1$. For solid solutions, $c_s < 1$ for separable components of the "parent" mineral (such as soluble Al ion impurities in an Fe oxide mineral). In both pure solids and trace contaminant situations, the corresponding γ_s values cannot be predicted *a priori* because they depend significantly on the crystallographic details of each mineral.

For this reason, the denominators of solubility equations are factored out (that is, they are numerically combined with the K values defined above). This, in turn, results in a new constant defined as the solubility constant (K_{so}) of the specific solid phase, such as:

$$K_{so} = (A^{z+})^x (B^{z-})^y \qquad [2-95]$$

for dissolution reactions that can be described by Equation [2-92], or

$$K_{so} = (A_xB_y(aq)) \qquad [2-96]$$

for dissolution reactions that can be described by Equation [2-82]. As with "w" in Equation

[2-3], the "so" subscript serves as a reminder that the solid phase activity has been factored into the K_{so} value; that is $K_{[2-94]} = K_{so}/(A_xB_y(s))$. Inevitably, the constant K_{so} value discussed here is specific to the mineral in question. For example, the K_{so} value for goethite (α-FeOOH) is different from the K_{so} value for akaganeite (β-FeOOH), and amorphous $SiO_2(am,s)$ K_{so} differs from crystalline $SiO_2(cr,s)$ K_{so}, although the mole fraction of each of these pure solid phases is unity. Clearly, if a solid solution is present, the mole fraction would change. But here, too, the solid solution should be considered a new phase with its own K_{so} value, or else we would need to introduce a solid activity coefficient (γ_s) in the calculations. By convention, the activity of the standard state of water and solids at standard temperature (25°C) and pressure (1 atm) are defined equal to one, which yields the same conclusions.

The product of the actual activities of the aqueous electrolytes present in the liquid is called the ion activity product (IAP). The IAP is not a constant, but a number that reflects the current conditions of the liquid portion of a sample. It is useful to compare the IAP number of a specific liquid sample with the solubility constants of various different minerals, expressed in terms of the saturation index (SI):

$$\text{SI} = \frac{\text{IAP}}{K_{so}} \qquad\qquad [2\text{-}97].$$

If SI < 1, then the solution is said to be *undersaturated* with respect to that particular mineral (remember, each particular mineral has its own K_{so} constant). That is, there are not enough ions present in the solution to cause the formation and subsequent precipitation of that particular mineral. It is also possible, however, to induce precipitation in a mixture with SI < 1 in the bulk solution if regions exist within the mixture with SI > 1(such as in highly concentrated thin film regions near surface adsorption processes, resulting in adsorbed solid precipitates).

If SI < 1 and a solid phase is present (one that corresponds to the K_{so} value that is being compared to the IAP value), then the solid phase will very probably dissolve because there are not enough ions present in the solution to allow that particular solid phase to remain stable. The word "probably" is used here because it is possible for the solid phase present to resist dissolution, such as samples in a metastable state. Typically, however, the dissolution of the mineral will continue until IAP = K_{so}.

If SI = 1, then the solution is said to be *saturated* with respect to the particular mineral whose K_{so} constant is being used. No dissolution of the solid mineral (if present) will occur. Also, no solid precipitation of that particular mineral should occur. If the soil sample contains the mineral, but for some reason the liquid and solid components have been separated, then the separated liquid portion continues to be saturated with respect to that particular mineral as long as its IAP = K_{so}.

If SI > 1, then the solution is said to be *supersaturated*. If the mineral is present in the mixture, then it will grow larger in size. If the mineral is not present, then it probably will form and precipitate out of solution until the IAP in the liquid phase drops down to IAP = K_{so}. The word "probably" is used here because it is possible that the mixture is in a metastable state with respect to that particular mineral. It is also possible that while the specified mineral should form, another mineral whose K_{so} constant is also less than the IAP value forms and precipitates instead. Which mineral forms in a supersaturated solution depends in part on the K_{so} values of the list of possible candidate minerals. However, it is incorrect to assume that the mineral with the lowest K_{so} constant will be the mineral that is most likely to precipitate. Some minerals have very complex structures, resulting in their rate of formation being very slow in various environments. Other minerals may form instead, as long as their $K_{so} <$ IAP.

The solubility of the mineral is also affected by the presence of other ions in the solution. The presence of common ions in the solution will affect the IAP value of the solution. For example, the presence of a NaCl salt solution will decrease the solubility of $FeCl_3(s)$ because an increase in the Cl^- anion concentration will inflate the IAP value. This is known as the common ion effect. Note that the presence of uncommon ions will also affect the IAP value because a change in the ionic strength will affect the activity values of the aqueous electrolytes.

The solubility of the mineral is significantly affected if the solubility product further reacts with the solvent. For example, if $Fe(OH)_3(s)$ dissolves to form $Fe(OH)_3(aq)$, the product may react with water to form numerous new aqueous Fe(III) species, such as: $Fe(OH)_4^-$, $Fe(OH)_2^+$, $FeOH^{2+}$, and Fe^{3+}. Note the presence of hydroxyl groups in this example, and, hence, the pH dependence of iron solubility in water. As these new products are formed, the amount of $Fe(OH)_3(aq)$ is briefly depleted, the IAP value is lowered, and more $Fe(OH)_3(s)$ dissolves to form more $Fe(OH)_3(aq)$. The solid dissolution and formation of all the aqueous products continues until either (1) the solid totally dissolves, or (2) the solution becomes fully saturated and can no longer hold any of the various secondary products following the formation of the initial aqueous compound. Additional examples of the impact of the acidity of the solution and solution reactivity with the soluble products are discussed in Sections 2.15 and 2.16.

[2.13.4] Other Factors Affecting Solubility

Published solubility data that differ by a few orders of magnitude are not uncommon. Stumm and Morgan (1981)[54] noted that differences as large as 10^{13} have been published on the solubility constants of $FePO_4(s)$. These differences may be due to various reasons. The most likely reasons are: (1) the presence of other ions in solution that can influence the solubility of the mineral (e.g., by surface or aqueous complex formation or by interfering with the reactivity of the mineral with water molecules), or (2) the minerals studied vary in their active forms or phases.

The presence of ions in the reactor will affect solubility, and it is easy to inadvertently overlook the presence of even high levels of interfering ions. A very common example of this is carbon dioxide impurities. It is very easy, unfortunately, to overlook the impact of inorganic-C ions on soil chemistry studies, namely $CO_2(aq)$ and its various pH-dependent species, which are ubiquitous in the environment. Laboratory solubility studies often seek to study the solubility of a mineral in the absence of impurities. Although they are generally successful at removing most impurities, high levels of inorganic-C impurities are probably often present in the mixtures studied. Inorganic C is easy to remove from the samples by purging with purified air, but this detail in the pretreatment of the samples is often missed. The inorganic C may originate from the mineral surface due to a prior exposure of the mineral to the air (hence, it can be an autochthonous source), or from absorption by the liquid sample of high levels of CO_2 in the lab or even the technician's breath (hence, it can be an allochthonous source). It may also originate from the pH-adjusting solutions, such as from the addition of NaOH solutions, which can contain high levels of soluble inorganic C.

The size and crystallinity of the mineral can affect the solubility of the mineral. A precipitate from a supersaturated solution may form a very fine crystalline precipitate with a disordered lattice. As it ages in solution, the precipitate may convert to a mineral with a more ordered lattice. A mineral with a disordered lattice will be more soluble (and reactive) than one with an ordered lattice. Also, a smaller precipitate is more reactive than one consisting of large particles.

When a mineral ages in water, the mineral phase may change entirely. Over time, a ferrihydrite may change to goethite or hematite, or a δ-Al_2O_3 can change to bayerite (α-$Al(OH)_3$). Each of these phase changes will result in different solubility patterns of the mineral studied. As a result of this, if one studies the solubility of a mineral in water over long equilibration times, then the original mineral present in the reaction vessel may change. If this crystallographic change is not detected, then the reported solubility properties of the starting material may have some errors.

Figure 2-16 illustrates the general pH-dependent patterns of precipitation and solubility studies. A supersaturated solution will precipitate, while an undersaturated solution will remain soluble. A supersaturated solution may form an intermediate, more soluble solid product. Some solutions may be only slightly supersaturated and can remain without forming a precipitate for a relatively long period. Conversely, some solid suspensions may have solutions that are only slightly undersaturated and can remain this way for a relatively long period. These conditions are referred to as metastable solutions. If these metastable conditions occur during solubility studies, then the accuracy of the determined solubility constants will be affected.

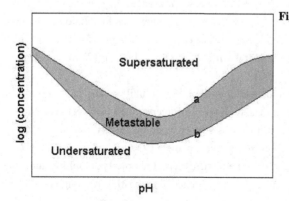

Figure 2-16: Schematic diagram of solubility. As a precipitate forms in a supersaturated solution, an intermediate product (often amorphous) will typically form first, noted by line "a". It will then slowly change into a more stable product (often crystalline), noted by line "b". *After Stumm and Morgan (1981).*[54]

[2.14] Solubility of Gases & Henry's Law

William Henry (1774–1836) studied the solubility of gases in liquids and is well known today for his gas–liquid partitioning equation of 1801. At equilibrium conditions between air and water, the dissolution of gases in the water can be predicted from Henry's Law:

$$a_A = K_H P_A \qquad [2\text{-}98],$$

where a_A = the activity of gas A in the soil solution (mol L^{-1}), P_A = the partial pressure of gas A in the air (atm), and K_H = Henry's Law constant for gas A. The "H" subscript serves as a reminder that this constant is defined based on Henry's Law. Note the difference on how a solid–liquid phase equilibrium (Equation [2-95] or [2-96]) is expressed with how an air–liquid phase equilibrium (Equation [2-98]) is expressed. This is because the partial pressure of the gas is not a constant and, therefore, remains a separate, easily quantifiable parameter of the K_H equilibrium constant definition. Conversely, the solid-phase activity is not easy to quantify and is, therefore, factored into the K_{so} equilibrium constant definition.

Equation [2-98] is a special case of the partitioning law, where the ratio of the activity of a compound in Liquid 1 and Liquid 2 is constant:

$$\text{partition coefficient, } K_p = \frac{(\text{activity in Liquid 1})}{(\text{activity in Liquid 2})} \qquad [2\text{-}99].$$

Table 2-9 lists the K_H values for numerous gases. The direction of motion of the molecules and the limit of mass transfer can be predicted based on Henry's Law. For example, when the soil air $CO_2(g)$ content is too high (> C_A/K_H), the excess molecules migrate into the soil solution. Conversely, if the soil solution $CO_2(aq)$ content is too high (> $K_H P_A$), the excess molecules migrate into the soil air.

Note that the rate of gas exchange between the two phases (air and water) may be slow due to poor mixing or lack of adequate agitation in the matrix. As a result of this, the hour-to-hour and day-to-day changes in the $CO_2(g)$ concentration in the soil air can be faster than the rate of gas transfer between the two phases. Consequently, true instantaneous equilibrium will not be present in a given field most of the time.

Table 2-9: Henry's Law constant for various gases at 25°C. Values decrease as temperature increases. *Data from Lide (2002).*[2]

Gas	K_H, mol L^{-1} atm^{-1}	Gas	K_H, mol L^{-1} atm^{-1}
He	3.88×10^{-4}	N_2	6.57×10^{-4}
H_2	7.83×10^{-4}	N_2O	2.43×10^{-2}
Rn	9.28×10^{-3}	NO	1.93×10^{-3}
Cl_2	9.23×10^{-2}	CO_2	3.42×10^{-2}
O_2	1.27×10^{-3}	CO	9.85×10^{-4}
O_3	6.57×10^{-4}	CH_4	1.42×10^{-3}
SO_2	1.40	C_2H_6	1.89×10^{-3}
H_2S	1.03×10^{-1}	C_3H_8	1.52×10^{-3}

[2.14.1] Impact of Reaction with Solvent on K_H

Equation [2-98] works well for gases that do not react with the solvent because these reactions will strongly enhance the apparent solubility of the gas. When reactions with the solvent occur, such as pH-dependent speciation reactions, then Equation [2-98] may be replaced with an effective Henry's Law constant (K_H^{eff}):

$$C_{A,Total} = K_H^{eff} P_A \tag{2-100},$$

where $C_{A,Total}$ = the total concentration of gas A in the soil solution (mol L^{-1}). Correlating K_H^{eff} to K_H will depend on the specifics of the reaction(s) of the dissolved gas with the solvent.

For example, let HL represent a gaseous compound that is soluble in water. Now assume that it can also dissociate in water to form H$^+$ (hydronium ion) and L$^-$ species with an acidity constant $K_a = (H^+)(L^-)/(HL)$. From conservation of mass

$$C_{HL,Total} = C_{HL} + C_L \tag{2-101},$$

and since HL is in equilibrium with L$^-$

$$C_{HL,Total} = C_{HL} + \frac{K_a C_{HL} \gamma_{HL}}{C_H \gamma_H \gamma_L} = C_{HL}\left(1 + \frac{K_a \gamma_{HL}}{C_H \gamma_H \gamma_L}\right) \tag{2-102},$$

or

$$C_{HL} = \frac{C_{HL,Total}}{1 + \dfrac{K_a \gamma_{HL}}{C_H \gamma_H \gamma_L}} = C_{HL,Total}\left(\frac{C_H \gamma_H \gamma_L}{C_H \gamma_H \gamma_L + K_a \gamma_{HL}}\right) \tag{2-103}.$$

From Equation [2-98]

$$K_H P_{HL} = a_{HL} = C_{HL} \gamma_{HL} = C_{HL,Total}\left(\frac{C_H \gamma_H \gamma_L \gamma_{HL}}{C_H \gamma_H \gamma_L + K_a \gamma_{HL}}\right) \tag{2-104},$$

$$\left(\frac{C_H \gamma_H \gamma_L + K_a \gamma_{HL}}{C_H \gamma_H \gamma_L \gamma_{HL}}\right) K_H P_{HL} = \left(\frac{1}{\gamma_{HL}} + \frac{K_a}{C_H \gamma_H \gamma_L}\right) K_H P_{HL} = C_{HL,Total} \tag{2-105},$$

or

$$K_H^{eff} P_{HL} = C_{HL,Total} \tag{2-106}$$

where

$$K_H^{eff} = K_H\left(\frac{1}{\gamma_{HL}} + \frac{K_a}{C_H \gamma_H \gamma_L}\right) \tag{2-107}.$$

Typically, it is easier or more convenient to measure the total concentration of a dissolved gas in water than it is to measure the concentration of a particular gaseous species (namely, HL(aq) in the example above). Accordingly, Equation [2-98] would not be practical to use if $C_{HL,Total}$ is known instead of C_{HL} (or a_{HL} or $C_{HL}\gamma_{HL}$). Equation [2-106] would be used instead. Note that in the example above, the effective gas–liquid equilibrium constant is a function of the pH of the solution (notice the inclusion of $C_H\gamma_H$ in Equation [2-107]). The additional terms in Equation [2-107] reflect the impact of the pH-dependent reaction of the gas with the water on the gas–liquid equilibrium.

[2.14.2] Impact of the Interface Curvature on K_H

Henry's Law (Equation [2-98]) assumes that the gas–liquid phase boundary is a flat surface. However, in moist soil environments, we have many different shapes present for water, ranging from the concave surfaces of droplets on hydrophobic organic matter (which is particularly common in soils following a forest fire) to the convex surfaces in the small hydrophilic capillary pores of soils.

The curvature of a liquid surface affects the ability of molecules at the gas–liquid interface to escape the attractive forces of the neighboring molecules in the liquid. The attractive forces between the liquid molecules are changed by the curvature at the interface. This phenomenon is described by the Kelvin equation[55]:

$$P_A^r = P_A^\infty \exp\left(\frac{-2\gamma V_L}{rRT} \cos\theta\right) \qquad [2\text{-}108],$$

where P_A^r = true partial pressure (p/p_o) over a curved liquid surface with radius r_m, P_A^∞ = partial pressure measured over a flat liquid surface, γ = surface tension, V_L = molar volume of the liquid (= molar mass / density), R = gas constant, T = absolute temperature, θ = contact angle, and r = radius of the pore. Note that r_m = radius of curvature of the meniscus, and $r_m \cos\theta = r$ (see Figure 5-17). The partial pressure of water vapor above the liquid is equal to the relative humidity (RH).

If the liquid does not wet the sides of the pore, then a concave meniscus is formed in the pore at the gas–liquid interface, $\theta > 90°$, and the right-hand side of Equation [2-108] is positive. For droplets sitting on a hydrophobic material or for aerosols, let $\theta = 180°$ and $r = r_m$, which again means that the right-hand side of Equation [2-108] is positive.

Henry's Law (Equation [2-98]) uses the partial pressure of gas A above a flat liquid surface because Henry's constant K_H for gas A is based on measurements across a flat liquid surface. With $P_A = P_A^\infty$, combining Equation [2-98] with [2-108] yields

$$a_A = K_H P_A^r \exp\left(\frac{2\gamma V_L}{rRT} \cos\theta\right) \qquad [2\text{-}109].$$

From Equation [2-108] we see that, for a droplet on a surface, or an aerosol or raindrop in the air, as the radius decreases (as $r \to 0$ and $\cos\theta \approx -1$), the concentration of A in the gaseous phase increases. From Equation [2-109], we see that the smaller the droplet, the lower the concentration of volatile species (a_A) in the droplet.

The same is also true in pore spaces when the liquid does not wet the pore walls and the meniscus is concave. Here, the amount of volatile species in the liquid-filled regions of the pore decreases as the pore radius decreases.

The opposite is true when the liquid wets the pore walls and the meniscus is convex. In these pore spaces, the amount of volatile species in the liquid-filled regions of the pore increases as the pore radius decreases (as $r \to 0$ and $\cos\theta \approx 1$). Accordingly, for a given partial pressure in the soil air spaces, the dissolved volatile gases will be more concentrated in the smallest of the liquid-filled pore channels and less concentrated in the larger ones.

[2.15] Aqueous Ion Speciation: Closed Systems

Quantifying the concentration of aqueous compounds and their numerous species as a function of pH will be illustrated below by way of two examples: phosphates and carbonates. We assume here a closed system, which means that the total amount of each element is constant. That is, there is no external source of any kind (e.g., solid-phase dissolution or atmospheric inputs) that would result in an essentially limitless supply of an element or compound.

[2.15.1] Phosphate Speciation

When phosphate dissolves in water, various aqueous species are formed. With each of the corresponding reactions with water, the law of mass action must be obeyed, which means that a specific balance between the reactants and the products must be maintained at equilibrium. To predict the relative concentration of each of the aqueous species, first construct a list of all the possible reactions of phosphate in water:

$$H_3PO_4 \rightleftharpoons H_2PO_4^- + H^+ \qquad\qquad [2\text{-}110],$$

$$H_2PO_4^- \rightleftharpoons HPO_4^{2-} + H^+ \qquad\qquad [2\text{-}111],$$

$$HPO_4^{2-} \rightleftharpoons PO_4^{3-} + H^+ \qquad\qquad [2\text{-}112].$$

Next, an equilibrium acidity constant is assigned to each of these reactions:

$$K_1 = \frac{(H_2PO_4^-)(H^+)}{(H_3PO_4)} = 10^{-2.12}, \quad pK_1 = 2.12 \qquad\qquad [2\text{-}113],$$

$$K_2 = \frac{(HPO_4^{2-})(H^+)}{(H_2PO_4^-)} = 10^{-7.20}, \quad pK_2 = 7.20 \qquad\qquad [2\text{-}114],$$

$$K_3 = \frac{(PO_4^{3-})(H^+)}{(HPO_4^{2-})} = 10^{-12.33}, \quad pK_3 = 12.33 \qquad\qquad [2\text{-}115].$$

Since the sum of all the phosphate species is constant, we obtain from the principle of mass balance (which is similar to the law of conservation of mass):

$$C_T = (H_3PO_4) + (H_2PO_4^-) + (HPO_4^{2-}) + (PO_4^{3-}) \qquad\qquad [2\text{-}116],$$

where C_T = total concentration. Note that we have four unknowns (i.e., the four phosphate species) and four equation (Equations [2-113] to [2-116]). The independent variable is pH and C_T is known or, as we shall see shortly, it can be ignored if the results are expressed in terms of percentages or fractions of total concentrations. Accordingly, the problem becomes strictly algebraic from this point onward. For example, to determine the relative concentration of PO_4^{3-}, all the parameters in Equation [2-116] are converted with the use of Equations [2-113] to [2-115] so that it is expressed in PO_4^{3-} concentration terms:

$$C_T = \frac{(PO_4^{3-})(H^+)^3}{K_1K_2K_3} + \frac{(PO_4^{3-})(H^+)^2}{K_2K_3} + \frac{(PO_4^{3-})(H^+)}{K_3} + (PO_4^{3-}) \qquad [2\text{-}117].$$

Next, rearrange and simplify the numerical expression:

$$\frac{C_T}{(PO_4^{3-})} = \frac{(H^+)^3}{K_1K_2K_3} + \frac{(H^+)^2}{K_2K_3} + \frac{(H^+)}{K_3} + 1 \qquad [2\text{-}118],$$

$$\frac{C_T}{(PO_4^{3-})} = \frac{(H^+)^3 + K_1(H^+)^2 + K_1K_2(H^+) + K_1K_2K_3}{K_1K_2K_3} \qquad [2\text{-}119],$$

$$\frac{(PO_4^{3-})}{C_T} = \frac{K_1K_2K_3}{(H^+)^3 + K_1(H^+)^2 + K_1K_2(H^+) + K_1K_2K_3} \qquad [2\text{-}120].$$

Equation [2-120] describes the relative concentration, or fraction, of the phosphate anion in water as a function of the pH of the liquid solution. Similar expressions can be obtained with the other phosphate species when the substitutions for PO_4^{3-} made in Equation [2-117] are changed for the appropriate substitutions of HPO_4^{2-}, $H_2PO_4^-$, or H_3PO_4. Accordingly, the fraction of each phosphate species is quantified as follows:

$$\alpha_1 = \frac{(H_3PO_4)}{C_T} = \frac{(H^+)^3}{D} \qquad [2\text{-}121],$$

$$\alpha_2 = \frac{(H_2PO_4^-)}{C_T} = \frac{K_1(H^+)^2}{D} \qquad [2\text{-}122],$$

$$\alpha_3 = \frac{(HPO_4^{2-})}{C_T} = \frac{K_1K_2(H^+)}{D} \qquad [2\text{-}123],$$

$$\alpha_4 = \frac{(PO_4^{3-})}{C_T} = \frac{K_1K_2K_3}{D} \qquad [2\text{-}124],$$

where $D = (H^+)^3 + K_1(H^+)^2 + K_1K_2(H^+) + K_1K_2K_3$. Figure 2-17A plots the fraction (α_i) of the various phosphate species in solution as a function of pH. The logarithm of each of the fractions can also be plotted as a function of pH. It is easy to see that the logarithm of Equations [2-121] to [2-124] will yield pH-dependent slopes with ratios of 3:1, 2:1, and 1:1 (Figure 2-17B). To determine the actual concentration of each of the phosphate fractions, the total phosphate concentration (C_T) must be known, which is often very easy to determine experimentally.

[2.15.2] Carbonate Speciation

As was discussed earlier in Section 2.1, inorganic C is ubiquitous in the soil environment. Depending on the pH, various carbonate species exist in water. The gaseous CO_2 tries to be in chemical equilibrium with the aqueous CO_2 (see Section 2.14):

$$CO_2(g) \rightleftharpoons CO_2(aq), \qquad K_H = 10^{-1.5} \qquad [2\text{-}125].$$

The partial pressure of $CO_2(g)$ in the atmosphere is 0.03%, or $10^{-3.45}$ atm. Using Henry's Law (Equation [2-98]) the concentration of $CO_2(aq)$ in equilibrium with the atmosphere is 10^{-5} M. Note that this aqueous CO_2 is a different species from carbonic acid, $H_2CO_3(aq)$. Both

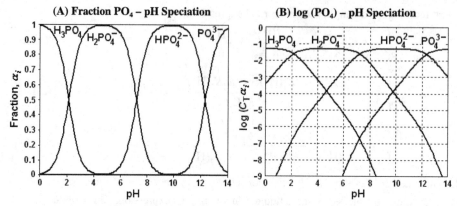

Figure 2-17: Speciation diagrams of phosphate as a function of pH, based on pK_i values shown in Equations [2-113] to [2-115] and assuming $(PO_4)_{total} = C_T = 0.05\ M$. (A) Fraction of phosphate species (α_i) versus pH. The sum of the lines must equal 100%. The lines cross at the pK_i values. (B) Log of phosphate species $(\log C_T \alpha_i)$ versus pH. Note the slope change from a horizontal line to a 1:1 slope, or from a 1:1 slope to a 2:1 slope, at each pK_i value. The change in slope is gradual over a ± 1 pH range. The sum of all of the lines must equal $\log C_T$. Note that $\log 0.5 = -0.30 =$ vertical drop shown on the log–log diagram whenever two lines cross at their corresponding pK_i values.

molecules are in solution due to the H bonds that they form with the surrounding water molecules. The central carbon in $CO_2(aq)$ forms covalent bonds with only two oxygens, while the central carbon in $H_2CO_3(aq)$ forms covalent bonds with three oxygens. These two species are in equilibrium with each other in water:

$$CO_2(aq) + H_2O \rightleftharpoons H_2CO_3(aq)\,, \quad K = \frac{(H_2CO_3(aq))}{(CO_2(aq))} = 10^{-2.5} \qquad [2\text{-}126].$$

Again, note that $a_{H_2O} = 1$ by convention. The concentration of $H_2CO_3(aq)$ is therefore always around 0.3% the concentration of $CO_2(aq)$. Because both of these species coexist in water at the same ratio for all pH conditions, they are sometimes referred to jointly as $H_2CO_3^*(aq)$, where $[H_2CO_3^*(aq)] = [H_2CO_3(aq)] + [CO_2(aq)]$. The asterisk serves as a convenient shorthand notation when both species are implied.

To study the pH-dependent speciation of carbonates, we first write down the reactions of inorganic C with water:

$$H_2CO_3^* \rightleftharpoons HCO_3^- + H^+ \qquad\qquad [2\text{-}127],$$

$$HCO_3^- \rightleftharpoons CO_3^{2-} + H^+ \qquad\qquad [2\text{-}128],$$

and their corresponding equilibrium acidity constants are

$$K_1 = \frac{(HCO_3^-)(H^+)}{(H_2CO_3^*)} = 10^{-6.35}\,, \quad pK_1 = 6.35 \qquad\qquad [2\text{-}129],$$

$$K_2 = \frac{(CO_3^{2-})(H^+)}{(HCO_3^-)} = 10^{-10.33}\,, \quad pK_2 = 10.33 \qquad\qquad [2\text{-}130].$$

Note that the acidity constant K_1 is a composite of the hydration reaction (Equation [2-126]) and the protolysis of true $H_2CO_3(aq)$:

$$H_2CO_3 \rightleftharpoons HCO_3^- + H^+ \qquad [2\text{-}131],$$

where

$$K_{H_2CO_3} = \frac{(HCO_3^-)(H^+)}{(H_2CO_3)} \qquad [2\text{-}132].$$

Expanding the denominator of Equation [2-129], and combining with Equations [2-126] and [2-132] yields

$$K_1 = \frac{K_{H_2CO_3}}{1 + K} \qquad [2\text{-}133].$$

In a closed system, the total amount of inorganic C is constant and is defined in a similar fashion as Equation [2-116]:

$$C_T = (H_2CO_3^*) + (HCO_3^-) + (CO_3^{2-}) \qquad [2\text{-}134].$$

Once again, determining the pH-dependent fraction of each species is an algebraic exercise from this point onward (similar to Equations [2-117] to [2-120]). The fractions of each carbonate species are quantified as follows:

$$\alpha_1 = \frac{(H_2CO_3^*)}{C_T} = \frac{(H^+)^2}{D} \qquad [2\text{-}135],$$

$$\alpha_2 = \frac{(HCO_3^-)}{C_T} = \frac{K_1(H^+)}{D} \qquad [2\text{-}136],$$

$$\alpha_3 = \frac{(CO_3^{2-})}{C_T} = \frac{K_1 K_2}{D} \qquad [2\text{-}137],$$

where $D = (H^+)^2 + K_1(H^+) + K_1 K_2$. Figure 2-18A plots the fraction (α_i) of the various carbonate species in solution as a function of pH. Figure 2-18B plots the logarithm of the concentration of each carbonate fraction as a function of pH while assuming a total inorganic-C (TIC) concentration of $10^{-3}\ M$. This concentration is close to the TIC values in

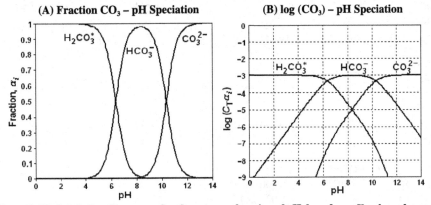

(A) Fraction CO₃ – pH Speciation   **(B) log (CO₃) – pH Speciation**

Figure 2-18: Speciation diagrams of carbonate as a function of pH, based on pK_i values shown in Equations [2-129] and [2-130] and assuming $(CO_3)_{Total} = C_T = 0.001\ M$. (A) Fraction of carbonate species (α_i) versus pH. (B) Log of carbonate species ($\log C_T \alpha_i$) versus pH.

nearly all river waters, and it is reasonably close to the TIC concentration of most soil surface runoff waters, but for rainwater TIC $\approx 10^{-5}$ M. Remember, however, that the underground levels of TIC can reach very high concentrations, as was discussed earlier in Section 2.1.

The actual concentration of each carbonate fraction depends on the total concentration of carbonate present (C_T). If the liquid is clean water at low pH (< 6), and it is in equilibrium with the atmosphere, then from Equation [2-98]: $C_T = 10^{-1.5} \times 10^{-3.45} \approx 10^{-5}$ M (note $\gamma = 1$ and $c = a$ at these low concentrations). In nature, however, the water is not very clean, but rather contains significant amounts of soluble ions including soluble carbonate minerals. The sum of the cations present equals the sum of the anions present, and the electroneutrality of the solution is maintained. Since the H^+ and OH^- ions are included in this charge balance, the pH of the system is also affected. Section 2.1 discussed the average total carbonate concentrations in various natural waters, varying significantly as a result of the seasonal and spatial biological activity.

The average composition of river water and seawater is described in Table 2-10. As a first approximation, the average composition of soil solutions can be presumed to be similar to the composition of the water leaving the soil and entering the nearby streams and rivers. Accordingly, the composition of the nearby streams mimics the composition of the soil solution. The problem with this initial assumption, however, is that the soil solution can be concentrated by evapotranspiration processes and freezing fronts, or it can be diluted with

Table 2-10: Composition of normal seawater, world average river water, and rainfall in the northeastern United States. Note that the concentration of some of the components will vary considerably with time and sampling location due to biological activity (particularly with carbonates, silicates, and phosphates). *Adapted from J.N. Butler (1982)[5], Carbon dioxide equilibria and their applications. Copyright © 1982. Adapted by permission of Pearson Education, Inc., Upper Saddle River, NJ.*

Component	Seawater, mmol kg^{-1}	River water, mmol kg^{-1}	Rainfall, µmol L^{-1}
H^+			72.3 ± 12.3
K^+	10.00	0.059	
Na^+	468.04	0.274	9.3 ± 12.2
NH_4^+			16.0 ± 4.6
Ca^{2+}	10.33	0.375	
Mg^{2+}	53.27	0.169	
Sr^{2+}	0.10		
Br^-	0.83		
Cl^-	545.88	0.220	12.1 ± 12.7
F^-	0.07	0.0053	
NO_3^-	0.0001 to 0.05	0.017	25.9 ± 5.0
SO_4^{2-}	28.20	0.117	28.1 ± 3.9
$B(OH)_3 + B(OH)_4^-$	0.43		
$HCO_3^- (+ CO_2 + CO_3^{2-})$	2.2 to 2.5	0.957	
$H_2PO_4^- + HPO_4^{2-} + PO_4^{3-}$	0.0001 to 0.005		
$Si(OH)_4 + SiO(OH)_3^-$	0.001 to 0.1	0.218	
Ionic strength	700	2.09	
pH	7.4 to 8.3	6.0 to 8.5	4.14 ± 0.07

rainwater, irrigation water, or melting snow and ice. Furthermore, these changes occur in a confined, poorly vented environment that can experience very high concentrations of trapped gases, particularly $CO_2(g)$ during the summer months (see Figures 2-1 and 2-2).

[2.16] Aqueous Ion Speciation: Open Systems

[2.16.1] Carbonate Speciation

The speciation of ions in the closed systems illustrated in the previous section will exist in nature wherever the total concentration of the ions present is limited. When the amount of a particular compound is essentially limitless, then the resulting pH-dependent concentration of its various aqueous species will change drastically. With the previous $CO_2(aq)$ speciation example (Section 2.15.2), a closed environment can be found in capped containers in the laboratory or in highly restricted porous cavities in soils. However, the exchange of gases through soils, albeit slow at times, often makes them open systems.

The mathematical predictions of aqueous carbonate speciation in an open system is modified as follows. Equation [2-134] is no longer helpful because the total concentration of inorganic C is not constant. The concentration of $H_2CO_3^*$ is pH independent due to its equilibrium with the atmosphere (Equation [2-125]) and, in pure water, is calculated based on Henry's Law:

$$[H_2CO_3^*] = K_H P_{CO_2(g)} \qquad [2\text{-}138].$$

Substituting Equation [2-138] into Equations [2-129] and [2-130], we obtain

$$[HCO_3^-] = \frac{K_1 K_H P_{CO_2(g)}}{(H^+)} \qquad [2\text{-}139],$$

$$[CO_3^{2-}] = \frac{K_1 K_2 K_H P_{CO_2(g)}}{(H^+)^2} \qquad [2\text{-}140].$$

Figure 2-19 plots the logarithmic concentration of each of these inorganic-C species in pure water in equilibrium with a moderately high partial pressure of $CO_2(g)$. In soil

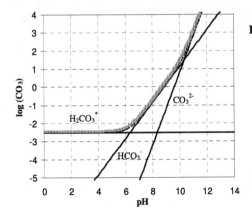

Figure 2-19: Speciation diagram of carbonate as a function of pH in equilibrium with an atmosphere having a $CO_2(g)$ partial pressure of 0.1 atm. The slope of each line is: 1:0 (horizontal) for $H_2CO_3^*$, 1:1 for HCO_3^-, and 1:2 for CO_3^{2-}. The total inorganic-C concentration is the sum of the three aqueous carbonate species, which is also shown.

environments, the partial pressure of $CO_2(g)$ is usually much higher than what is present in the atmosphere (see Section 2.1). Keep in mind that the high aqueous concentrations of inorganic C at high pH (such as the high values shown in Figure 2-19) will only be present in the soil solution if the soil air is proportionally large enough to supply a "limitless" quantity of gaseous CO_2. If the supply of gaseous inorganic C is depleted, then the lines in Figure 2-19 will change and mimic the slopes shown in Figure 2-18B.

If a solution of inorganic C is in equilibrium with the gaseous phase and the partial pressure of the $CO_2(g)$ is forced to decrease (e.g., it is purged or vented), then the liquid will degas. The pH will always increase because protons are consumed when the bicarbonate and carbonate ions are removed from the liquid. That is, the irreversible process of $CO_3^{2-} \rightarrow HCO_3^-$ $\rightarrow H_2CO_3^*(aq) \rightarrow CO_2(g)$, which is pulled to the right, will remove protons in the first two steps. Conversely, if additional $CO_2(g)$ is added so as to raise the partial pressure of this gas, then the liquid phase will absorb inorganic C and the pH will decrease. Hence, the large fluctuations in the partial pressures of $CO_2(g)$ in soil environments will influence the equilibrium pH of the soil solution. This impact is often minimized, however, by the buffering capacity of the soil.

[2.16.2] Aluminum Speciation

Another example of an open system is one where a solid mineral phase is present. The supply of a soluble component has the same effect that atmospheric $CO_2(g)$ had in the previous example. This is illustrated here with a soluble Al oxide mineral (α-$Al(OH)_3$), whose pH-dependent solubility affects the concentration of the various aqueous Al species.

The aqueous Al species are quantified as before: (1) describe the reaction, (2) construct the definition of the equilibrium constant, (3) insert an appropriate value for the equilibrium constant (resort to published data for this), and (4) solve the equation for the aqueous species in question. Note that the concentration of aqueous $Al(OH)_3$ is constant when the solid phase is present. Note also that the concentration of H_2O, which is essentially a constant ($\approx 55.56\ M$), is factored into the equilibrium constants (K_i values). As discussed earlier, the convention is to set the activity of water equal to unity whenever it appears in an equilbrium constant definition, $(H_2O) = 1.0$, which yields the same conclusions. The following reactions and equations are used to construct the speciation diagram for aqueous aluminum:

$$\alpha\text{-}Al(OH)_3(s) \rightleftharpoons Al(OH)_3(aq) \qquad\qquad\qquad [2\text{-}141],$$

$$K_{so} = (Al(OH)_3(aq)) = 10^{-6.5} \qquad\qquad [2\text{-}142],$$

$$\log (Al(OH)_3(aq)) = -6.5 \qquad\qquad [2\text{-}143],$$

$$Al(OH)_3(aq) + H^+ \rightleftharpoons Al(OH)_2^+ + H_2O \qquad\qquad [2\text{-}144],$$

$$K_1 = \frac{(Al(OH)_2^+)}{(Al(OH)_3(aq))(H^+)} = 10^{5.7} \qquad\qquad [2\text{-}145],$$

$$\log (Al(OH)_2^+) - \log (Al(OH)_3(aq)) - \log (H^+) = 5.7$$
$$\log (Al(OH)_2^+) = 5.7 + \log (Al(OH)_3(aq)) + \log (H^+)$$
$$\log (Al(OH)_2^+) = 5.7 - 6.5 - pH$$
$$\log (Al(OH)_2^+) = -0.8 - pH \qquad\qquad \text{[2-146]},$$

$$Al(OH)_3(aq) + 2H^+ \rightleftharpoons AlOH^{2+} + 2H_2O \qquad\qquad \text{[2-147]},$$

$$K_2 = \frac{(AlOH^{2+})}{(Al(OH)_3(aq))(H^+)^2} = 10^{10.03} \qquad\qquad \text{[2-148]},$$

$$\log (AlOH^{2+}) - \log (Al(OH)_3(aq)) - 2\log (H^+) = 10.03$$
$$\log (AlOH^{2+}) = 10.03 + \log (Al(OH)_3(aq)) + 2\log (H^+)$$
$$\log (AlOH^{2+}) = 10.03 - 6.5 - 2pH$$
$$\log (AlOH^{2+}) = 3.53 - 2pH \qquad\qquad \text{[2-149]},$$

$$Al(OH)_3(aq) + 3H^+ \rightleftharpoons Al^{3+} + 3H_2O \qquad\qquad \text{[2-150]},$$

$$K_3 = \frac{(Al^{3+})}{(Al(OH)_3(aq))(H^+)^3} = 10^{15.0} \qquad\qquad \text{[2-151]},$$

$$\log (Al^{3+}) - \log (Al(OH)_3(aq)) - 3\log (H^+) = 15.0$$
$$\log (Al^{3+}) = 15.0 + \log (Al(OH)_3(aq)) + 3\log (H^+)$$
$$\log (Al^{3+}) = 15.0 - 6.5 - 3pH$$
$$\log (Al^{3+}) = 8.5 - 3pH \qquad\qquad \text{[2-152]},$$

$$Al(OH)_3(aq) + H_2O \rightleftharpoons Al(OH)_4^- + H^+ \qquad\qquad \text{[2-153]},$$

$$K_4 = \frac{(Al(OH)_4^-)(H^+)}{(Al(OH)_3(aq))} = 10^{-8.0} \qquad\qquad \text{[2-154]},$$

$$\log (Al(OH)_4^-) - \log (Al(OH)_3(aq)) + \log (H^+) = -8.0$$
$$\log (Al(OH)_4^-) = -8.0 + \log [Al(OH)_3(aq)] - \log (H^+)$$
$$\log (Al(OH)_4^-) = -8.0 - 6.5 + pH$$
$$\log (Al(OH)_4^-) = -14.5 + pH \qquad\qquad \text{[2-155]}.$$

The lines drawn in Figure 2-20 correspond to Equations [2-143], [2-146], [2-149], [2-152], and [2-155]. Note that the slope of each line (e.g., 1:0, 1:1, 1:2, and 1:3) reflects the number of protons involved in each reaction. At a specific pH value, if a solution contains a total Al concentration that is above the sum of all of these lines, then the solution is supersaturated and a precipitate of α-Al(OH)$_3$(s) is expected to form. The precipitation process will shift the pH of the solution because each of the aqueous Al ions will release or consume hydroxyl anions as the α-Al(OH)$_3$(s) is formed. The pH will shift away from 6.85 (($pK_4 - pK_1$)/2 = (8.0 + 5.7)/2 = 6.85); it will lower if the supersaturated solution pH is less then 6.85, or rise if the supersaturated solution pH is greater than 6.85. Conversely, if a solution contains an Al concentration that is below the sum of all of these lines, then the solution is undersaturated and no precipitate will form, or the α-Al(OH)$_3$(s), if present, will continue to dissolve so as to bring the total aqueous Al concentration to a value that equals the sum of the lines. The dissolution process will affect the pH of the solution, but this time

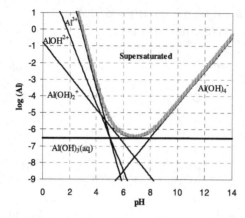

Figure 2-20: Speciation diagram of aqueous Al as a function of pH in equilibrium with α-Al(OH)$_3$(s). The solution is supersaturated if the total Al concentration in solution falls above the sum of these lines.

pulling the undersaturated solution pH toward 6.85. Hence, the processes of precipitation and dissolution affect the pH of the matrix. Similarly, the pH of soil samples are strongly affected by the minerals present.

[2.17] Speciation Modeling

It is easy to see that the longhand calculations of ion speciation discussed in Sections 2.15 and 2.16 can quickly grow into an overwhelmingly large algebraic exercise if multiple reactions are present. The soil aqueous environment is a highly complex environment that will often render a speciation exercise into one that is too difficult to solve by hand. Fortunately, there are numerous computer software programs available that were specifically written to solve these kinds of equations, and several of these are listed in Table 2-11. Instead of generating exact solutions, such as those illustrated by Equations [2-135] to [2-137], computer programs use numerical methods to solve the problems. Some of these computer programs will also model surface adsorption reactions.

The most common approach by computer programs to solve the equilibrium composition is the equilibrium constant approach. A series of equilibrium, or mass action, relationships are linked with mass balance equations for all the elements in the solution. A set of linear equations are generated, which are then solved by a series of iterative loops that seek the solution with the least amount of error. The error estimate is generally based on the conservation of mass; that is, the resultant total mass of the matrix (sum of all the final elements) is compared with the initial total mass (sum of all the initial elements).

A key component of all of these computer software programs is the availability of reliable equilibrium stability constants. Table 2-12 is a short list of various aqueous stability constants. A more comprehensive list can be found in published lists, such as in the six volumes of data compiled by Smith and Martell (1974 to 1989).[56–61] Unfortunately, there is often a wide variation (sometimes as large as an order of magnitude) in the stability constants reported in the literature, and this can result in a problem for the end user who may not have the time or the skill to evaluate which stability constant is most appropriate for his or her situation. These errors in the literature arise from the presence of errors in ligand purity and experimental measurements or calculations. A more detailed discussion of how to select the "most correct" equilibrium constant from the literature is given by Smith and Martell (1995).[62] Validation of an aqueous species concentration predicted by a computer program is often necessary, which typically involves comparing the predicted results with measured experimental results.

In these speciation models, it is presupposed that equilibrium exists for all the reactions in the system. Fortunately, many aqueous reactions occur very rapidly. In well stirred systems, adsorption and desorption reactions are also fast, often limited only by the kinetics of diffusion through the thin films that surround all particles or through the porous solid–liquid environment. In systems that cannot be stirred, such as natural soil environments, the diffusion rates are much slower and equilibrium conditions may take several hours or days. Note also that in natural soil environments, the total concentration of the compounds

Table 2-11: Some computer software programs used for speciation modeling in dilute aqueous systems ($I \leq 0.5\ M$). The input of data to some of the older programs (particularly those that did not run on Windows) can be quite complicated, resulting in the need for additional user interactive software programs that assist in the proper construction of the input data. Most of these models can also be used to predict surface adsorption and solid precipitation reactions.

VMINTEQ: Windows version of MINTEQA2, developed by Gustafsson (2002)[63] with funding from the Swedish research councils VR and MISTRA. Free software is available.[63]

MINTEQA2: It has an extensive thermodynamic database. First developed by Allison et al. (1990)[64] of the USEPA, it was last modified in 2000 (ver. 4.02). Free MINTEQA2 software available at http://www.epa.gov/ceampubl/mmedia/minteq/index.htm. Various other useful related software programs may be found at http://www.epa.gov/ceampubl/.

    MINTEQ: predecessor to MINTEQA2, originally developed by Felmy et al. (1984)[65] at Battelle Pacific Northwest Laboratory (PNL). It had the mathematical structure of MINEQL and the thermodynamic database of WATEQ3.[66]

MINEQL+: This Windows version uses a similar database to the one supported by the USEPA's MINTEQA2 database. MINEQL+ is distributed and sold by Environmental Research Software of Hallowell, ME, USA (http://www.mineql.com/).

    MINEQL: developed by Westall et al. (1976)[67] at the Massachusetts Institute of Technology based on the earlier REDEQL model of Morel and Morgan (1972).[68] MINEQL introduced the fundamental algorithms that were subsequently incorporated in numerous other software programs for speciation modeling.

WATEQ4F: Although the User's Manual to this USGS software program was last released by Ball and Nordstrom in 1991[69], the program continues to be updated. The most recent release was in 2002 (ver. 2.46). A free copy of WATEQ4F is available at http://wwwbrr.cr.usgs.gov/projects/GWC_chemtherm/index.htm.

    WATEQ: developed by Truesdell and Jones (1974)[70], it used a simple iteration scheme to solve the aqueous speciation problems. Numerous modifications followed, namely WATEQF, WATEQ2, WATEQ3[66], and WATEQ4F.

PHREEQC: developed by Parkhurst in 1995[71], replacing its predecessor PHREEQE by Parkhurst et al. (1980).[72] The most recent release was in 2002 (ver. 2.6)[73], and a free copy is available at http://wwwbrr.cr.usgs.gov/projects/GWC_coupled/phreeqc/.

present may change prior to the system achieving equilibrium conditions. One example of this is the rapidly and constantly changing CO_2 concentrations present in soils. Accordingly, predicting the products formed from the dissolution of solid phases or from the redox transformation of elements in the system can be very uncertain due to the strong dependence on reaction kinetics.

Precipitation reactions will also affect the balance of ions remaining in the liquid phase. Yet accurate predictions on which is the solid phase that will actually form are often impossible without prior knowledge of various other details about the mixture (such as how fast each of the components in the mixture were added or produced).

Table 2-12: Equilibrium constants of some compounds commonly found in aqueous soil environments. Values assume $a_{solids} = 1.0$, and $a_{H_2O} = 1.0$.

Equilibrium Reaction	pK_i at 25°C, $I = 0$
$Al(OH)_3(s, \alpha\text{-phase}) \rightleftharpoons Al(OH)_3(aq)$	6.5
$Al(OH)_3(aq) + H_2O \rightleftharpoons Al(OH)_4^- + H^+$	8.0
$Al(OH)_3(aq) + H^+ \rightleftharpoons Al(OH)_2^+ + H_2O$	−5.7
$Al(OH)_2^+ + H^+ \rightleftharpoons AlOH^{2+} + H_2O$	−4.31
$AlOH^{2+} + H^+ \rightleftharpoons Al^{3+} + H_2O$	−4.99
$Fe(OH)_3(s, am) + 3H^+ \rightleftharpoons Fe^{3+} + 3H_2O$	−3.2
$FeOOH(s, \alpha\text{-phase}) + 3H^+ \rightleftharpoons Fe^{3+} + 3H_2O$	−0.5
$\frac{1}{2}Fe_2O_3(s, \alpha\text{-phase}) + 3H^+ \rightleftharpoons Fe^{3+} + 3H_2O$	0.7
$Fe(OH)_2^+ + 2H_2O \rightleftharpoons Fe(OH)_4^- + 2H^+$	15.9
$Fe(OH)_2^+ + H^+ \rightleftharpoons FeOH^{2+} + H_2O$	−3.51
$FeOH^{2+} + H^+ \rightleftharpoons Fe^{3+} + H_2O$	−2.19
$Fe(OH)_2(s) \rightleftharpoons Fe(OH)_2(aq)$	7.7
$Fe(OH)_2(aq) + H_2O \rightleftharpoons Fe(OH)_3^- + H^+$	11.4
$Fe(OH)_2(aq) + H^+ \rightleftharpoons FeOH^+ + H_2O$	−11.1
$FeOH^+ + H^+ \rightleftharpoons Fe^{2+} + H_2O$	−9.5
$SiO_2(s, am) \rightleftharpoons Si(OH)_4(aq)$	2.74
$Si(OH)_4(aq) \rightleftharpoons Si(OH)_3O^- + H^+$	9.86
$Si(OH)_3O^- \rightleftharpoons Si(OH)_2O_2^{2-} + H^+$	13.1
$NH_3(aq) + H^+ \rightleftharpoons NH_4^+$	−9.244
$HNO_2 \rightleftharpoons NO_2^- + H^+$	3.15
$H_3PO_4 + H_2O \rightleftharpoons H_2PO_4^- + H^+$	2.12
$H_2PO_4^- + H_2O \rightleftharpoons HPO_4^{2-} + H^+$	7.20
$HPO_4^{2-} + H_2O \rightleftharpoons PO_4^{3-} + H^+$	12.33
$HSO_4^- + H_2O \rightleftharpoons SO_4^{2-} + H^+$	1.99
$HSeO_4^- + H_2O \rightleftharpoons SeO_4^{2-} + H^+$	1.70
$H_3AsO_4 + H_2O \rightleftharpoons H_2AsO_4^- + H^+$	2.24
$H_2AsO_4^- + H_2O \rightleftharpoons HAsO_4^{2-} + H^+$	6.96
$HAsO_4^{2-} + H_2O \rightleftharpoons AsO_4^{3-} + H^+$	11.50
$H_2CO_3^* + H_2O \rightleftharpoons HCO_3^- + H^+$	6.35
$HCO_3^- + H_2O \rightleftharpoons CO_3^{2-} + H^+$	10.33
formic acid, $HCOOH + H_2O \rightleftharpoons HCOO^- + H^+$	3.745
acetic acid, $CH_3COOH + H_2O \rightleftharpoons CH_3COO^- + H^+$	4.757
phenol, $C_6H_5\text{-}OH + H_2O \rightleftharpoons C_6H_5\text{-}O^- + H^+$	9.98

[2.18] Induction Time & Precipitation Processes

Supersaturated solutions occur often in soils, particularly when the soil water is cooled in winter or when the salt solution is concentrated through evaporation of the excess water. The solubility diagram shown in Figure 2-20 predicts that if an Al solution is supersaturated, then a precipitate will form. The calculations that produced Figure 2-20 assume that the precipitate is an α-phase Al hydroxide. However, numerous conditions in the supersaturated liqueur (for example: redox conditions, lack of interfering compounds, temperature, mixing rate) must be just right for the anticipated solid product to form. In a particular laboratory or field condition, the precipitate may be a different mineral than the predicted α-$Al(OH)_3(s)$. It is prudent, therefore, to confirm the crystallography of the precipitated mineral, such as with x-ray diffraction.

If a solid phase must be tentatively identified, then an initial but rough approach that can be useful is to assume that the mineral formed is the mineral with the smallest solubility constant. Having the lowest solubility constant also means that said mineral is probably the most stable mineral, and it will ultimately affect the availability of its elements in the liquid phase. But the most stable mineral might not form immediately because alternate minerals or intermediate minerals (minerals that will eventually be replaced by the formation of a more stable mineral) may easily form instead. Alternate minerals can form when (1) the differences in the saturation indices are small for several minerals, (2) the precipitation process is rapid, which does not favor the formation of structurally complex minerals, or (3) "poisons" are present in the solution that hinder the formation of a particular mineral.

In a supersaturated solution, the formation of a crystal nucleus is thermodynamically favored and results in the net release of energy. However, an energy barrier is present that must be overcome due to the fact that the ions must dehydrate in order to come close enough to each other to form the crystal bonds. It must also do this in a particular orientation depending on the mineral being formed. If the energy barrier is small, as it is for some mixtures, a light tap on the container may suffice to initiate the precipitation process.

Rapid cooling typically will raise the saturation index of the solution relative to a large number of mineral candidates that could form. If this occurs, there might not be any significant difference in the saturation index of the solution relative to any of the various mineral candidates. With rapid cooling, any one of (or mixture of) the mineral candidates could form. Rapid cooling of the mixture favors the formation of amorphous solids, simple mineral structures, or mixture of minerals, and disfavors the formation of complex mineral structures. For the formation of complex mineral structures, warm environments and lots of time is typically needed.

The time needed for the first crystal nucleus to form (that is, for nucleation to occur) is called the *induction time*. It may be very fast or slow, depending on the complexity and stability of the mineral structure, as well as various characteristics of the supersaturated solution. Nucleation is a complex process that involves the probability of getting all the right elements in the right place, all at nearly the same time for the first crystal to form. The process must continue before this initial crystal dissolves back into the liquid. An excellent way to guarantee and accelerate the formation of a particular mineral is to seed the supersaturated solution with the desired mineral, or with any mineral that has the desired structure (an isomorphic mineral). Crystal growth will generally follow the structural patterns laid out by the seed, and the energy barrier needed for nucleation to occur is circumvented by the addition of a seed to the mixture. The energy barriers for ion adsorption and crystal growth remain. The addition of a seed to the supernatant solution enhances the quality of the crystals formed in the mixture, and decreases the probability of forming unwanted crystal structures (e.g., amorphous solids, simple minerals, or a mixture of minerals). Even if the desired mineral is a structurally simple one, the addition of a seed will still greatly accelerate the formation and crystal growth of the desired mineral.

Figure 2-21 illustrates the induction time for the crystal growth of MgF_2 while using TiO_2 as the seed. No precipitates were observed in the absence of the seed. Although the induction time is easily reproduced for a specific set of conditions, it is very sensitive to various parameters, such as pH, SI, and the presence of "poisons."

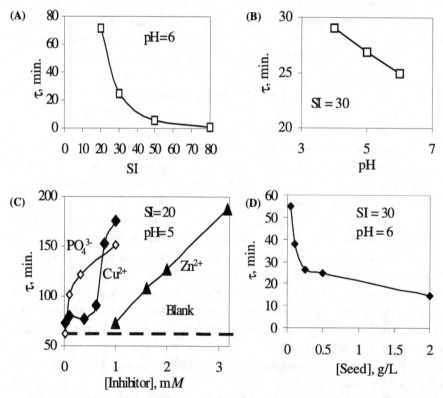

Figure 2-21: Induction time (τ) for nucleation of MgF_2(s) seeded with TiO_2(s) as a function of (A) saturation index (SI), (B) pH, (C) presence of metal-$(NO_3)_2$ or NaH_2PO_4 inhibitors or "poisons", and (D) seed concentration. TiO_2(s) concentration = 0.5 g L^{-1} in (A), (B), and (C). All data were obtained in 0.1 M $NaNO_3$ solutions at 25°C. For (C), the "blank" line at 62 minutes is induction time in the absence of added nucleation inhibitors. *Data from Schulthess and Hohl (1982).*[74]

[2.19] Acid–Base Titrations

[2.19.1] Principle of Electroneutrality

As was observed in Sections 2.15 and 2.16, the pH of a solution plays an important role in the speciation of a compound in water. The reactivity of the proton (H^+) is easily observed in acid–base titrations, as is illustrated in Figure 2-22 for the titration of a boric acid solution; the titration of plain water is also shown for comparison. The concentration of the base added is known, and this is obtained by titration of the strong base with a very carefully prepared acid standard solution (typically potassium hydrogen phthalate, KHP). The base thus calibrated can then also be used to calibrate other strong acid solutions if needed. Without these calibrations, our interpretations of the titration results will be limited. The *x*-axis can also be expressed in terms of moles of base added rather than volume of base added.

A titration reaction is considered completed at the *equivalence point*. More precisely, the equivalence point is reached when the amount of titrant added satisfies a specific zero-level proton condition in the reacting mixture, which is discussed in more detail in Section 2.20. A color change occurs at the *end point* when a colorimetric indicator is present. However, when using pH electrodes, which are needed to generate detailed titration curves such as those presented in this chapter, our interest is on the identification of the equivalence point.

In Figure 2-22, it is easy to notice that the pH of the boric acid solution does not rise the same way as ordinary water. There is, in fact, an area that resists a rise in pH when the alkaline titrant is added (pH 8 to 10). This area, or plateau, is due to the buffering effect of the boric acid on the mixture. Around pH 10.4, an equivalence point is reached, which marks the end of the buffering effect. Above pH 10, the pH of the solution gradually approaches the pH of the titrant, taking into consideration the dilution effect on the concentration of the added base.

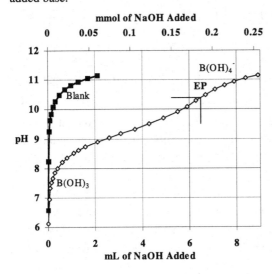

Figure 2-22: Titration of boric acid, H_3BO_3(aq), with 0.0286 *M* NaOH at 25°C, *I* = 0.01 *M* (using NaCl as background electrolyte), and an initial volume of 35 mL. Blank run was NaCl solution. The equivalence point (EP) is at 6.503 mL (or 0.186 mmol) and pH 10.39. *Data from Schulthess (1986).* [75]

The pH value at the center of the buffered area in the titration curve is equal to the acidity constant (pK_a) of a specific reaction of the aqueous compound with the water's proton or hydroxyl ions. All pH-dependent reactions will have a corresponding pK_a value. However, although it is highly narrowed in scope, the identity of the reaction is not generally explicitly identified by the extrapolated pK_a value. Because of the simplicity of the sample titrated in Figure 2-22, it is very easy to identify the speciation reaction that is causing this particular buffered zone around pH 9.2. The reaction of boric acid with base is as follows:

$$B(OH)_3(aq) + H_2O \rightleftharpoons B(OH)_4^- + H^+ \qquad [2\text{-}156],$$

and its acidity constant (K_a) is defined as

$$K_a = \frac{(B(OH)_4^-)(H^+)}{(B(OH)_3(aq))} \qquad [2\text{-}157].$$

Other commonly used acidity constants are the conditional acidity constant (K_a^c), where concentration units are used rather than activity units, and the mixed acidity constant (K_a^m), where both concentration and activity units are used:

$$K_a^c = \frac{[B(OH)_4^-][H^+]}{[B(OH)_3(aq)]} \qquad [2\text{-}158],$$

$$K_a^m = \frac{[B(OH)_4^-](H^+)}{[B(OH)_3(aq)]} = 10^{-9.2}, \; pK_a^m = 9.2 \qquad [2\text{-}159].$$

Determining the amount of product that has formed is best done using the *principle of electroneutrality*. Namely, at any given time in the titration process, the sum of the negative charges must equal the sum of the positive charges:

$$\sum \text{negative charges} = \sum \text{positive charges} \qquad [2\text{-}160],$$

where, for the present exercise,

$$[OH^-] + [Cl^-] + [B(OH)_4^-] = [H^+] + [Na^+] \qquad [2\text{-}161].$$

Be sure to use concentration rather than activity terms. Rearranging,

$$[B(OH)_4^-] = [Na^+] - [Cl^-] + [H^+] - [OH^-] \qquad [2\text{-}162],$$

or

$$[B(OH)_4^-] = (C_B V_B - C_A V_A)/V_T + ([H^+] - [OH^-]) \qquad [2\text{-}163],$$

and

$$[B(OH)_4^-]_T = [B(OH)_4^-]V_T = (C_B V_B - C_A V_A) + ([H^+] - [OH^-])V_T \quad [2\text{-}164],$$

where V_T = total volume present with each increment of acid or base added, $V_{A,B}$ = volume of acid or base added, $C_{A,B}$ = concentration of acid or base added, $[Na^+]$ = concentration of sodium in the mixture = ([NaCl added] + [NaOH added])/V_T, $[Cl^-]$ = concentration of chloride in the mixture = ([NaCl added] + [HCl added])/V_T. Note that $[Na^+] - [Cl^-] =$ ([NaOH added] − [HCl added])/$V_T = (C_B V_B - C_A V_A)/V_T$.

Multiplying Equation [2-163] by V_T yields Equation [2-164], where $[B(OH)_4^-]_T$ is the total amount of the $B(OH)_4^-$ product present with each increment of base added. These results are plotted in Figure 2-23. Experience shows that there is usually a significant improvement in the understanding of Figure 2-23 when it is obtained in an alternate way, as will now be illustrated below.

Electroneutrality must also be obeyed by the blank solution, where from Equation [2-160] we now obtain:

$$[OH^-] + [Cl^-] = [H^+] + [Na^+] \qquad [2\text{-}165],$$

$$0 = [Na^+] - [Cl^-] + [H^+] - [OH^-] \qquad [2\text{-}166],$$

$$0 = (C_B V_B - C_A V_A)/V_T + ([H^+] - [OH^-]) \qquad [2\text{-}167].$$

Subtracting Equation [2-167] from [2-163],

$$[B(OH)_4^-] = (C_B V_B - C_A V_A)_{sample}/V_{Ts} + ([H^+] - [OH^-])_{sample}$$

$$\underline{- \ 0 \qquad = - (C_B V_B - C_A V_A)_{ref}/V_{Tr} - ([H^+] - [OH^-])_{ref}}$$

$$[B(OH)_4^-] = (C_B V_B - C_A V_A)_{sample}/V_{Ts} - (C_B V_B - C_A V_A)_{ref}/V_{Tr}$$
$$+ ([H^+] - [OH^-])_{sample} - ([H^+] - [OH^-])_{ref} \qquad [2\text{-}168].$$

The total volume parameters V_{Ts} and V_{Tr} have the same meaning as V_T in Equation [2-163], but they are calculated separately here for the sample and the reference solutions, respectively. Now if $([H^+]-[OH^-])_{ref} = ([H^+]-[OH^-])_{sample}$, such as by subtracting the reference titration curve from the sample titration curve at the same pH value, then Equation [2-168] becomes

$$[B(OH)_4^-] = (C_B V_B - C_A V_A)_{sample}/V_{Ts} - (C_B V_B - C_A V_A)_{ref}/V_{Tr} \qquad [2\text{-}169].$$

Multiplying by the total volume again yields the total amount of product present in the sample:

$$[B(OH)_4^-]_T = [B(OH)_4^-]V_{Ts} = (C_B V_B - C_A V_A)_{sample} - (C_B V_B - C_A V_A)_{ref}\left(\frac{V_{Ts}}{V_{Tr}}\right) \qquad [2\text{-}170].$$

A horizontal subtraction of the titration curve of water (adjusted by the factor V_{Ts}/V_{Tr}) from the titration curve of the boric acid solution subtracts the dilution effect, and what remains is only the boric acid buffering effect on the matrix (Figure 2-23). In other

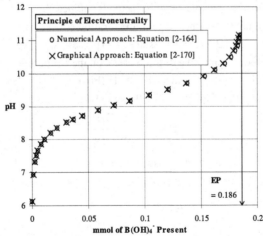

Figure 2-23: Principle of electroneutrality analysis of boric acid titration data with 0.0286 *M* NaOH (same as Figure 2-22). Equivalence point (EP = 0.186 mmol, or 6.503 mL NaOH) is based on modeling shown in Figure 2-26. Amount of B(OH)$_4^-$ based on the numerical approach (Equation [2-164]), where total B(OH)$_4^-$ = 0.0286V_B + [(H$^+$)/γ_H – (OH$^-$)/γ_{OH}]V_T, and on the graphical approach (Equation [2-170]). Predicted values from Figure 2-26 and both measured values in this graph indicate that Reaction [2-156] is 99% complete at the maximum attained pH value of 11.15.

words, if we horizontally subtract the reference blank data in Figure 2-22 from the sample data in the same figure, we would also obtain the data as that plotted based on Equation [2-164]. Note, of course, that when a reference datum point does not have the same pH as the sample datum point, then you must first estimate the reference value that is expected at the pH value that exactly matches the pH of the sample datum.

In summary, we have two different ways of obtaining the amount of product formed: (1) a numerical approach based on the principle of electroneutrality, such as via Equation [2-163], or (2) a graphical approach, where an appropriate blank titration curve is horizontally subtracted from the sample titration curve. The numerical approach is fast and easy to do, particularly if all of the aqueous species in the liquid phase are known. The graphical approach does not require an *a priori* knowledge of all of the aqueous species in the liquid phase, but the choice of the blank solution must accurately, and exactly, complement the sample solution.

[2.19.2] Gran Titrations

The numerical approach described above for determining the amount of compound present in the liquid can be modified so as to have the titration data presented as linearized (Gran) plots. There are numerous variations to the preparation of Gran plots, each depending on what part of the titration is being linearized (such as, regions before or after the equivalence point) and on what is being titrated (such as, strong acid with strong base or weak acid with strong base). These linearized plots are known as Gran titrations in honor of Gunnar Gran, who introduced this approach in the early 1950s.[76] Although Gran titrations are not particularly easier to construct than the methods outlined above with Equation [2-164] or [2-170], they do highlight that the shape of the titration curve near the equivalence point (such as that shown in Figure 2-22) is the result of two processes: (1) the consumption of acid or base to form the product, and (2) a dilution effect on the pH of the solution.

For the titration of boric acid with NaOH (Figure 2-22), a Gran titration can be constructed as follows. From Equation [2-159] we obtain

$$(H^+) = K_a^m \frac{[B(OH)_3]}{[B(OH)_4^-]}$$
[2-171].

Prior to the equivalence point, we can assume that most of the NaOH base added will be consumed to produce the product. Accordingly,

$$[B(OH)_4^-] \approx \frac{C_B V_B}{V_o + V_B}$$
[2-172],

where V_o = initial volume, V_B = volume of base added, C_B = concentration of base added, and $V_o + V_B = V_T$ = total volume present following each increment of base added. At the start of the titration, the amount of $B(OH)_3$ present equals $C_T V_o / V_T$, which diminishes with each increment of product formed:

$$[B(OH)_3] \approx \frac{C_T V_o}{V_o + V_B} - \frac{C_B V_B}{V_o + V_B}$$
[2-173],

where C_T = total concentration of B present = $[B(OH)_3] + [B(OH)_4^-]$. At the equivalence point, the total amount of $B(OH)_4^-$ present is approximately equal to $C_B V_e$, where V_e = volume of base added to reach the equivalence point, and $C_B V_e$ should equal the total amount of $B(OH)_3$ that was present at the beginning of the titration. Accordingly, Equation [2-173] becomes

$$[B(OH)_3] \approx \frac{C_B V_e}{V_o + V_B} - \frac{C_B V_B}{V_o + V_B} = \frac{C_B (V_e - V_B)}{V_o + V_B}$$

[2-174].

Combining Equations [2-171], [2-172] and [2-174]

$$(H^+) = K_a^m \frac{(V_e - V_B)}{V_B}$$

[2-175],

and

$$f_1 = (H^+) V_B = K_a^m (V_e - V_B)$$

[2-176].

A plot of f_1 versus the amount of base added (V_B) should yield a straight line with a slope of K_a^m and an intercept on the x-axis at the equivalence point (that is, $V_B = V_e$ when $f_1 = 0$).

Beyond the equivalence point, we can assume that the NaOH base added will merely result in an increase in pH. Accounting for the dilution effect, we obtain

$$f_2 = [OH^-] = \frac{(H^+)}{K_w \gamma_{OH^-}} = C_B \left(\frac{V_B - V_e}{V_o + V_B} \right)$$

[2-177].

A plot of f_2 versus the amount of base added (V_B) should yield a straight line with an intercept on the x axis at the equivalence point (that is, $V_B = V_e$ when $f_2 = 0$), as shown in Figure 2-24.

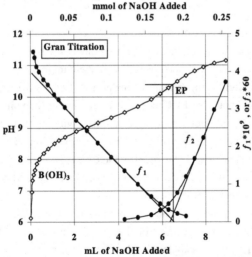

Figure 2-24: Gran titration analysis of boric acid titration data with 0.0286 M NaOH (same as Figure 2-22). Equivalence point (EP = 0.186 mmol, or 6.503 mL NaOH) is based on modeling shown in Figure 2-26. Linearization of the approach to EP, f_1, is based on Equation [2-176] and has an x-intercept at 6.524 mL (0.187 mmol) and a slope of $10^{-9.21}$ (= K_a^m). Linearization after the EP, f_2, is based on Equation [2-177] and has an x-intercept at 6.353 mL (0.182 mmol). Averaging the f_1 and f_2 results, the equivalence point is at 6.439 mL (0.185 mmol).

[2.19.3] Buffering Capacity

The buffering capacity (β) of a solution is a measure of the resistance of the solution to pH changes following the addition of acid (V_A) or base (V_B) (Figure 2-25):

$$\beta = \frac{dV_B}{dpH} = -\frac{dV_A}{dpH} \qquad [2\text{-}178].$$

A mixture of 50% acid (HL) and 50% of its conjugate base (L$^-$) will always result in a solution that is highly buffered at pH equal to the pK_a of the reaction HL $\rightleftharpoons$ L$^-$ + H$^+$. For example, a mixture of 50% H_3BO_3(aq) and 50% Na$^+$B(OH)$_4^-$ forms a solution buffered at pH 9.2.

In Figure 2-22, as NaOH is added, Reaction [2-156] proceeds to the right and releases a proton that, in balance with all the other components of the mixture, keeps the pH of the solution from rising too fast. The strongest buffering capacity in this process is found at the center of the buffered region (Figure 2-25). The center of the buffered region always corresponds to the condition where the concentration of acid (in this case, B(OH)$_3$) and the concentration of its conjugate base (in this case, B(OH)$_4^-$) are equal. The value of pK_a^m = 9.2 in Equation [2-159] can be obtained from the pH value at the center of the buffered zone in Figure 2-23 or 2-25, where the concentration of the acid and its conjugate base were equal and canceled out in the equation. This value agrees well with published values[58], as well it should.

The inverse of the buffering capacity is equal to the first derivative (β^{-1}) analysis of the titration data:

$$\beta^{-1} = \frac{dpH}{dV_B} = -\frac{dpH}{dV_A} \qquad [2\text{-}179].$$

This kind of analysis, which is also illustrated in Figure 2-25, is the most commonly used method for determining the value of the equivalence point (EP). The equivalence point is the location on the titration graph that corresponds with the completion of the reaction being studied. In effect, it calibrates the amount of material present in the sample solution that exhibits a pH-dependent reaction. Plotting β^{-1} as a function of the volume of acid or base added is very easy to do and, not surprisingly, is done routinely. It is also nearly always offered as an analytical option in the accompanying software of autotitrators.

Note, however, that using β^{-1} to identify the EP value is not always accurate. There are no problems, normally, when the pH value of the equivalence point is far away from the pH value of the center of the buffered zone (that is, far from the pK_a of the reaction). However, if the pH value of the equivalence point is very close to the buffering pH value, then the estimated EP value may be offset or skewed. This is a result of the characteristics of the processes occurring in the sample after the EP value has been reached. For example, in the titration illustrated in Figure 2-25, the β^{-1} analysis underestimates the EP value. The concentration of the base was weak (namely, 0.0286 M), which allowed us to have high detail and precision in the changes in pH as a function of the amount of base added. The down side of titrating with a weak base is that the dilution of the base in the reaction vessel will be observed sooner, rather than later. The maximum pH value that can be reached in the example given is 12.5 (= 14 − log 0.0286) and, as can be seen in Figure 2-25, it will be a

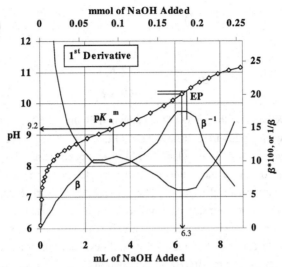

Figure 2-25: Buffering capacity (β) and first derivative analysis (β^{-1}) of boric acid titration data with 0.0286 M NaOH (same as Figure 2-22). Equivalence point (EP = 0.186 mmol, or 6.503 mL NaOH) is based on modeling shown in Figure 2-26. The pK_a^m of Reaction [2-156], defined by Equation [2-159], is equal to the midpoint of the buffered region, which is also equal to the midpoint of the reaction process and [B(OH)$_4^-$]=[B(OH)$_3$]. Typically, the EP based on β^{-1} will equal the EP based on modeling estimations. A discrepancy in the EPs between these two methods is observed here (6.3 versus 6.503 mL) because of the nearness of the dilution effect to the pK_a^m of the reaction. In other words, β^{-1} of the data has its maximum too close to the buffered region, which, in turn, skews the expression of the first derivative maximum. At V_B = 6.3 mL, pH = 10.29 and Reaction [2-156] is only 92.5% complete.

slow, long climb toward that upper pH limit. Note also that the dilution effect is accompanied by an increase in the apparent buffering capacity of the sample solution.

[2.19.4] Modeling the Titration Data

The data in Figure 2-22 or 2-23 can be easily described mathematically, and this is referred to as "modeling the data." Based on simple rules, this is the most accurate method available for quantifying the pK_a (or pK_a^m) and C_T (or EP) values from titration data.

From conservation of mass laws, the total concentration (C_T) of boron is constant. Using concentration rather than activity terms, the mass balance of the boric acid mixture is defined as

$$C_T = [B(OH)_3(aq)] + [B(OH)_4^-] \qquad [2\text{-}180].$$

Equations [2-159] and [2-180] are combined to predict the boron anion concentration as follows:

$$C_T = \frac{[B(OH)_4^-](H^+)}{K_a^m} + [B(OH)_4^-] \qquad [2\text{-}181],$$

$$C_T = [B(OH)_4^-] \left(\frac{(H^+)}{K_a^m} + 1 \right) = [B(OH)_4^-] \left(\frac{(H^+) + K_a^m}{K_a^m} \right) \qquad [2\text{-}182],$$

$$[B(OH)_4^-] = \frac{C_T K_a^m}{(H^+) + K_a^m} \qquad [2\text{-}183],$$

where $pK_a^m = -\log K_a^m = pH$ at the center of the plateau in Figure 2-26A, and C_T = maximum value reached by the data in Figure 2-26A. Note that Equations [2-181] to [2-183] can be expressed slightly more precisely if we replace the mixed acidity constant (K_a^m) with the acidity constant (K_a), adding, of course, the activity coefficients of the negative ion product ($\gamma_{B(OH)4}$) and the neutral starting material ($\gamma_{B(OH)3}$). That is, replace K_a^m with $K_a(\gamma_{B(OH)3})/(\gamma_{B(OH)4})$. Nevertheless, at low ionic strengths, Equation [2-183] predicts the data in Figure 2-26A very well and, therefore, improving the model prediction is not warranted here.

The raw data can also be predicted as follows. The boron anion predicted by Equation [2-183] is equivalent to $[B(OH)_4^-]_T$ in Equation [2-164]. Accordingly, combining Equations [2-183] with [2-164], and solving for V_B as a function of pH (remember, $V_T = V_o + V_B$), yields the prediction of the titration curve plotted in Figure 2-26B. The titration results of the blank solution are also predicted using Equation [2-167] and solving for V_B is a function of pH.

The C_T value could be based on the experimental details of the preparation of the solution, if known (such as, in the preparation of the boric acid solution illustrated in Figure 2-26). It is better, however, to extrapolate the C_T value from the titration data because the titrating solutions (such as NaOH or HCl) are usually very well calibrated and added,

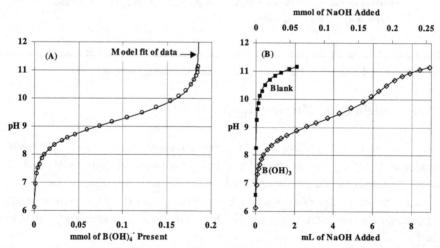

Figure 2-26: Modeling the titration of boric acid. All lines drawn are the result of the model predictions. Solid symbols are raw data, which are the same as those plotted in Figures 2-22 and 2-23 (based on the numerical approach). (A) Modeling the amount of boric acid formed as a function of pH based on Equation [2-183], where $pK_a^m = 9.2$ and $C_T = 0.186$ mmol. (B) Modeling the volume of base needed as a function of pH for the blank solution and the boric acid sample.

typically making use of precise autotitrators. Conversely, the concentration of the sample solution may be unknown or known only approximately. Hence, the titration of the sample with well-calibrated titrating solutions serves to calibrate the sample solution.

Note that whenever a pH-buffered region is encountered, then a pH-dependent reaction *must* be present in the matrix. More than one reaction may be involved in generating a buffered region; but if two separate regions are observed, then at least two separate reactions are involved. Knowing the pH value of the buffered region will not identify the cause of the buffering effect or the reaction involved. More information is needed to positively identify the species involved (such as knowledge of what was added to the mixture, or spectroscopic information of the sample). Nevertheless, it is possible to proceed with a model or mechanistic prediction of the data even when the actual chemical nature of the aqueous species involved is only guessed at initially. This "blind" approach allows one to subsequently select an analytical method that tries to definitively identify the presumed aqueous species present. That is, it offers an idea of what to look for when no other information is available. The down side of this approach is that one often forgets that the model was blind and, therefore, might not see the need for an independent confirmation of the model. One might even subsequently insist (sometimes erroneously) that the species presumed by the model must be correct.

[2.20] The Proton Condition & Equivalence Point (EP)

As noted in the previous section, we must be able to recognize the equivalence point in a titration experiment in order to determine if the acid–base reactions in the titration experiment are completed. The pH value of the equivalence point (pH_{EP}) is typically based on the first-derivative analysis of the data. It is sometimes helpful, however, to know in advance what the approximate pH will be at the equivalence point. This is particularly useful when a titration curve yields numerous inflection points, where an advance estimate of the theoretical equivalence point will help you immediately recognize the correct experimental EP in the graph. Obviously, the more details we have about the solution, the more accurate will be our predictions. A numerical approach and a graphical approach for estimating the pH of the equivalence point (pH_{EP}) using various simple sample solutions are illustrated below.

A. Titration of Boric Acid:

We seek here to estimate pH_{EP} of the sample illustrated in Figure 2-22. We begin first with describing the proton condition of the solution at the pH value of the equivalence point. The *proton condition* is a mathematical relation that balances the distribution of free or complexed protons around a "zero level" condition. The "zero level" condition is equal to the condition present at the equivalence point. For example, adding base to an acid solution of $B(OH)_3$ will reach a zero level condition with just $B(OH)_4^-$ in water. Conversely, adding acid to an alkaline solution of $B(OH)_4^-$ will reach a zero level condition of just $B(OH)_3$ in water. As we overshoot and pull away from the zero condition, we will have either excess protons (free or complexed) or hydroxyls (free or complexed) in the solution:

	Acidic EP		Alkaline EP	
Increase in pH:	$B(OH)_4^-$	OH^-		OH^-
Zero level:	$B(OH)_3$	H_2O	$B(OH)_4^-$	H_2O
Decrease in pH:		H^+	$B(OH)_3$	H^+

To obtain the proton condition, we assume that the concentration of species that are formed by the release of protons must equal the concentration of species that were formed by the consumption of protons. In other words, we generate mass balance equations that track the protons in the solution:

For alkaline EP: $$[OH^-] = [B(OH)_3] + [H^+]$$ [2-184],

For acidic EP: $$[B(OH)_4^-] + [OH^-] = [H^+]$$ [2-185].

Note that once again everything is defined in concentration units rather than activity units. Equations [2-184] and [2-185] are actually similar to Equation [2-163]. We can define the zero level as if it were a condition that exists prior to the addition of acids. Hence, Equation [2-185] is equivalent to Equation [2-163] with the volume of acid added set equal to zero. For the alkaline EP, if $[B(OH)_4^-] = C_T - [B(OH)_3]$ (from Equation [2-180]) and $V_B = V_e$ (where V_e = volume of base added to reach the equivalence point), then substituting these into Equation [2-163] yields Equation [2-184] with $C_B V_e / V_T$ canceling out C_T. Note, however, that it is much easier to construct Equation [2-184] and [2-185] by tabulating the proton condition, as was just illustrated at the top of this page.

Solving for the pH value of the alkaline EP, we begin by defining the boric acid concentration based on a derivation similar to the one used to obtain Equation [2-183]:

$$[B(OH)_3] = \frac{C_T [H^+]}{[H^+] + K_a^c}$$ [2-186].

Combine Equations [2-184] and [2-186], and solve for pH

$$[OH^-] = \frac{C_T [H^+]}{[H^+] + K_a^c} + [H^+]$$ [2-187].

Multiplying by $[H^+]$ and $([H^+] + K_a^c)$, and rearranging yields

$$[H^+]^3 + (C_T + K_a^c)[H^+]^2 - K_w^c[H^+] - K_w^c K_a^c = 0$$ [2-188],

where $[OH^-] = K_w^c/[H^+]$, $K_w^c = K_w/(\gamma_H \gamma_{OH})$, $K_w = 10^{-14}$, and C_T = total concentration of boron present. Solving Equation [2-188] for the titration illustrated in Figure 2-22, where $K_a^m = 10^{-9.2} = \gamma_H K_a^c$, $C_T = 0.0053\ M$, and $(H^+) = [H^+]\gamma_H$, yields $(H^+) = 10^{-10.42}$. The alkaline $pH_{EP} = 10.42$, which is 6.56 mL (or 0.188 mmol) in Figure 2-22, versus $pH_{EP} = 10.39$ at 6.503 mL (or 0.186 mmol) obtained by analyzing the experimental data as discussed in Section 2.19.4. This is a very close match considering that the pH reading given by most autotitrators will trail slightly behind the true equilibrium pH value of each datum point, particularly in the poorly buffered regions of the titration. That is, autotitrators do not really wait long enough for true equilibrium conditions (or pH stability) prior to adding the next incremental volume of acid or base. The C_T calculation illustrated in Figure 2-26A is not affected by this small pH trailing error, but the estimated pK_a value may have some error if the buffered region is too small (which would happen if the concentration of the acidic or basic titrant is too high). Starting with Equation [2-185], the pH of the acidic equivalence point when titrating a 0.0053 M solution of $B(OH)_4^-$ with acid can also be predicted: acidic $pH_{EP} = 5.76$.

The close match between the estimated and experimental pH_{EP} values is rarely observed because this procedure of estimating the pH_{EP} value does not normally have such precise advance knowledge of the C_T value. It sometimes also lacks precise information about the pK_a values. Instead, we generally start with a rough estimate of the C_T value followed by a rough theoretical calculation of pH_{EP}, which then helps us to find it in the experimental data. Knowing where to look for the equivalence point in the titration curve (which then allows you to calculate C_T accurately) is the purpose of this procedure.

For most titrations, solving the proton condition numerically can become a very difficult and time consuming exercise. A quick and easy way to solve the proton condition equations is a graphical approach.[77] Again, this approach is first illustrated here with the very simple boric acid titration example. Figure 2-27 illustrates a log-pH plot of all the aqueous species present in the boric acid titration mixture. This graph is easy to generate since all the lines are horizontal or follow a 1:1 slope with an intercept at $pH = pK_a$.

Based on Equation [2-185], the acidic EP is the pH where the $[H^+]$ line equals the sum of the $[B(OH)_4^-]$ and $[OH^-]$ lines. We quickly see in the graph that the acidic EP is near pH 5.7. If the total concentration of boron increases, then the acidic EP would occur at lower pH values. Conversely, if the total concentration of boron decreases, then the acidic EP would occur at higher pH values, but not higher than pH 7.

Based on Equation [2-184], the alkaline EP is the pH where the $[OH^-]$ line equals the sum of the $[B(OH)_3]$ and $[H^+]$ lines. We quickly see in the graph that the alkaline EP is near pH 10.5. If the total concentration of boron increases, then the alkaline EP would occur at higher pH values. Conversely, if the total concentration of boron decreases, then the alkaline EP would occur at lower pH values, but not lower than pH 7.

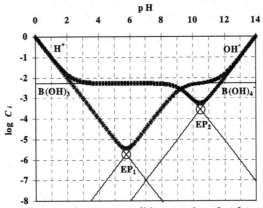

Figure 2-27: Graphical solution to the proton condition equations that determine the acidic (EP_1) and alkaline (EP_2) equivalence points of boric acid titration (Equations [2-184] and [2-185]). Note the 1:1 slopes of the H^+ and OH^- lines, and the 1:1 slope of the boron species on either side of $pH = pK_a$. Changing the total concentration of boron, C_T, moves the two boron lines up or down, and shifts the pH_{EP} values.

Clearly, identifying the equivalence point is particularly valuable when the titration curve has numerous inflections and buffered regions, yielding numerous equivalence points with all appearing to be equally valid for the desired analysis of the sample. Solving the proton condition for a very simple titration experiment (e.g., titration of boric acid) is not normally necessary because confusion on which equivalence point is the correct one is rare when there is only one to chose from (one for an acidic titration, or one for an alkaline titration). But do keep in mind that the titration of most natural samples involve far more complicated mixtures — fortunately, the electroneutrality rules that they all must follow do not change, and the analytical procedure for a complicated sample is fundamentally the same as the simple one illustrated here. Also, it is best to learn the procedure by starting with a simple example. The next two examples below are slightly more difficult.

B. Titration of Inorganic C:

Since most aqueous solutions contain inorganic C, we should understand how the presence of these compounds affect the equivalence point. We again begin with defining the zero level condition:

	EP$_1$		**EP$_2$**		**EP$_3$**	
	CO_3^{2-}					
Proton loss:	HCO_3^-	OH^-	CO_3^{2-}	OH^-		OH^-
Zero level:	$H_2CO_3^*$	H_2O	HCO_3^-	H_2O	CO_3^{2-}	H_2O
Proton gain:		H^+	$H_2CO_3^*$	H^+	HCO_3^-	H^+
					$H_2CO_3^*$	

The proton conditions are then defined as follows:

For EP$_1$: $\quad 2[CO_3^{2-}] + [HCO_3^-] + [OH^-] = [H^+]$ $\qquad$ [2-189],

For EP$_2$: $\quad [CO_3^{2-}] + [OH^-] = [H_2CO_3^*] + [H^+]$ $\qquad$ [2-190],

For EP$_3$: $\quad [OH^-] = [HCO_3^-] + 2[H_2CO_3^*] + [H^+]$ $\qquad$ [2-191].

Using a graphical approach, these equations are solved in Figure 2-28, where pH$_{EP1}$ = 4.3, pH$_{EP2}$ = 8.3, and pH$_{EP3}$ = 11.0. Note that the pH$_{EP1}$ and pH$_{EP3}$ values will change as a

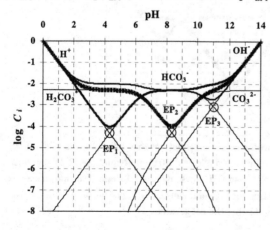

Figure 2-28: Graphical solution to the proton condition equations that determine the various equivalence points of inorganic C (Equations [2-189] to [2-191]). Let C_T = 0.005 M.

function of the total inorganic C (TIC) present in the solution, whereas the pH_{EP2} value does not change unless the TIC present is very low ($< 0.2mM$).

C. Titration of Inorganic C & Al hydroxide:

For this last example, assume a solution of 0.002 M inorganic C and 0.001 M Al hydroxide. Also assume for now that the desired equivalence point is the one with the following zero level condition: HCO_3^-, $Al(OH)_3$, and H_2O. The proton condition equation for this titration experiment is defined as follows:

Proton loss:	CO_3^{2-}	$Al(OH)_4^-$	OH^-
Zero level:	HCO_3^-	$Al(OH)_3$	H_2O
Proton gain:	$H_2CO_3^*$	$Al(OH)_2^+$	H^+
		$AlOH^{2+}$	
		Al^{3+}	

$$[CO_3^{2-}] + [Al(OH)_4^-] + [OH^-] = [H_2CO_3^*] + [Al(OH)_2^+]$$
$$+ \; 2[AlOH^{2+}] + 3[Al^{3+}] + [H^+] \qquad [2\text{-}192].$$

The graphical solution to Equation [2-192] yields $pH_{EP} = 7.35$ (marked "a" in Figure 2-29). Notice that this pH_{EP} value will vary with the relative concentrations of total Al and total inorganic C (TIC) concentrations, rather than with their actual concentrations. As the total Al concentrations decrease relative to the TIC concentrations, the pH_{EP} increases toward pH 8.33, where $[CO_3^{2-}] = [H_2CO_3^*]$ (marked "b" in Figure 2-29).

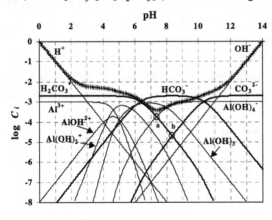

Figure 2-29: Graphical solution to proton condition expressed by Equation [2-192]. The equivalence point marked "a" assumes TIC = 0.002 M and $[Al]_T = 0.001$ M, while the point marked "b" assumes TIC $\gg [Al]_T$.

[2.21] Total Alkalinity & Total Acidity

The pH of a solution tells you the activity of free hydrogen cations (H^+) and free hydroxyl anions (OH^-) present. These ions are free because they are complexed only with water molecules, forming, for example, hydronium ions (H_3O^+). These ions do not contribute to the buffering capacity of the medium.

Based on the earlier discussions on the pH-dependent speciation of aqueous ions, it should be clear by now that there are many protons and hydroxyl ions in liquid solutions that are not "free"; that is, H^+ and OH^- ions can be held by molecules other than H_2O. These pH-dependent compounds contribute to the buffering capacity of the mixture, as well as to the total mass balance of available protons and hydroxyls present in the mixture. In order to release and measure these complexed protons or hydroxyls, it is necessary to titrate the solution with a strong base (or acid). Accordingly, the alkalinity of a water is equal to the sum of all the titratable bases present in the liquid. Alkalinity is a measure of the capacity of the solution to neutralize added acids. Similarly, the acidity of a water is equal to the sum of all the titratable acids present in the liquid. Acidity is a measure of the capacity of the solution to neutralize added bases.

There are various subsets of acidity and alkalinity, all of which depend on the equivalence point (EP) chosen in the titration procedure. Since the liquid phase of nearly all natural samples contains some inorganic C, the EP chosen is typically based on the aqueous carbonate reactivity. These EP values were estimated in Figure 2-28 for a sample that contains only inorganic C in solution. Figure 2-30 illustrates the resultant titration curve of a sample that is high in dissolved inorganic-C concentration, and this figure will be used to describe the various subsets of acidity and alkalinity commonly used with natural samples.

If we begin in Figure 2-30 with an acidic solution and titrate with a strong base, the amount of base added to reach the EP value prior to any buffering in the matrix (typically around pH 4.5) is a measure of the *mineral acidity*. As we continue adding base, we often pass a buffered region that is caused by the presence of various inorganic-C compounds and complexes, such as $H_2CO_3^* \rightleftharpoons HCO_3^- + H^+$, and reactions with $NaHCO_3$, $MgHCO_3^+$, and many others. The amount of base added to reach the next EP value (typically around pH 8) is a measure of the *CO_2 acidity*. As we again continue adding base, the amount added to reach the last equivalence point (typically around pH 10.3) is a measure of the *total acidity*. Note that the designation of the equivalence point and the corresponding name of the acidity measurement may be modified based on the specific characteristics of the sample being titrated (for example, high concentrations of boric acid or natural organic matter may be present).

Now, in reverse order, if we begin in Figure 2-30 with an alkaline solution and titrate with a strong acid, the amount of acid added to reach the first EP value prior to any

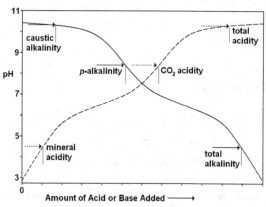

Figure 2-30: Illustration of various alkalinity and acidity quantities for a given solution based on its acid-neutralizing and base-neutralizing capacity. *After Stumm and Morgan (1981).*[54]

buffering in the matrix is a measure of the *caustic alkalinity*. This is typically around pH 10.3, but it may be higher since alkaline solutions will have very high concentrations of inorganic-C species dissolved in them, such as CO_3^{2-}, $MgCO_3$, $NaCO_3^-$ and many others. As we continue adding acid, we reach an equivalence point (typically around pH 8), which gives us a measure of the *carbonate alkalinity* or *p-alkalinity* (the letter "*p*" here is in reference to phenolphthalein, a common colorimetric EP indicator). As we again continue adding acid, the amount added to reach the last equivalence point (typically around pH 4.5) is a measure of the *total alkalinity*. Again note that the designation of the equivalence point and the corresponding name of the alkalinity measurement may be modified based on the specific characteristics of the sample being titrated.

Acidity and alkalinity are typically expressed in moles of charge per liter ($mol_c\ L^{-1}$, which is sometimes still referred to as equivalents per liter, eq L^{-1}). Alkalinity is also sometimes expressed as mg equivalents of $CaCO_3$ per liter (or ppm $CaCO_3$):

$$\text{Alkalinity, } \frac{\text{mg } CaCO_3}{L} = \frac{V_A C_A}{V_o} \times MW_{CaCO_3} \times \frac{1000 \text{ mg}}{L} \times \frac{mol}{2\ mol_c}$$

$$= \frac{V_A C_A}{V_o} \times 50,000 \qquad\qquad \text{[2-193]},$$

where V_o = initial volume of sample, V_A = volume of acid added to reach the equivalence point, C_A = concentration of acid added (mol L^{-1}), MW_{CaCO3} = molecular weight of $CaCO_3$ = 100 g mol^{-1}, and 1 mol $CaCO_3$ neutralizes two moles of acid (= mol/2 mol_c). Measuring the acidity or alkalinity of a liquid sample is based on the volume of base or acid needed to change the current, natural pH of the sample to the appropriate equivalence point pH. Determining the pH value of the appropriate equivalence point condition is based on the proton condition equations discussed in the previous section.

[2.22] Interstitial Water

The concentration of ions or neutral compounds dissolved in soil water is often relatively easy to measure. Complications arise when the soil sample is not saturated with water and complete extraction of the water is countered by retentive capillary forces. Various methods that can be employed to extract these interstitial waters are summarized in Table 2-13. It is generally preferable to use those methods that collect only the ions dissolved in the water that is passing through the soil, rather than using those methods that also collect some of the ions that are attached to the soil.

Use of tensiometers is the least intrusive method, but the centrifugation method is the most effective for extraction of water from very small capillary regions. The faster the spin rate in the centrifugation method, the smaller the diameter of the capillary regions that are also contributing to the total volume of extracted water. Only large-diameter pore regions will release their entrained water at the low spin rates. Note, however, that there may be a shearing potential present, and if this is occurring, then there is some induced desorption from the solid phase into the liquid extract.

Table 2-13: Summary of unsaturated soil water extraction methods.

Method	Comments
Squeezer	Hydraulic or gas operated; works well for unconsolidated sediments
Lysimeter	Tensiometric *in situ* extraction of fluids; works well for continuous monitoring of unsaturated zones
Elutriation or leaching	The addition of water to the column may result in contamination or a shift in the adsorption–desorption equilibria
Vacuum displacement	The addition of the displacing solution (e.g., saturated $CaSO_4$ with 4% K-thiocyanate, KCNS) may result in contamination or a shift in the adsorption–desorption equilibria
Centrifugation with an immiscible liquid	Works well, but the addition of liquids to the sample may result in contamination; common immiscible liquids used are: carbon tetrachloride (CCl_4), tetrachloroethylene ($Cl_2C=CCl_2$), and 1,1,1-trichloroethane (Cl_3C-CH_3)
Centrifugation with a trap cup assembly	Has a high extraction efficiency and a low potential for contamination; see Figure 2-31 for assembly

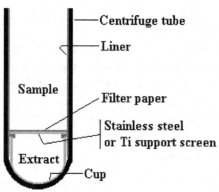

Figure 2-31: Centrifuge assembly for extraction of interstitial waters. The inner liner is needed to help remove the solid sample after centrifugation. The cup at the bottom holds the extracted liquid, but note that a small assembly will still work fine without the cup. *After Edmonds and Bath (1976).*[78]

The volume of water displaced by centrifugation at a set speed is affected by the water tension, which is, in turn, related to the percentage of silt and clay present in the sample (Figure 2-32). It is interesting to note an observation made by Edmonds and Bath (1976)[78], which may have significant implications on how and where porous media retain, adsorb, or entrap dissolved ions and compounds. They reported that an analysis of the concentration of the elements in the displaced water may vary with the centrifugation speed and the amount of pore fluid removed, which is illustrated in Figure 2-33.

[2.23] pH of Soils

One of the most important parameters that controls the reactivity of soils is the concentration of protons present in the soil water. The hydrogen activity in soil water directly affects the quality of our groundwaters and the health of the vegetation that grows on it. Measurements of soil pH are done routinely worldwide by soil testing laboratories as one of

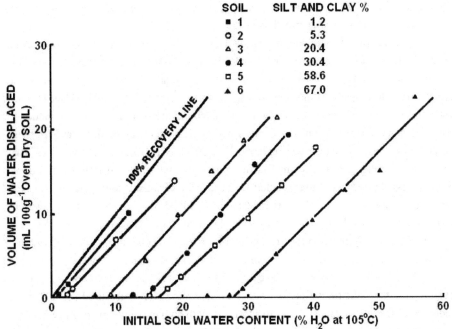

Figure 2-32: Recovery of water from various soils at different initial water tensions. Displacent is CCl_4, centrifuge speed is $265,000 \ m \ s^{-2}$, centrifuge time is 2 h. *From Whelan and Barrow (1980)[79], with permission.*

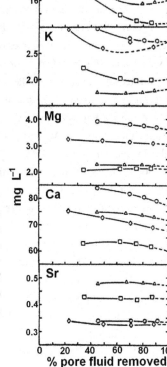

Figure 2-33: Concentration of various cations in water extracted from various chalk samples using a centrifuge extraction with a trap cup assembly. Precision = ± 2–3%. An increase in the percentage of pore fluid removed is obtained by increasing the centrifugation speed. The change in the concentration of some of the cations with centrifuge speeds is interpreted to mean that the small-diameter pore regions contain fewer dissolved ions in their entrained water. *Reprinted from Edmonds and Bath (1976)[78], copyright © 1976, with persmission from the American Chemical Society.*

the criteria needed for determining how much lime needs to be added to a particular soil for a particular crop for optimum growth conditions.

The measurement of soil pH measures the pH of the soil water; it does not measure the pH of the soil's solid phase *per se*. As was described in the previous section, extraction of interstitial soil water is laborious and difficult. Since this process is time consuming and costly, all soil testing laboratories routinely seek only a representative soil pH value. The true soil pH is not precisely pinpointed by soil testing laboratories because liquids are added to the dried soil samples to create a soil slurry whose pH is then measured. Because of these added liquids, the method measures representative pH values rather than true pH values. That is, the soil testing labs are not measuring native, undisturbed soil liquid extractions. Nevertheless, this is not a serious problem because the recommendations generated by these laboratories are based on numerous field studies and calibrations. Since the pH values determined by the calibration field studies and the unknown sample soils were all measured by the same procedure, the inherent errors are presumed to be nearly constant within a narrow range of soils and soil textures, and the field calibrations render most errors present in the methods to be self correcting. The objectives of the soil testing laboratories are not to quantify the true soil pH value, but rather to determine if the measured soil pH value indicates a high or low value relative to a test soil, with both measured by the same method. They also apply the same objectives with respect to the measurement of a field's nutrient status. We do not need, however, to overstate the problem because the systematic errors of representative pH versus true pH values are actually not that large, particularly in soils with a large buffering capacity (such as soils with highly soluble minerals).

There are two liquids commonly used by soil testing laboratories for measuring soil pH: (1) pure water, and (2) a salt solution, such as 10 mM $CaCl_2$. The soil-to-liquid ratio is often 1:2 by volume. The pH values obtained when water is added are always higher than the pH values obtained when $CaCl_2$ (or other salt solution) is added. This is because the aqueous Ca^{2+} cations will exchange with the protons adsorbed to the soil's solid phase. As the protons are released or exchanged, the pH of the liquid phase is lowered. There is also aqueous Ca^{2+} cation exchange with adsorbed Al^{3+} cations, which easily hydrolyze in the liquid phase and consume OH^- anions as $Al(OH)_x$ species are formed — and, once again, the pH is lowered. Soil pH values taken in 10 mM $CaCl_2$ are probably very close to true soil pH values because these solutions mimic more closely the ionic strength of the natural soil liquid solutions.

For purposes of lime recommendations, many soil fertility researchers might also evaluate the soil's pH buffering capacity. This value can be inherently integrated with the field calibrations of the various particular soil textural classes. In many soil testing laboratories, the region covered by their customers tends to cover a small diversity in soil textural classes anyway. Alternatively, the soil's pH buffering capacity can also be measured by noting the degree of pH shift a buffered solution exerts on the soil pH value. Sometimes, two measurements are made with buffers at two different pH values. The shift in the soil pH reflects the strength of the soil's pH buffering capacity relative to the pH buffering capacity of the added liquid buffers.

When a soil is no longer limed, then the soil pH will gradually return to the pH value that is maintained by various factors influencing the soil environment. The dissolution of the mineralogy present does not seem to affect the pH of a soil's A horizon[80], but it may have a significant role in the soil's B horizon pH or in soil pH in mountainous areas. Areas that are dominated by carbonate parent materials will have higher soil pH values. Areas downstream from mine deposits, such as coal, will generally have very low pH values. Other factors that may influence soil pH include: the chemical composition of the plant and animal litter on the surface, the nature of the surviving soil biota, the aboveground vegetation (forest versus grasslands), the chemical composition of the rainfall on the soil, and anthropogenic activity (such as liming or cultivation).

The Natural Resources Conservation Service of the United States Department of Agriculture (USDA-NRCS) has compiled various soil chemical and physical properties in the State Soil Geographic Database (STATSGO). From this database, Miller and White (1998)[81] developed a second database (CONUS-SOIL) for large-scale studies. Studies of pH values of the USA soils show that the dry regions in the west tend to have alkaline soils, while the humid regions in the east and pacific coast tend to have acidic soils (Figure 2-34).

Folkoff et al. (1981)[80] were able to predict the A horizon pH of soils reasonably well using climatic parameters. The most important influence on the pH of A horizon soils is the soil moisture deficit minus surplus ($D-S$). In fact, excellent general pH predictions can be achieved with just this value (Figure 2-35):

$$pH = 3.34 + 0.0028 (D - S) \qquad [2\text{-}194],$$

where the moisture deficit and surplus are given in inches (1 inch = 2.54 cm); $R^2 = 0.79$. The

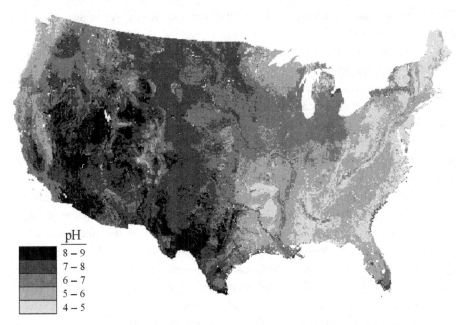

Figure 2-34: Soil pH values of the conterminous United States of America. The pHs shown are for the top 5 cm of soil. *Image is courtesy of Dr. D.A. Miller of The Pennsylvania State University, University Park, PA. See also Miller and White (1998)*[81] *for images showing various other chemical and physical properties of soils of the USA.*

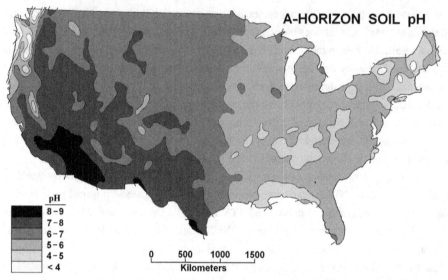

Figure 2-35: General values of A horizon soil pH values predicted based on soil moisture deficit minus surplus ($D - S$). Forest soil pH values should be about 0.5 units lower. Predicted values for mountainous areas are questionable. Model responds to urban heat areas, such as around New York City. *Adapted from Folkoff et al. (1981)[80]; adapted with permission from Physical Geography, 1981, Vol. 2, No. 2, pp. 116–124. © V.H. Winston & Sons, Inc., 360 South Ocean Boulevard, Palm Beach, FL 33480. All rights reserved.*

pH 7 isoline runs diagonally from the gulf coast of Texas to eastern Washington. The pH 6 isoline runs north-south.

The pH range of mineral soils varies worldwide from 3 to 10. Arid areas have high pH values, while humid areas have low pH values. This is expected because there is an accumulation of soil cations in arid environments, resulting in a high alkalinity in these soils. Conversely, in moist environments the soils are generally acidic due to the increased leaching of soil cations.

Average pH values of the soil orders are listed in Table 2-14. Note that the pH values of the representative pedons used in soil taxonomy may differ significantly from the average pH values. We can draw a few generalities here. First, low pH values are found in wet soils, while high pH values are found in dry soils. Again, this suggests that soil cations are accumulating in arid environments. Second, the pH values in the lower horizons are often a little higher than in the A horizons. It is easy to see that in most soils, the rainfall events will leach the cations from the A horizon, leaving an acidic A horizon behind. The cations are deposited in the lower horizons, where they accumulate and increase the pH. Note that the depth of water infiltration in soils depends on the amount of rainfall and hydraulic conductivity. The higher organic content in the A horizons also contributes to their lower pH values.

The soil pH values listed in Table 2-14 correlate poorly with the typical soil temperature regime or the degree of soil weathering (that is, relative age of the soil). Note that the ranking used in Table 2-14 for the degree of soil weathering is not well quantified, and there can be much variability present within any of the soil orders.

Table 2-14: Average pH values of the soil orders. (A) *Data from Manrique et al. (1991)*[82]; **these are averages for the soil horizons of soils from around the world. (B)** *Data from Holmgren et al. (1993)*[83]; **these are averages for the soil horizons of soils sampled in the United States of America. (C)** *Data from the USDA Soil Survey Staff (1999)*[84]; **these are from the "Representative Pedon" description for each soil order; first pH value is for the A horizon, second pH value is for the B horizon. All pH values are of 1:1 soil/water mixtures. For degree of weathering: 1 = least weathered, 12 = most weathered.**

Soil Order	pH (A)	pH (B)	pH (C) A	B	Degree of Weathering	Typical Temperature Regime	Typical Moisture Regime [a]
Aridisol	8.1 ± 0.7	7.26	8.2	8.7	6	varies	dry
Vertisol	7.5 ± 1.0	6.72	7.2	7–8[b]	7	warm – hot	dry
Mollisol	7.0 ± 1.0	6.51	6.2	7.2	8	warm – hot	moist
Entisol	6.9 ± 1.4	7.32	6.3	7.0[c]	1	varies	dry – moist
Inceptisol	6.0 ± 1.1	6.08	6.5	5.9	2	varies	varies
Alfisol	6.0 ± 1.1	6.00	7.7	7.8	9	cool – warm	moist
Andisol	—	—	5.3	5.7	3	varies	varies
Gelisol	—	—	5.2	6.8	4	cold	moist – wet
Histosol	—	5.50	—	—	5	cool	moist – wet
Oxisol	5.5 ± 0.8	—	4.2	5.1	11	hot	very wet
Spodosol	5.1 ± 0.7	4.93	3.9[d]	4.5	10	cool – cold	wet – very wet
Ultisol	5.0 ± 0.6	5.60	5.1	5.0	12	warm – hot	wet

[a] This combines rainfall frequency and intensity with soil drainage characteristics.
[b] pH 7.0 at 15 to 122 cm depth; pH 8.0 to 8.2 at 122 to 234 cm depth.
[c] C horizon pH value; typically, Entisols do not have B horizons.
[d] E horizon pH value.

References Cited

[1] Smith, W.O., H.W. Olsen, R.A. Bognold, and J.C. Rice. 1966. Certain flows of air and water in sands during infiltration. Soil Sci. 101:441–449.

[2] Lide, D.R. (Editor-in-Chief). 2002. CRC Handbook of chemistry and physics. 83rd ed. CRC Press, Boca Raton, FL.

[3] Buyanovsky, G.A., and G.H. Wagner. 1983. Annual cycles of carbon dioxide level in soil air. Soil Sci. Soc. Am. J. 47:1139–1145.

[4] Castelle, A.J., and J.N. Galloway. 1990. Carbon dioxide dynamics in acid forest soils in Shenandoah National Park, Virginia. Soil Sci. Soc. Am. J. 54:252–257.

[5] Butler, J.N. 1982. Carbon dioxide equilibria and their applications. Addison-Wesley, Reading, MA.

[6] Fernandez, I.J., and P.A. Kosian. 1987. Soil air carbon dioxide concentrations in a New England spruce–fir forest. Soil Sci. Soc. Am. J. 51:261–263.

[7] Robbins, G.A., B.G. Deyo, M.R. Temple, J.D. Stuart, and M.J. Lacy. 1990. Soil-gas surveying for subsurface gasoline contamination using total organic vapor detection instruments: Part II. Field experimentation. Ground Water Monit. Rev. 10(4):110–117.

[8] Wijnja, H., and C.P. Schulthess. 2002. Effect of carbonate on the adsorption of selenate and sulfate on goethite. Soil Sci. Soc. Am. J. 66:1190–1197.

[9] Wijnja, H., and C.P. Schulthess. 2000. Interaction of carbonate and organic anions with sulfate and selenate adsorption on an aluminum oxide. Soil Sci. Soc. Am. J. 64:898–908.

[10] Bernstein, S.J. 1948. The effect of catalyst poisons on sorption isotherms. Master's thesis. Montana State Univ. Library. Bozeman, MT.

[11] de Grotthuss, C.J.T. 1806. Sur la décomposition de l'eau et des corps qu'elle tient en dissolution à l'aide de l'électricité galvanique. Ann. Chim. 58:54–74.

[12] Barrow, G.M. 1973. Physical chemistry. 3rd ed. McGraw-Hill, New York, NY.

[13] Anderson, D.M., and P.F. Low. 1958. The density of water adsorbed by lithium-, sodium-, and potassium-bentonite. Soil Sci. Soc. Am. Proc. 22:99–103.

[14] Pauling, L. 1935. The structure and entropy of ice and of other crystals with some randomness of atomic arrangement. J. Am. Chem. Soc. 57:2680–2684.

[15] Kamb, B. 1968. Ice polymorphism and the structure of water. p. 507–542 *In* A. Rich and N. Davidson (ed.) Structural chemistry and molecular biology. Freeman and Company, San Francisco, CA.

[16] Marion, G.M. 1995. Freeze–thaw processes and soil chemistry. Cold Regions Res. Eng. Lab. (CRREL) Spec. Rep. 95-12. Hanover, NH.

[17] Workman, E.J., and S.E. Reynolds. 1950. Electrical phenomena occurring during the freezing of dilute aqueous solutions and their possible relationship to thunderstorm electricity. Phys. Rev. 78:254–259.

[18] Debye, P., and E. Hückel. 1923. Zur theorie der electrolyte. I. Gefrierpunktserniedrigung und verwandte Erscheinungen. Phys. Z. 24:185–206.

[19] Kielland, J. 1937. Individual activity coefficients of ions in aqueous solutions. J. Am. Chem. Soc. 59:1675–1678.

[20] Pitzer, K.S. 1991. Ion interaction approach: Theory and data correlation. p. 75–153 *In* K.S. Pitzer (ed.) Activity coefficients in electrolyte solutions. CRC Press, Boca Raton, FL.

[21] Carslaw, K.S., T. Peter, and S. Clegg. 1997. Modeling the composition of liquid stratospheric aerosols. Rev. Geophys. 35:125–154.

[22] Felmy, A.R., and J.H. Weare. 1995. The development and application of aqueous thermodynamic models: The specific ion-interaction approach. p. 31–52 *In* R.H. Loeppert, A.P. Schwab, and S. Goldberg (ed.) Chemical equilibrium and reaction models. Soil Sci. Soc. Am. Spec. Publ. no. 42. Soil Sci. Soc. Am., Madison, WI.

[23] He, S., and J.W. Morse. 1993. The carbonic acid system and calcite solubility in aqueous Na–K–Ca–Mg–Cl–SO$_4$ solutions from 0 to 90°C. Geochim. Cosmochim. Acta 57:3533–3554.

[24] Pitzer, K.S. (Ed.) 1991. Activity coefficients in electrolyte solutions. 2nd ed. CRC Press, Boca Raton, Fl.

[25] Walther, J.V. 1997. Determination of activity coefficients of neutral species in supercritical H$_2$O solutions. Geochim. Cosmochim. Acta 61:3311–3318.

[26] Martin, L.R. 1984. Atmospheric liquid water as a reaction medium. p.23(1)–23(6). *In* L. Newman (ed.) Abstracts of conference on gas-liquid chemistry of natural waters. Vol. 1. April 1984. Brookhaven National Laboratory, Upton, NY.

[27] Seinfeld, J.H. 1986. Atmospheric chemistry and physics of air pollution. John Wiley & Sons, New York, NY.

[28] Wolt, J.D. 1994. Soil solution chemistry: Applications to environmental science and agriculture. John Wiley & Sons, New York, NY.

[29] Rhoades, J.D. 1996. Salinity: Electrical conductivity and total dissolved solids. p. 417–435 *In* J.M. Bigham (ed.-in-chief) Methods of soil analysis. Part 3. Chemical methods. SSSA Book Ser. no. 5. Soil Sci. Soc. Am., Madison, WI.

[30] Maas, E.V. 1990. Crop salt tolerance. p. 262–304 *In* K.K. Tanji (ed.) Agricultural salinity assessment and management. ASCE Manuals Prac. no. 71. Am. Soc. Civ. Eng., New York, NY.

[31] Bigham, J.M. (Editor-in-Chief). 1997. Glossary of soil science terms. Soil Sci. Soc. Am., Madison, WI.

[32] Kavanau, J.L. 1964. Water and solute–water interactions. Holden-Day, Inc., San Francisco, CA.

[33] Samoilov, O.Ya. 1957. Structure of aqueous electrolyte solutions and the hydration of ions. USSR Academy of Sciences, Moscow. Translated from Russian by D.J.G. Ives, Consultants Bureau, NY, 1965.

[34] Hoffmann, M.M., J.G. Darab, and J.L. Fulton. 2001. An infrared and x-ray absorption study of the equilibria and structures of chromate, bichromate, and dichromate in ambient aqueous solutions. J. Phys. Chem. A105:1772–1782.

[35] Marcus, Y. 1985. Ion solvation. John Wiley & Sons Ltd., New York, NY.

[36] Brownlow, A.H. 1979. Geochemistry. Prentice-Hall, Englewood Cliffs, NJ.

[37] Jenkins, H.D.B., and Y. Marcus. 1995. Viscosity B-coefficients of ions in solution. Chem. Rev. 95:2695–2724.

[38] Murrell, J.N., and A.D. Jenkins. 1994. Properties of liquids and solutions. 2nd ed. John Wiley & Sons, Chichester, England.

[39] Hofmeister, F. 1888. On the understanding of the effects of salts. Arch. Exp. Pathol. Pharmakol. (Leipzig) 24:247–260.

[40] Hajdu, F. 1977. A model of liquid water tetragonal clusters: Description and determination of parameters. Acta Chim. (Budapest) 93:371–394.

[41] Chaplin, M.F. 1999. A proposal for the structure of water. Biophysical Chem. 83:211–221.

[42] Atwood, J.L., L.J. Barbour, T.J. Ness, C.L. Raston, and P.L. Raston. 2001. A well-resolved ice-like $(H_2O)_8$ cluster in an organic supramolecular complex. J. Am. Chem. Soc. 123: 7192–7193.

[43] Tanaka, H. 2000. Simple physical model of liquid water. J. Chem. Phys. 112:799–809.

[44] Dore, J.C., M.A.M. Sufi, and M.-C. Bellissent-Funel. 2000. Structural change in D_2O water as a function of temperature; The isochoric temperature derivative function for neutron diffraction. Phys. Chem. Chem. Phys. 2:1599–1602.

[45] Wiggins, P.M. 1995. High and low-density water in gels. Progr. Polymer. Sci. 20:1121–1163.

[46] Elliott, S.R. 1995. Interpretation of the principal diffraction peak of liquid and amorphous water. J. Chem. Phys. 103:2758–2761.

[47] Mishima, O., L.D. Calvert, E. Whalley. 1985. An apparently first-order transition between two amorphous phases of ice induced by pressure. Nature 314:76–78.

[48] Davy, H. 1811. On some of the combinations of oxymuriatic gas and oxygene, and on the chemical relations of these principles, to inflammable bodies. Phil. Trans. Roy. Soc. (London) 101:1–35.

[49] Miller, S.L. 1973. The clathrate hydrates — Their nature and occurrence. p. 42–50 *In* E. Whalley, S.J. Jones, and L.W. Gold (ed.) Physics and chemistry of ice. Royal Soc. Canada, Ottawa, Canada.

[50] Kleinberg, R.L., and P.G. Brewer. 2001. Probing gas hydrate deposits. American Scientist 89:244–251.

[51] Head-Gordon, T. 1995. Is water structure around hydrophobic groups clathrate-like? Proc. Natl. Acad. Sci. USA 92:8308–8312.

[52] Harvey, L.D.D., and Z. Huang. 1995. Evaluation of the potential impact of methane clathrate destabilization on future global warming. J. Geophys. Res. 100:2905–2926.

[53] Robbins, O., Jr. 1967. Ionic reactions and equilibria. Macmillan, New York, NY.

[54] Stumm, W., and J.J. Morgan. 1981. Aquatic chemistry. 2nd ed. John Wiley & Sons, New York, NY.

[55] Seinfeld, J.H., and S.N. Pandis. 1998. Atmospheric chemistry and physics. John Wiley & Sons, New York, NY.

[56] Smith, R.M., and A.E. Martell. 1974. Critical stability constants, Vol. 1: Amino acids. Plenum Press, New York, NY.

[57] Smith, R.M., and A.E. Martell. 1975. Critical stability constants, Vol. 2: Amines. Plenum Press, New York, NY.

[58] Smith, R.M., and A.E. Martell. 1976. Critical stability constants, Vol. 4: Inorganic complexes. Plenum Press, New York, NY.

[59] Martell, A.E., and R.M. Smith. 1977. Critical stability constants, Vol. 3: Other organic ligands. Plenum Press, New York, NY.

[60] Smith, R.M., and A.E. Martell. 1982. Critical stability constants, Vol. 5: First supplement. Plenum Press, New York, NY.

[61] Smith, R.M., and A.E. Martell. 1989. Critical stability constants, Vol. 6: Second supplement. Plenum Press, New York, NY.

[62] Smith, R.M., and A.E. Martell. 1995. The selection of critical stability constants. p. 7–29. *In* R.H. Loeppert, A.P. Schwab, and S. Goldberg (ed.) Chemical equilibrium and reaction models. SSSA Spec. Publ. no. 42. Soil Sci. Soc. Am., Madison, WI.

[63] Gustafsson, J.P. 2002. VMINTEQ. Windows version of MINTEQ2A. http://www.lwr.kth.se/english/OurSoftware/Vminteq/.

[64] Allison, J.D., D.S. Brown, and K.J. Novo-Gradac. 1990. MINTEQA2/PRODEFA2, a geochemical assessment model for environmental systems: Version 3.00 user's manual. EPA-600/3-91-021. USEPA, Athens, GA.

[65] Felmy, A.R., D.C. Girvin, and E.A. Jenne. 1984. MINTEQ — A computer program for calculating aqueous geochemical equilibria. EPA-600/3-84-032. USEPA, Athens, GA.

[66] Ball, J.W., E.A. Jenne, and M.W. Cantrell. 1981. WATEQ3: A geochemical model with uranium added. Open File Rep. 81-1183. U.S. Geol. Surv., Washington, D.C.

[67] Westall, J.C., J.L. Zachary, and F.M.M. Morel. 1976. MINEQL, a computer program for the calculation of chemical equilibrium conpositions of aqueous systems. Tech. Note 18. Dept. Civil Eng., Massachusetts Inst. Technol., Cambridge, MA.

[68] Morel, F., and J. Morgan. 1972. A numerical method for computing equilibria in aqueous chemical systems. Environ. Sci. Technol. 6:58–67.

[69] Ball, J.W., and D.K. Nordstrom. 1991. User's manual for WATEQ4F, with revised thermodynamic data base and test cases for calculating speciation of major, trace, and redox elements in natural waters. Open-File Rep. 91-183. U.S. Geol Surv, Washington, D.C.

[70] Truesdell, A.H., and B.F. Jones. 1974. WATEQ, A computer program for calculating chemical equilibria of natural waters. J. Res. U.S. Geol. Surv. 2:233–274.

[71] Parkhurst, D.L. 1995. User's guide to PHREEQC — A computer program for speciation, reaction-path, advective-transport, and inverse geochemical calculations. Water-Resour. Invest. Rep. 95-4227. U.S. Geol. Surv., Washington, D.C.

[72] Parkhurst, D.L., D.C. Thorstenson, and L.N. Plummer. 1980. PHREEQE — A computer program for geochemical calculations. Water-Resour. Invest. Rep. 80-96 (Revised, 1990). U.S. Geol. Surv., Washington, D.C.

[73] Parkhurst, D.L., and C.A.J. Appelo. 1999. User's guide to PHREEQC (Version 2) — A computer program for speciation, batch-reaction, one-dimensional transport, and inverse geochemical calculations. Water-Resour. Invest. Rep. 99-4259. U.S. Geol. Surv., Washington, D.C.

[74] Schulthess, C.P., and H. Hohl. 1982. Laboratory notes from EAWAG, Dübendorf, ZH, Switzerland. (Unpublished data.)

[75] Schulthess, C.P. 1986. Laboratory notes from the University of Delaware, Newark, DE. (Unpublished data.)

[76] Gran, G. 1952. Determination of the equivalence point in potentiometric titrations. Part II. Analyst 77:661–671.

[77] Bard, A.J. 1966. Chemical equilibrium. Harper & Row, New York, NY.

[78] Edmonds, W.M., and A.H. Bath. 1976. Centrifuge extraction and chemical analysis of interstitial waters. Environ. Sci. Technol. 10:467–472.

[79] Whelan, B.R., and N.J. Barrow. 1980. A study of a method for displacing soil solution by centrifuging with an immiscible liquid. J. Environ. Qual. 9:315–319.

[80] Folkoff, M.E., V. Meentemeyer, and E.O. Box. 1981. Climatic control of soil acidity. Phys. Geogr. 2:116–124.

[81] Miller, D.A., and R.A. White. 1998. A conterminous United States multi-layer soil characteristics data set for regional climate and hydrology modeling. Earth Interactions 2(2):1–26. (See also http://www.essc.psu.edu/soil_info/ and web pages therein.)

[82] Manrique, L.A., C.A. Jones, and P.T. Dyke. 1991. Predicting cation-exchange capacity from soil physical and chemical properties. Soil Sci. Soc. Am. J. 55:787–794.

[83] Holmgren, G.G.S., M.W. Meyer, R.L. Chaney, and R.B. Daniels. 1993. Cadmium, lead, zinc, copper, and nickel in agricultural soils of the United States of America. J. Environ. Qual. 22:335–348.

[84] USDA Soil Survey Staff. 1999. Soil taxonomy: A basic system of soil classification for making and interpreting soil surveys. 2nd ed. Agriculture Handbook no. 436. U.S. Department of Agriculture, Washington, D.C.

Questions

1. Assume that a soil contains the following compounds in the aqueous phase (M_c = moles of charge per L):

    Cations: 0.405 M Mg^{2+}, 0.032 M NH_4^+, 0.013 M K^+, 0.008 M_c sum of other cations.

    Anions: 0.335 M SO_4^{2-}, 0.009 M Cl^-, 0.104 M_c sum of other anions.

    (A) What should the pH be so as to maintain electroneutrality?

    (B) What should the pH be if $CO_2(g)$ is 0.03, 0.3, or 7.5% in the soil air?

2A. Add HCl to water so that the solution is pure $H_3O^+Cl^-$, and that the density of the liquid is 1.19 g cm^{-3}. Since this hypothetical example assumes that every molecule of water is protonated, this essentially results in the most acidic solution ever imagined. What is the pH^c of this hypothetical medium? The pH^c measures $-\log [H^+]$, where $[H^+]$ is the hydrogen ion concentration rather than its activity (that is, assume $\gamma_H = 1.0$).

    Answer: 1.0 liter of liquid will contain 1190 g of solution. The formula weight of each molecule is 54.5 g mol^{-1}. Since 1190/54.5 = 21.8 M, the pH^c of this liquid is $-\log(21.8) = -1.34$. Although pH^c ≠ pH, we do get a sense that the pH may be less then zero, and that the traditional statement that the pH limits of water are 0 to 14 is not clear. In reality, concentrated HCl solutions are 12.1 M (pH$^c = -1.08$ if fully dissociated, $\rho = 1.19$ g cm^{-3}).

2B. Saturated NaOH in water is 10.5 M. What is the pH^c of this saturated solution?

    Answer: pH^c = 15.0 if fully dissociated. Again note that the pH range of water does not have to be "0 to 14" as is often stated in introductory texts. This phrase is convenient because pK_w = 14, pH 7 is the neutral pH value, and a range of 0 to 14 places the neutral pH 7 value in the middle. Your pH meter, however, will probably not measure the hydrogen ion activity at these extreme values. This is not due to a pH range limitation of the water, but rather to a limitation of the instrument and methodology used. You should also always measure *within* the range of your calibrating buffer solutions, which are typically 4, 7, and 10 for measurement of most soil solutions. Fortunately, the pH of natural environments are generally not anywhere close to these extreme values. One exception noted here is the extremely low pH (< 0) of stratospheric aerosols consisting of saturated acid solutions[21]; but these liquids are not likely to naturally come in contact with soils. However, accidental spillage of strong acids and bases, particularly during overland transport of these chemicals, does occur sometimes.

3.   In Section 2.12.4, the presence of IC in NaOH was mentioned. Why is inorganic C (IC) present in NaOH solutions? How can you make NaOH solutions that are CO_2-free?

Answer: As can be seen in Figure 2-19, alkaline solutions have a strong affinity for IC. CO_2-free solutions are difficult to prepare. Boiling water will release CO_2(aq), but it will readily re-absorb it as it begins to cool down. NaOH pellets used to prepare NaOH solutions will already have some IC impurities in it, and the pellets will very easily sorb more IC while they are being weighed in preparation for the alkaline solution. Purging with pure air (CO_2-free air) or N_2(g) will remove dissolved carbonates from acidic solutions, but the degassing rate is too slow for it to be practical with alkaline solutions. The best method is as follows: (1) Prepare a solution of saturated NaOH ($\approx 10\ M$) making sure that excess NaOH pellets are present in the sealed container. (2) If there is any IC present, it will precipitate out of solution as Na_2CO_3(s). (3) Pipette a small aliquot of this saturated NaOH solution, avoiding any solids suspended in the liquid, and immediate dilute it in freshly deionized water (distill the water before it is deionized). (4) Keep the container protected from contacting atmospheric CO_2 gases. (5) If necessary, calibrate the basic solution with KH-phthalic acid standard and, if available, confirm that the basic solution is IC-free using an IC autoanalyzer.

4.   Using Table 2-12, draw pH-dependent speciation diagrams for 0.001 M arsenate and 0.0005 M aluminum hydroxide species in a 0.01 M NaCl background electrolyte solution. Using numerical and graphical methods, what is the equivalence point of this mixture if we assume the zero level condition to represent the addition of $Al(OH)_3$ and NaH_2AsO_4 in pure water?

5.   A conductivity titration monitors the electrical conductivity of the solution as a function of amount of acid or base added. Although the slopes will differ, the general shape of the conductivity titration data resembles the f_1 and f_2 data illustrated in Figure 2-24. Why do the results of conductivity titrations resemble those of Gran titrations?

6.   Figure 2-35 shows predicted A horizon soil pH values for the USA. If you are measuring soil pH values anywhere in the area shown, how likely is it for your soil measurement to differ from the predicted values? Discuss the variability of soils in general, and in your present region in particular.

Partial answer: Note that the smallest unit of area that can be shaded and recognized in Figure 2-35 covers many, many km^2.

7.   Is the charge of an acid solution (pH < 7) positive or negative? Is the charge of an alkaline solution (pH > 7) positive or negative?

Answer: Although they are not pH neutral, they are both charge neutral. Say that you add

HCl acid to water. The liquid has excess free hydronium cations and a pH < 7; but it also has an equal number of free Cl⁻ anions, and the solution has no net charge. Now assume that you add NaOH base to water. The liquid has excess free hydroxide anions and a pH > 7; but it also has an equal number of free Na⁺ cations, and the solution has no net charge.

Note that when you say that a liquid is "neutral" you are probably talking about it being *pH neutral* because it is always *charge neutral* regardless of its acidity or alkalinity. An exception would be if the liquid is subjected to an applied electric field; but, if the ions were not selectively removed, then here, too, the solution would again be charge neutral when the applied electric field is turned off. Note also that a battery's cathode and anode components are technically charge neutral at all times, but charge travels in and out of them if redox reactions are allowed to proceed at the moment of closing the circuit.

Chapter 3:
Soil Organic Matter

Soil organic matter (SOM) is all dead organic matter in soil. It may be highly decomposed or not at all. *Humus* material is decomposed organic matter. The *nonhumus* organic matter fraction refers to the soil organic matter that has not yet decomposed. Historically, we have known for a long time about the importance of soil organic matter to the quality of our soils, but its impact on the physical and chemical properties of soils still continues to fascinate us.

Much of our understanding of SOM depends on the extraction procedures used and, as will be seen, our nomenclature of SOM components are based on the extraction method used rather than on the presence of any particular chemical structure or functional group. The first 6 sections of this chapter focus on the methods and problems encountered with various SOM extraction procedures. This is followed by 9 sections that discuss some of the characteristics and chemical reactivity of SOM, as well as some of the natural processes that affect the composition of SOM. The final 4 sections discuss selected topics that highlight the impact of SOM on the geologic, biologic and anthropogenic activities on the earth's surface.

[3.1] Nomenclature & Extraction of SOM

The first extraction of humic acid was by F.K. Achard, in 1786, who treated peat with alkali and obtained a dark amorphous precipitate upon acidification. The term humus is credited to T. de Saussure (1804) to describe the dark-colored organic material in soil. In 1822, J.W. Dobereiner introduced the term *humus acid*. In 1862, G.J. Mulder classified humic substances on the basis of solubility and color as follows: (i) *ulmin* and *humin*, insoluble in alkali, (ii) *ulmic acid* (brown) and *humic acid* (black), soluble in alkali, and (iii) *crenic* and *apocrenic acids*, soluble in water. The term *hymatomelanic acid* was proposed by F. Hoppe-Seyler in 1889. Note that each of these names depends on the extraction methods or the solubility properties of the soil organic material. Black and Christman (1963)[1] noted that 43 different names for soil organic matter were introduced by 1919. In these early years, many investigators repeated the mistake of regarding the isolated substances as chemically pure compounds of specific compositions.

In 1919, S. Oden classified humic substances into four groups: (1) *humus coal* (same as Mulder's humin), (2) *fulvic acid* (similar to crenic and apocrenic acids), (3) *humic acid* (dark-brown to black material, soluble in alkali, insoluble in acid), and (4) *hymatomelanic acid*. Page (1930)[2] modified Oden's scheme of classification by suggesting Mulder's term *humin* for *humus coal*, and agreed with the term *humic acid* in place of *humus acid*. A final distinction came when Waksman in 1936 specified *humus acid* to be an inclusive term of all soil organic acids, and *humic acid* to be the precipitate obtained from acidification of an alkali extract.

The terms used today to identify the fractions of soil organic matter continue to focus on their acid–base behavior. Figure 3-1 shows this mechanistic scheme for the fractionation of soil organic matter, or humus. Humin is the alkali-insoluble fraction of humus. Fulvic acid (FA) is the fraction that remains in solution after the alkali humus solution has been acidified. Humic acid (HA) is the fraction that precipitates when the alkali humus

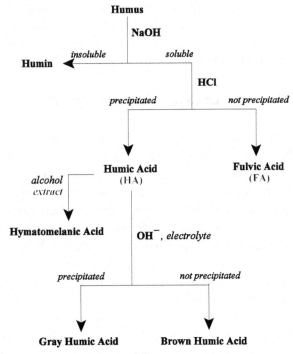

Figure 3-1: Fractionation of humus. The NaOH concentration in the first step is typically around 0.1 to 0.5 *M* with a soil/extractant ratio of 1:5 to 1:10 (g mL⁻¹) resulting in an initial pH ≈ 13, which is often maintained for 24 h at room temperature. The mixture is often N₂-purged in an attempt to minimize autoxidation of the organic matter. The supernatant liquid is separated from the insoluble portion and acidified to around pH 1 with HCl. The dark brown to black precipitate that immediately forms is labeled HA, while the fulvous (meaning dull reddish-yellow) colored supernatant liquid is labeled FA. Additional extraction steps may be pursued, as indicated, but are rarely done.

solution is acidified. Hymatomelanic acid is the alcohol-soluble fraction of humic acid. Since FA is the most soluble fraction of humus, it generally dominates the organic component in soil solutions and nearby streams, while the typically insoluble HA dominates the "immobile" organic component of soils.

Humic acid may be further divided into two groups.[3] In the presence of an electrolyte under alkaline conditions, humic acids characteristic of some Mollisols (such as Rendolls and cold-region Mollisols) will coagulate and are termed *gray humic acids*. Humic acids characteristic of Histosols and Alfisols do not coagulate and are termed *brown humic acids*. The HA fraction may also be fractionated based on the strength of the acidifying solution (Step 2 in Figure 3-1). For example, the HA that precipitates at pH 4 may be differentiated from the HA that precipitates at pH 3, and so forth.

Natural organic matter will complex easily with various metals. Accordingly, the humic acid extracts will probably have a significant amount of metals in them. If we eliminate them, we obtain an "ash-free" humic acid product. This is accomplished by shaking the HA solid extract in a strongly acidic solution consisting of 0.3 M HF and 0.1 M HCl.[3] This acid washing procedure is repeated three times, followed by dialysis to remove the dissolved salts and freeze-drying the ash-free HA product. If most of the ash present is due to entrained clays, we might be able to avoid the HF–HCl treatment by simply filtering the NaOH extract (Step 1 in Figure 3-1) through a 0.2-μm filter.

[3.2]   Are Our Soil Organic Extractions Too Harsh?

Ideally, the soil organic extraction procedure should result in a product whose physical structure and chemical composition are similar to (or, if possible, identical to) the structure and composition of the original organic material present in the undisturbed soil sample. Hence, do the HA and FA products obtained by alkali extraction (first step in Figure 3-1) truly represent the organic components present in the original humus sample?

Strong alkali is known to react strongly with organic matter. It causes the fatty acids in your finger to dissolve, resulting in a false impression that the NaOH feels "oily". It will sterilize a soil medium. An alkali extraction procedure will dissolve protoplasmic and structural components from fresh organic tissues, and these dissolved compounds will become mixed with the humified organic matter. Alkali extractions will also dissolve the mineral components of soils (e.g., Si, Fe, and Al dissolution will be high), which will contaminate the organic fractions separated from the extract. Accordingly, a red flag should be raised if the humic acid was obtained by treating a humus sample with alkali. It is very likely that the humus was chemically modified (perhaps even significantly destroyed) by the initial alkali treatment. That is, was the HA truly "extracted" or was it "derived" from soil organic matter when strong alkali was used? Similarly, if SOM is treated with strong acids, are the HA and FA products truly representative of the chemical components existing in the original untreated sample?

Some researchers feel that minimal chemical change will occur if dissolved oxygen is missing in the NaOH solution. Accordingly, the alkali solution is often purged with N_2 to remove the dissolved oxygen and reduce the likelihood of autoxidation. However, this precaution may not be enough. As Figure 3-2 shows, there is a very rapid increase in the production of FA from HA when it is in contact with an alkali solution, even in the absence of dissolved oxygen. The longer the contact time, the greater will be the chemical change. That is, an "extraction" of HA and FA fractions from humus using the method illustrated in Figure 3-1 will probably (1) yield a false distribution of the two fractions because the FA content is inflated by the degradation of HA to FA in the NaOH treatment step, and (2) yield products that are physically and chemically different from the compounds that actually existed in the original humus material. Schematically,

$$HA + Na^+OH^- \rightarrow \text{FA-like products} \qquad [3\text{-}1],$$

where the products do not revert to the original HA material in the presence of acids. That is, Reaction 3-1 is irreversible.

While the mechanistic scheme illustrated in Figure 3-1 is very useful *for describing* an organic sample, it is not a useful method *for extracting* soil humus fractions. For example, a soil scientist may confirm that a given organic sample is completely humic acid because it is not soluble in acid but soluble in base. Hopefully, the soil scientist did not obtain the organic sample by exposing it to harsh chemical environments. If any physical or chemical damage has occurred, or is suspected to have occurred, then the ensuing discussion about the organic product needs to be carefully framed with this in mind. Experimental analysis and results obtained with a harshly extracted organic product may or may not properly represent the expected behavior and properties of the original, undisturbed soil organic material.

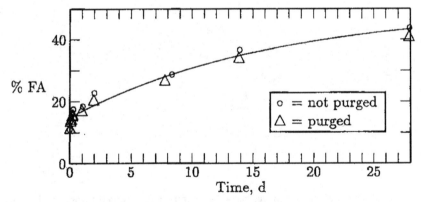

Figure 3-2: Kinetics of fulvic acid (FA) production from a pure humic acid (HA) source (2.0 g L^{-1}) while dissolved at pH 13.35 (0.5 M NaOH that was purged or not purged with He gas). At time $t = 0$, FA present was zero percent. The organic matter source was from a prior HA extraction of a Carlisle muck consisting of about 91% SOM using the loss-on-ignition method. The solid line drawn is based on first-order kinetics assuming two reactions: an initial fast reaction ($k_1 = 40$ d^{-1}, max. = 15%), followed by a slow reaction ($k_2 = 0.06$ d^{-1}, max. = 35%). *From Schulthess et al. (1997)*.[4]

[3.3]   Impact of pH on SOM Extraction

Multiple extractions with NaOH recovers up to 80% of the soil organic matter from soils.[3] Accordingly, the use of strong alkaline solutions has become quite widespread in the literature for extraction of SOM. Figure 3-3 illustrates the enormous impact that pH plays on the extractability of organic compounds from the SOM of an organic soil, where plain water (equilibrium pH 4.3) yielded nearly 200 times less than the extraction in 0.5 M NaOH solution (pH 13.25). While only fulvic acids are extracted by acidic solutions, both humic and fulvic acids are extracted by alkaline solutions. Based on the discussion presented above with Figure 3-2, it is questionable if the FA alkaline extraction shown in Figure 3-3 is truly representative of these highly soluble organic compounds in soils (particularly, in the Carlisle soil for the case presented). By extension, it is also questionable if the HA alkaline extraction is truly representative of these acid-insoluble organic compounds in soils.

Considering the very extensive application of alkaline extractions of soil organic matter for obtaining HA and FA fractions, it may seem as if these concerns are being ignored by many researchers. Note, however, that many interesting generalities can still be made about soil organic matter even when the study material has been slightly altered. Over longer time frames, most of these alkaline-accelerated alterations may be occurring naturally in soils anyway. Nevertheless, sometimes it would be better to have unaltered soil organic matter samples to work with.

If only FA is needed, and small quantities are acceptable, then the extracts obtained with weakly acidic solutions will probably be ideal FA samples that are unaltered either

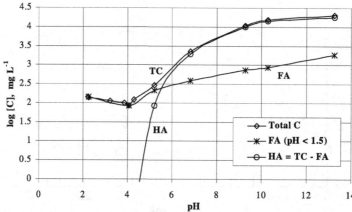

Figure 3-3: Extraction of soil organic matter with water at various pH values over a 24 h period (2.5 g of a dried Carlisle muck sample in a 35-mL solution of H_2O + NaOH or HCl). All extracting solutions were purged with He gas to remove the dissolved O_2 content. For total C, the organic C content of the extracted solution was filtered and analyzed on a total organic carbon autoanalyzer (TOC), expressed as mg C L^{-1}. For FA, an aliquot of the filtered extracting solution was then acidified with HCl (final pH < 1.5), left standing for 2 h, and the supernatant solution was then again analyzed on a TOC autoanalyzer. The HA content equals the difference of these two values. *From Schulthess (1996).*[5]

physically or chemically. If unaltered HA is needed in large or small quantities, or if unaltered FA is needed in large quantities, then perhaps alternate methods of obtaining these materials should be used.

[3.4] Mild SOM Extracting Solutions

In the hope of reducing the probability of autoxidation, milder extraction procedures can be used instead. Mild extractants include salts of complexing agents (such as $Na_4P_2O_7$ and EDTA), organic complexing agents in aqueous media (such as acetylacetone, cupferron, and 8-hydroxyquinoline), dilute acid mixtures containing HF, and acidified organic solvents of various types (such as acetone, dioxane, cyclopentanone, dimethylformamide, and dimethyl sulfoxide).[3] Table 3-1 illustrates the chemical structure of these compounds.

Sodium pyrophosphate, EDTA, ammonium oxalate and salts of weak organic acids are effective because they form soluble complexes with polyvalent cations (such as Ca^{2+}, Fe^{3+}, and Al^{3+}) that are responsible for flocculating the soil organic matter. Organic matter will generally be more soluble if these polyvalent cations are replaced with monovalent cations (such as Na^+, K^+, and NH_4^+). These complexing agents are used at pH 7, but sometimes they are used at higher pH values (such as pH 8 or 9).

Acetylacetone, cupferron, and 8-hydroxyquinoline will also complex with polyvalent cations, but more selectively with the Fe^{3+} and Al^{3+} cations. These organic complexing agents are generally only effective in solubilizing the SOM if the soil is rich in just these cations. Dilute acid mixtures with HF will increase the extraction of SOM presumably by acid dissolution of the minerals present followed by complexation of the F^- with the released iron and aluminum.

Mixtures of HCl with organic solvents will extract SOM presumably by having the acid break the salt bridges between the organic matter and the minerals, followed by the

Table 3-1: Formulas of some compounds used in mild SOM extracting solutions.

organic solvents aiding in keeping the SOM in solution. Solvents used for this include acetone, dioxane, cyclopentanone, dimethyl formamide, and dimethyl sulfoxide. A similar effect is seen with formic acid, which acts as both an acid and a solvent. Note also that, in general, extraction of the soil organic matter components can be increased by pretreating the soil with acids to remove carbonates or silicates, which tend to bind with the organics.

The down side of using many of these mild SOM extractants is that it can be difficult to remove the solvent or organic complexing agent from the extracted material. Some extracted or dissolved minerals may also be present in the product obtained.

These mild extracting solutions do not typically yield high concentrations of extracted humus material. The effectiveness of these solutions will vary. The general rule of thumb is "like dissolves like." Accordingly, the chemical properties of the extracting solution, particularly its polarity and relative dielectric constant, will dictate the type of SOM extracted.

[3.5]   Supercritical Fluids (SCFs)

[3.5.1] Overview & General Properties of SCFs

Temperature–pressure (T–P) phase diagrams of compounds, such as those shown in Figure 3-4, are useful for illustrating the transitions of solids, liquids, and gases. The solid phase forms at low temperatures, and it is generally favored by an increase in pressure. For example, we observe an increase in CO_2 solid-phase formation with increase in pressure because the solid phase is more dense (that is, it occupies less volume) than the liquid phase. The opposite, however, is observed with H_2O, which has the rare property of having a solid phase that is less dense than the liquid phase. Various different solid structures may form within this region of the phase diagram. There are, for example, 14 known structures of ice, which are designated as ice structures I to XIV. All of the solid H_2O shown in Figure 3-4A are ice I because the transition from ice I to ice II or III does not occur at these low pressures. Since only the ice I solid is less dense than the liquid phase, the ice–liquid phase boundaries for the other ice structures will slope to the right. There are also five known structures of solid CO_2 (I to V), with structures II to V forming at extremely high pressures (not shown).[6]

A gas phase forms at moderate to high temperatures and low pressures, while a liquid phase forms at moderate temperatures and moderate to high pressures. Except for autodecomposition at high temperatures, there are typically no chemically distinct phases observed in gas or liquid molecules within these two phases. Note that at ambient pressures (1 atm), H_2O can exist as either vapor (gas), water (liquid), or ice (solid), depending on the temperature. However, CO_2 can only exist as dry ice (solid) or gas.

We focus here on the properties of liquids and gases at high temperatures and pressures, known as supercritical fluids (SCFs). They are formed when the environment is above a critical temperature (T_c) and pressure (P_c). Do not confuse this with the triple point of a substance, which refers to the temperature and pressure where the gas, liquid, and solid phases of a substance are all stable and can coexist (see Table 3-2). Also, do not confuse this with the limit of superheat of a liquid, which refers to the highest temperature that a liquid

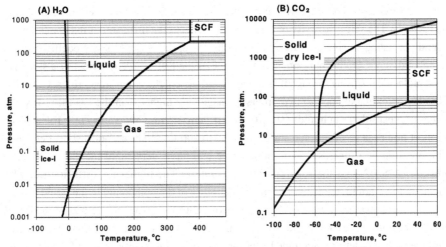

Figure 3-4: Temperature–pressure (T–P) phase diagrams for (A) H_2O and (B) CO_2. Note that the boundary lines are at the temperature and pressure where two phases can coexist: solid–liquid, solid–gas, or liquid–gas. Each boundary line describes the conditions for a particular phase change: melting (solid → liquid), freezing (liquid → solid), evaporation (liquid → gas), condensation (gas → liquid), sublimation (solid → gas), and deposition (gas → solid). The liquid–gas boundary line ends at the critical temperature and pressure, above which the compound exists as a supercritical fluid (SCF). *Data from Lide (2002)*[7], *and Span and Wagner (1996).*[8]

Table 3-2: Important temperature and pressure values for water and carbon dioxide. T_m = melting point at 1 atm, T_b = boiling point at 1 atm, T_{tp} = triple point temperature at the triple point pressure (P_{tp}), T_c = critical temperature at the critical pressure (P_c), and ρ_c = critical density at T_c and P_c, where °C = K − 273.15 and 1 atm = 101.325 kPa. *Data from Lide (2002).*[7]

Substance	T_m °C	T_b °C	T_{tp} °C	P_{tp} atm	T_c °C	P_c atm	ρ_c g cm⁻³
H_2O	0	100	0.1	0.006	373.99	217.75	0.322
CO_2	−78.45	sublimes	−56.57	5.112	30.99	72.786	0.468

can sustain in a metastable state at a constant pressure without undergoing a phase transition.

As the temperature of a liquid increases, thermal expansion causes its density to decrease. Conversely, the density of a gas increases as the pressure increases. There is a point reached, known as the critical temperature (T_c) and pressure (P_c), where the densities of the two phases become identical and the distinction between them disappears. The boundary curve that separates liquids from gases in a T–P phase diagram comes to an end at this point (Figure 3-4). The supercritical fluid that forms at $T \geq T_c$ and $P \geq P_c$ behaves as both a gas and a liquid. Of the four states of matter (solid, liquid, gas, and plasma), SCFs are generally classified as gases in spite of their liquid-like characteristics. Solid-phase research articles, however, will refer to solid–SCF boundaries in T–P phase diagrams as solid–liquid boundaries, or as melting points.

Supercritical fluids have high mass transfer rates due to their high diffusion coefficients. The self-diffusivity of a compound refers to the diffusion of the compound in a fluid consisting of similar compounds. Furthermore, the self-diffusivity coefficient of a compound, say CO_2, is approximately the same as the diffusion coefficient of a similarly sized molecule diffusing through it. The self-diffusivity coefficient of supercritical CO_2 (SC-CO_2) is approximately one to two orders of magnitude higher than the typical diffusivity values of most solutes in liquids ($D = 10^{-4}$ to 10^{-3} cm^2 s^{-1} for CO_2, versus $D \approx 10^{-5}$ cm^2 s^{-1} for most liquids at $P = 70$ to 200 atm and $T = 10$ to 100°C).[9] The self-diffusivity decreases with pressure, but increases with temperature and has a very rapid increase at a given pressure ($P > P_c$) near the supercritical temperature. The higher the diffusion rate of a compound in a liquid, the faster the rate of solubility of the compound in the liquid.

As shown in Figure 3-5, the viscosity of SC-CO_2 increases rapidly as a function of pressure near the critical point. Nevertheless, the high viscosity values at 300 to 400 bars (or atm) are still an order of magnitude lower than the typical viscosity values of most organic solvents. Note also that, as with all gases, the surface tension of SC-CO_2 is zero.[9] As a result of this low viscosity and zero surface tension, SC-CO_2 will flow very easily into hard to reach places, such as deep cavities in microporous materials.

Supercritical fluids have high solvating capacities due to their liquid-like densities. Figure 3-6 shows the sharp increase in density of SC-CO_2 near the critical point. As the density increases, the solubility of organic compounds also increases. For example, the SC-CO_2 solubility of polycyclic aromatic hydrocarbons (PAHs) at 50 and 70°C increases 20- to 300-fold as the CO_2 density increases from 0.4 to 0.8 g cm^{-3}, with the higher solubility observed at the higher temperature.[10]

As with the density of the gas, the relative dielectric constant (ϵ_r) of SC-CO_2 also increases as the pressure increases.[11] At about 50°C, a sharp increase is observed from $\epsilon_r = 1.0$ at ambient pressures to $\epsilon_r = 1.6$ at around 450 atm pressure, followed by a slower increase in ϵ_r with pressure (with $\epsilon_r \approx 1.8$ at 2000 atm). Even so, the relative dielectric

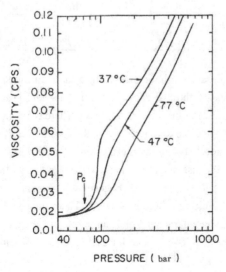

Figure 3-5: Viscosity of carbon dioxide as a function of temperature and pressure. Units are in centipoise (cps). *Reprinted from McHugh and Krukonis (1994)[9], "Supercritical fluid extraction", 2nd ed., Butterworth-Heinemann, copyright © 1994, with permission from Elsevier.*

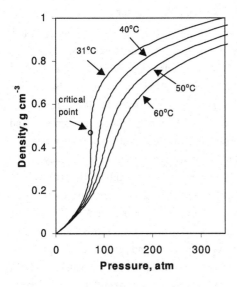

Figure 3-6: Density of carbon dioxide as a function of temperature and pressure based on the Peng–Robinson semi-empirical equation of state[12], where the pressure of a gas is the result of repulsive and attractive forces.

constant of SC-CO_2 remains low, and it is comparable to that of carbon tetrachloride (ϵ_r = 2.23 at 25°C, Table 2-8). This means that although the type of molecules that can dissolve in SC-CO_2 will change slightly with the applied pressure, using pure SC-CO_2 will only dissolve nonpolar molecules.

[3.5.2] Extracting SOM with SCFs

A supercritical fluid has good wetting characteristics, which makes it especially useful for extraction applications, particularly from porous materials. For example, supercritical CO_2 (SC-CO_2) is used in a wide range of applications, from extraction of caffeine from coffee and tea to cleaning of tools and equipment. When use of the SC-CO_2 is completed, it is simply allowed to evaporate, leaving behind the extracted compounds at ambient temperature and pressure. The solvent is completely removed upon depressurization, which, accordingly, also minimizes problems associated with contamination of the extracted compounds by the solvent. Use of SC-CO_2 also has minimal environmental impact because it is nontoxic at low concentrations or in well-ventilated areas, and because the compression and eventual decompression of the SC-CO_2 results in no net increase of CO_2 to the earth's atmosphere.

Supercritical CO_2 has recently been used to extract numerous organic toxic contaminants from soil, such as polycyclic aromatic hydrocarbons (PAHs) and polychlorinated biphenyls (PCBs).[10,13,14] If the organic compounds present in the soil are highly soluble chelating agents, then various metals will also be extracted from the soil.[15] Current use of SC-CO_2 for soil remediation has been limited, but the interest is clearly there mostly because, as noted earlier, it is an extracting solution that is easily disposed of when done (minimizing bulk waste problems) and has minimal environmental impact when evaporated back to the atmosphere.

Supercritical CO_2 can also be used to extract SOM from soil for subsequent research studies. $SC-CO_2$ is probably the best SCF choice for SOM extractions because of its low critical temperature (Table 3-2). Little, if any, thermal decomposition of SOM is expected to occur at around 31°C or even at slightly higher temperatures. Conveniently, when the SC-CO_2 is depressurized and removed, the extracted SOM will also be concentrated. In contrast, $SC-H_2O$ will expose the organic matter to very high temperatures (close to 400°C, Table 3-2), which will easily cause the organic matter to oxidize. Although $SC-H_2O$ is corrosive and costly to use, it could be potentially useful for the oxidative destruction of toxic waste.

The water content in the soil will not interfere with the extraction efficiency of SC-CO_2 as long as it is low (<10% by weight water content in soils with little or no clay).[16,17] Low yields are obtained when the water content is high, presumably because the water is acting as a barrier to carbon dioxide's penetration through the porous medium.[18]

Carbon dioxide is a linear, non-polar molecule (O=C=O). Following once again from the rule of thumb "like dissolves like," the organic compounds extracted with pure SC-CO_2 will be the hydrophobic fractions of SOM. Polar modifiers, such as methanol (CH_3OH), are typically added to the supercritical fluid when the extraction of polar molecules are desired.[19,20] Generally, 1 to 10% of the modifier is added. The more modifier added, the higher the polarity of the extractant. However, the more modifier added, the more we reintroduce the problems associated with these solvents — particularly the environmentally safe separation of the modifier from the extracted product.

In order to maintain better control of the extraction process, you should ensure that there is only one phase present when a modifier is used. That is to say, ensure that the modifier and the supercritical fluid are completely miscible with each other. Page et al. (1992)[21] give comprehensive phase data for various modifier–CO_2 mixtures. For example, at 50°C methanol–CO_2 might exist as two phases below 95 atm depending on the percentage of methanol present, but there is only one phase above 95 atm whatever the composition.[21]

[3.6]   Fractionation of DOC with Exchange Resins

Soil organic extraction with NaOH or other solvents is not necessary if the desired SOM is already dissolved in the soil liquid phase. Significant fractions of a soil's dissolved organic carbon (DOC) will also be present in nearby aquatic environments — for example ponds, streams, and groundwaters. When the desired fraction of SOM is the soil's DOC, then the experimental procedure to obtain it is basically concentration followed by fractionation of the dilute soluble components.

Concentration and fractionation of dissolved compounds may involve dialysis membranes and freeze-drying of large volumes of water. Freeze-drying is generally considered to be a safe and rapid method for the collection of dissolved humic matter in large volumes of water.[22] A much more common approach involves a preconcentration step, where compounds dissolved in large volumes of water are adsorbed onto a specific solid phase, followed by elution of the adsorbed compounds with a small volume of an extracting solution.

Figure 3-7 illustrates a commonly used procedure for obtaining DOC involving the use of exchange resins. Although there are numerous variations to this procedure found in the literature, Figure 3-7 serves to illustrate the general principles involved. The types of organic products thus concentrated depend on the reactivity of the dissolved organic compounds with the exchange resin. The resins shown, also known as exchangers, are commercially available organic polymers that selectively retain various types of organic compounds. Some additional information about the resins is given in Table 3-3.

There are, however, a few concerns about obtaining concentrated samples of DOC using exchange resins. Note in particular that the eluting solution is a strong base with a pH as high as 13, and this may cause some chemical changes in the sample similar to those discussed earlier in Section 3.2. Minimize this potential problem by quickly acidifying the eluted alkaline solution.

If the water sample is acidified to pH 2 prior to filtration, then all the fractions collected by these resins are, by definition, fulvic acids. If, instead, the sample is acidified after filtration, then both humic and fulvic acids are collected by the resins. These HA and FA fractions are obtained based on the pH-dependent solubility of the eluted solutions.

Note that sometimes the eluted products are operationally described as either "hydrophobic" or "hydrophilic" organic compounds. These terms are misleading because all the fractions can be quite hydrophilic, as can be expected since they were initially water soluble in the environment. These terms were introduced because it was believed that the retention of the organic compounds by the resins was either by hydrophobic interactions or by H-bonding, and the corresponding names of the eluted products reflected this belief. As it turns out, however, much of the retention of organic compounds by these resins is by H-bonding, and not entirely by hydrophobic interactions. An organic compound sorbing to an exchanger can also be ambiphilic (that is, it can have both hydrophobic and hydrophilic portions within the same molecule).

The porosity of the exchanger is as important to the efficiency of the exchanger as is the chemical structure of the matrix. This is known as a size-exclusion effect. Specifically, large humic molecules (particularly those with MW > 30,000 g mol^{-1}, but also those with

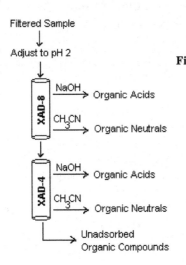

Figure 3-7: Example of a typical extraction and fractionation set-up for concentrating DOC using organic polymer resins. The organic acids are eluted from the resins with 0.1 M NaOH (pH ≈ 13) followed by immediate acidification to avoid oxidation of the humic substances. The neutral organic compounds are eluted from the resins with acetonitrile (CH$_3$CN). If only fulvic acids are desired, acidify to pH 2 first and then filter prior to passing the liquid sample through the columns.

Table 3-3: Description of some organic polymer resins commonly used for extraction and fractionation of dissolved organic carbon (DOC). These resins may need to be washed prior to use in exchanger columns. The washing of these resins is typically as follows: first wash the beads in 0.1 M NaOH, then rinse with distilled water, and finally place them in a soxhlet extractor and sequentially extract for 48 h each with methanol (CH_3OH) and acetonitrile (CH_3CN). The chemical structures shown are to be used only as a rough description of the matrix because this information is generally classified confidential by the manufacturers.

Resin Name	Matrix	Comments
Amberlite XAD-4:	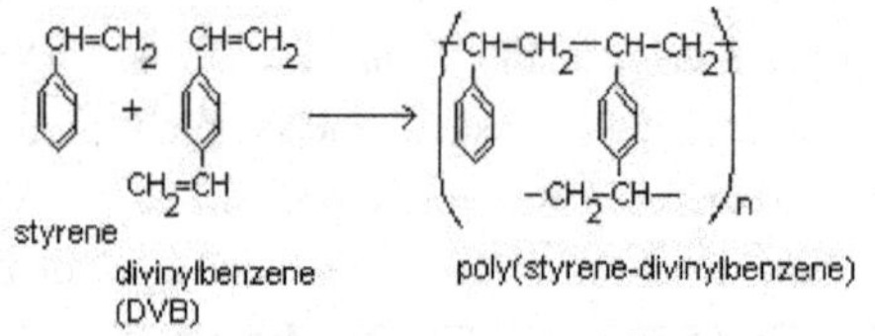 styrene divinylbenzene (DVB) poly(styrene-divinylbenzene)	Maximum pore volume contributed by pore diameters of 14.0 nm.
Amberlite XAD-8:	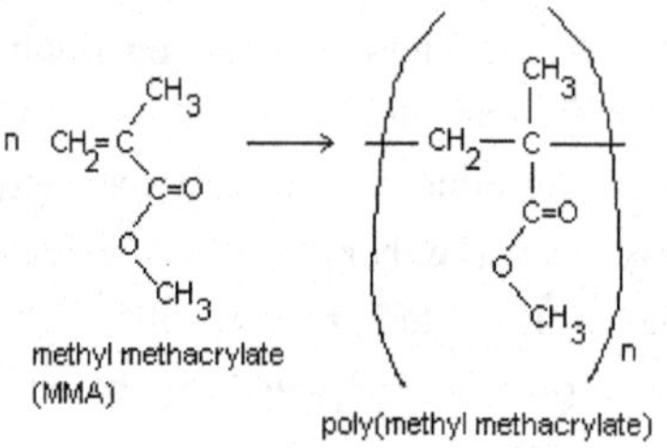 methyl methacrylate (MMA) poly(methyl methacrylate)	XAD7HP can replace XAD-8. In XAD7HP, the methyl ester group ($-COOCH_3$) is replaced with an unspecified aliphatic ester group ($-COOR$). Other agents are added to provide cross-linking, which occurs across these ester groups ($-COO-R-OOC-$). The maximum pore volume in XAD7HP is contributed by pore diameters of 36.0 nm.
Dowex MSC-1:	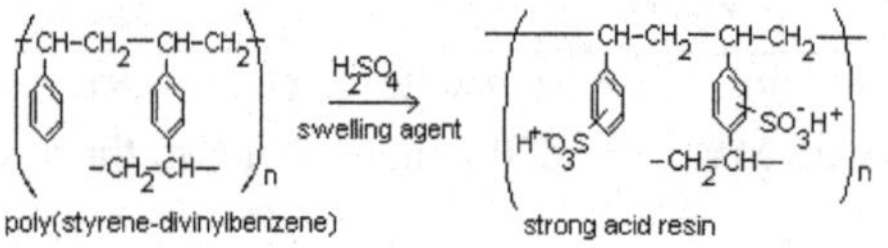 poly(styrene-divinylbenzene) strong acid resin	The percentage of monomers present varies with resin type, but it is typically 6 to 20% DVB. In MSC-1 the monomer is approximately 20% DVB. The ratio of sulfonates to benzene rings is nearly 1:1 in MSC-1.
Polyclar:	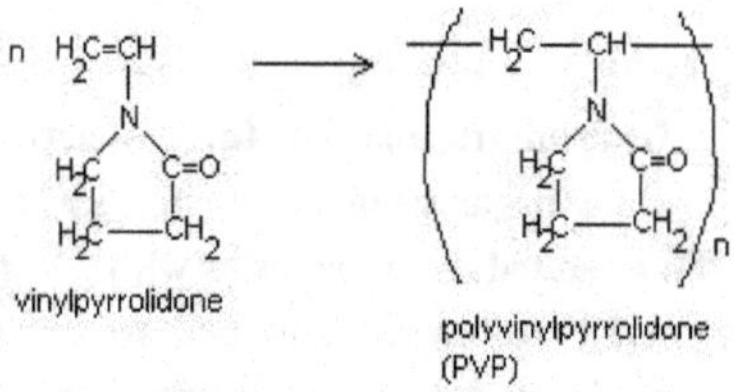 vinylpyrrolidone polyvinylpyrrolidone (PVP)	Polyvinylpolypyrrolidone (PVPP), also known as Polyclar, is a cross-linked PVP polymer (shown here). PVP is water soluble, while PVPP is an insoluble resin.
Duolite A-7:	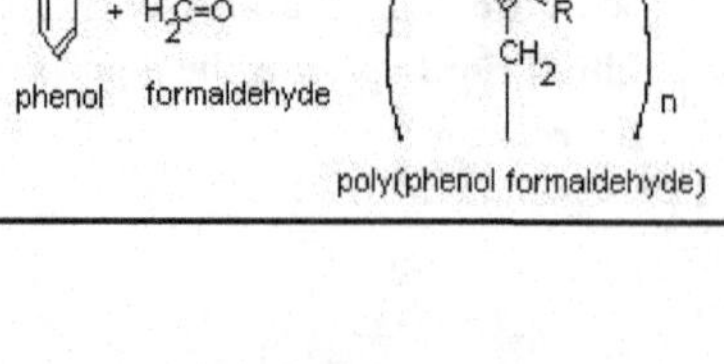 phenol formaldehyde poly(phenol formaldehyde)	This polyamine is a porous, weak base anion exchanger, with amine groups (e.g., $-CH_2-NH_2$) randomly distributed within the matrix. Effective pH range is 0–6. Pore sizes are 15 to 30 nm.

MW as low as 5,000 g mol^{-1}) are not significantly adsorbed by many of the commonly used resins, such as the XAD resins.[23] As a result of this, a potential problem with the use of exchangers is that the resins will not necessarily concentrate a representative sample of the dissolved humic material.

Finally, note that the fractions collected by exchangers may differ significantly from the fractions collected by other methods. For example, the humic acid (HA) and fulvic acid (FA) fractions collected by an XAD-8 column may be chemically different from the HA or FA fractions collected by freeze-drying followed by conventional fractionation methods (Figure 3-1).[24]

[3.7]   General Characteristics of Humic Substances

Soil organic matter has various lasting effects on soil, which are summarized in Table 3-4. A high content of organic matter is usually found in wet environments, such as wetlands, where the rate of degradation of organic matter is slower than its rate of accumulation or deposition. The organic matter of surface mineral soils typically ranges from 0.5 to 5% by weight, but may be nearly 100% for some peat soils.

Soil organic matter can hold a large amount of water by weight. It also has a high affinity for nutrients and trace elements essential to plant life. Clearly, from the general properties described in Table 3-4, SOM fully deserves its nickname of "Black Gold" by experienced gardeners. (Note that this nickname is also used for oil.)

Using the platinum–cobalt method, one standard unit of color is produced by 1 mg Pt L^{-1} in the form of chloroplatinate ion, $PtCl_6^{2-}$. Humic acid has a much stronger effect on color then fulvic acid. At pH 8, a solution of 1 mg L^{-1} of humic acid yields 26.5 color units (CU) while a solution of 1 mg L^{-1} of fulvic acid yields 2.8 CU. If the soil organic matter washes into nearby streams, then it darkens their color also. Surface waters are typically 5 to 200 CU. Note that drinking water standards permit 15 CU.[25]

Organic soils tend to be wet and the evaporation of water from the ground has a very strong cooling effect. The evaporation of water absorbs heat from its environment, absorbing 540 calories g^{-1} H$_2$O. Therefore, although the dark color of SOM absorbs sunlight very efficiently, organic soils are not necessarily warmer.

The dissolved organic carbon fractions in the soil solution include both the organic carbon that is dissolved in solution plus the large, colloidal suspensions of organic carbon compounds or mixtures. These colloids play an important role in the sequestering of other organic carbons from the soil solution. The actual biochemicals present in the soil solution vary spatially and temporally. The nature of these compounds reflects the nature and activity of the soil biota.

[3.8]   Chemical Characteristics of Humic Substances

The four humic substances classified by Oden in 1919 (humin, FA, HA, and hymatomelanic acid) differ in molecular weight, with fulvic acid having the lowest value.

Table 3-4: General properties of humus and associated effects in the soil. *From Stevenson (1994)*[3], *Humus chemistry: Genesis, composition, reactions. 2nd ed., copyright © 1994 by John Wiley & Sons. This material is used by permission of John Wiley & Sons, Inc.*

Property	Remarks	Effect on Soil
Color	The typical dark color of many soils is caused by organic matter	May facilitate warming
Water retention	Organic matter can hold up to 20 times its weight in water	Helps prevent drying and shrinking; improves moisture-retaining properties of sandy soils
Combination with clay minerals	Cements soil particles into structural units called aggregates	Permits exchange of gases; stabilizes structure; increases permeability
Chelation	Forms stable complexes with Cu^{2+}, Mn^{2+}, Zn^{2+}, and other polyvalent cations	Enhances availability of micronutrients to higher plants
Solubility in water	Insolubility of organic matter is due to its association with clay; also, salts of divalent and trivalent cations with organic matter are insoluble	Little organic matter is lost by leaching
Buffer action	Exhibits buffering in slightly acid, neutral, and alkaline ranges	Helps to maintain a uniform reaction in the soil
Cation exchange	Total acidities of isolated fractions of humus range from 300 to 1400 cmoles kg^{-1}	Increases cation exchange capacity (CEC) of the soil; from 20 to 70% of the CEC of many soils (e.g., Mollisols) is caused by organic matter
Mineralization	Decomposition of organic matter yields CO_2, NH_4^+, NO_3^-, PO_4^{3-}, and SO_4^{2-}	Source of nutrients for plant growth
Combines with xenobiotics	Affects bioactivity, persistence, and biodegradability of pesticides	Modifies application rate of pesticides for effective control

Their molecular weights can range from 700 to 200,000 g mol^{-1}. They are usually negatively charged under the pH conditions of most natural waters. They can also be considered to be colloids due to their colloidal dimensions of approximately 20 nm.[26] The concentrations of organic matter in surface waters are from 15 to 50 mg L^{-1} with the following general distribution: 87% fulvic acid, 11% hymatomelanic acid, and 2% humic acid. The concentrations of humic acid in water are low, ranging from 0.5 to 2.0 mg L^{-1}.[27] In soils, however, most of the soil humus is 50 to 80% humic acids and 10 to 30% polysaccharides. The nonhumus organic matter (that is, the undecomposed SOM) influences short-range effects (such as source of food and energy), while humic organic matter influences long-range conditions of the soil (such as increasing soil cation exchange, maintaining good soil structure, pH buffering, and water-holding capacity).

Degradation of humic substances (Table 3-5) show that their main functional groups are: alcoholic (ROH), phenolic (⬡-OH), carboxylic (RCOOH), methoxylic (ROCH$_3$), and chinoidic (or quinodic, O=⬡=O). Although the actual chemical composition of humic and fulvic acids in soils will vary significantly from soil to soil, the functional groups present will vary little over a range of soils. In general, HA contains about 10% more C than

Table 3-5: Distribution of oxygen-containing functional groups and elemental analysis of humic and fulvic acids isolated from soils of widely different climatic zones. *Adapted from Schnitzer (1977)[28], with permission.*

Functional Group & Elemental Analysis	Climatic Zone					Range	Average
	Arctic	Cool, Temperate Acid Soils	Cool, Temperate Neutral Soils	Subtropical	Tropical		
Humic Acids:							
Total Acidity	560	570–890	620–660	630–770	620–750	560–890	725 cmol kg^{-1}
COOH	320	150–570	390–450	420–520	380–450	150–570	360 cmol kg^{-1}
Phenolic OH	240	320–570	210–250	210–250	220–300	210–570	390 cmol kg^{-1}
Alcoholic OH	490	270–350	240–320	290	20–160	20–490	255 cmol kg^{-1}
Quinonoid C=O	230	{10–180}	{450–560}	{80–150}	140–260	{10–560}	{285 cmol kg^{-1}}
Ketonic C=O	170				30–140		
OCH$_3$	40	40	30	30–50	60–80	30–80	55 cmol kg^{-1}
C	56.2	53.8–58.7	55.7–56.7	53.6–55.0	54.4–54.9	53.6–58.7	56.2 % (w/w)
H	6.2	3.2–5.8	4.4–5.5	4.4–5.0	4.8–5.6	3.2–6.2	4.7 % (w/w)
N	4.3	0.8–2.4	4.5–5.0	3.3–4.6	4.1–5.5	0.8–5.5	3.2 % (w/w)
S	0.5	0.1–0.5	0.6–0.9	0.8–1.5	0.6–0.8	0.1–1.5	0.8 % (w/w)
O	32.8	35.4–38.3	32.7–34.7	34.8–36.3	34.1–35.2	32.7–38.3	35.5 % (w/w)
Fulvic Acids:							
Total Acidity	1100	890–1420		640–1230	820–1030	640–1420	1030 cmol kg^{-1}
COOH	880	610–850		520–960	720–1120	520–1120	820 cmol kg^{-1}
Phenolic OH	220	280–570		120–270	30–250	30–570	300 cmol kg^{-1}
Alcoholic OH	380	340–460		690–950	260–520	260–950	605 cmol kg^{-1}
Quinonoid C=O	200	{170–310}		{120–260}	30–150	{120–420}	{270 cmol kg^{-1}}
Ketonic C=O	200				160–270		
OCH$_3$	60	30–40		80–90	90–120	30–120	75 cmol kg^{-1}
C	47.7	47.6–49.9	40.7–42.5	42.2–44.3	42.8–50.6	40.7–50.6	45.7 % (w/w)
H	5.4	4.1–4.7	5.9–6.3	5.9–7.0	3.8–5.3	3.8–7.0	5.4 % (w/w)
N	1.1	0.9–1.3	2.3–2.8	3.1–3.2	2.0–3.3	0.9–3.3	2.1 % (w/w)
S	1.6	0.1–0.5	0.8–1.7	2.5	1.3–3.6	0.1–3.6	1.9 % (w/w)
O	44.2	43.6–47.0	47.1–49.8	43.1–46.2	39.7–47.8	39.7–49.8	44.8 % (w/w)

FA, while FA contains 10% more O and has a higher total acidity, more carboxylic groups, and more alcoholic groups than HA.

Amino acids and polypeptides are also isolated among decomposition by-products and, because of this, the nitrogenous groups are often considered to be a part of these compounds instead of the basic chain structures of SOM. Nitrogen, however, may well be part of the humic structure. Nitrogen content is 2 to 6% in humic acids and humin, and <1 to 3% in fulvic acids.[29]

Some basic chemical structures of SOM proposed in the literature are shown in Figure 3-8. Most of the structures proposed are polymeric units with a high density of reactive functional groups. Do not assume that SOM can be described by any particular chemical structure. The formulas depicted in Figure 3-8 are only illustrating functional groups that SOM *may* contain. Comparing FA with HA, the general trends are as follows: FA > HA in density of functional groups, FA < HA in molecular weights, and FA < HA in rigidity of the molecules.

Very little is known about the humus fractions humin and hymatomelanic acid. It is not known if hymatomelanic acid is a distinct chemical entity since it may be an artifact produced by the fractionation process. Stevenson (1985)[30] notes that the simple process of redissolving the alcohol-insoluble material (that is, HA) in alkali followed by reprecipitation with acid results in a further increase in alcohol-soluble material (that is, hymatomelanic acid). Comparing Stevenson's (1985)[30] observation with the discussion in Section 3.2, we note again that alkali extractions are too harsh.

Humin has similar elemental composition and functional groups as humic acids (HA). It is not known if humin is really a separate fraction, or if it consists of other fractions but with such a strong association with mineral matter that they are not solubilized by alkali extracting solutions.[30] Some nuclear magnetic resonance (NMR) studies, however, suggest that (1) humins are significantly more aliphatic than HA due to higher concentrations of polysaccharides and paraffinic components, (2) can have high concentrations of lignin and carbohydrates present, and (3) are not clay–humic acid complexes.[31] Humin is related to humic acids, but it is more likely that degradation of humin yields HA products rather than HA being precursors to humin. Over time and under anaerobic conditions, terrestrial humin will transform to coal, while aquatic humin will transform to kerogen and petroleum.[31]

There are many different kinds of low-molecular-weight organic compounds found in soils.[32] All compounds that exist inside living organisms will also exist in soil solutions when the organisms die and their cell walls lysate, albeit sometimes for brief periods of time. Many compounds are also excreted as waste by the living organisms. The list of compounds that may be present in soils expands when you also consider the chemical and photochemical degradation products of the compounds present. Atmospheric deposition of air-borne pollutants can be found in the topsoil of very large areas of land. In several localized areas, we also need to consider the vast array of agricultural chemicals that may have been added or the accidental spillage of anthropogenic compounds.

The relative abundance of low-molecular-weight organic compounds is strongly related to the biological activity present and the affinity of the organic compounds toward the

Humic Acids:

(A)

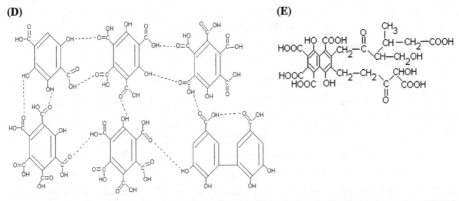

(B)

(C)

Fulvic Acids:

(D)

(E)

Figure 3-8: Some proposed humic acid (HA) and fulvic acid (FA) chemical formulas. Note that these formulas are to be interpreted as merely representative of what may be present in the soil as HA and FA fractions. Numerous other representative structures have been proposed. *The structures shown here were proposed by:* **(A)** *Dragunov et al. (1948)*[33]*,* **(B)** *Flaig (1960)*[34] *according to Stevenson (1994)*[3]*,* **(C)** *Fuchs (1931)*[35]*,* **(D)** *Schnitzer (1978)*[36]*, and* **(E)** *Buffle (1977).*[37]

mineral surfaces present in the soil. Accordingly, much higher concentrations of organic compounds (particularly organic acids) are found in the rhizosphere soil solution than in the ambient soil solution.

Low-molecular-weight organic acids are important complexing agents of metals in soils. They will also adsorb onto soil minerals and enhance chemical weathering processes. The more soluble aliphatic acids generally dominate the soil solution (concentration range in mM), while the less soluble and strongly sorbed aromatic acids generally have a much lower concentration in solution (concentration range in µM, usually).[38] In most soils, the relative abundance of low-molecular-weight organic acids in soil solutions is[38]:

volatile aliphatic acids > nonvolatile aliphatic acids > aromatic acids   [3-2].

[3.9]   Analysis of SOC & SOM

A generalized chemical formula of soil organic matter is often desired in environmental sciences. This is apparent, for example, whenever problems arise that require a conversion of the quantity of organic matter present to the quantity of organic carbon present in the soil. The percentage of organic matter (%OM) is not the same as the percentage of organic carbon (%OC); the former is organic material that includes carbon, while the latter is only concerned with the amount of carbon in the soil or organic material. The conversion formula is:

$$(\%OM) = f \times (\%OC) \tag{3-3},$$

where the conversion factor, f, is based on the typical amount of C in organic matter. The amount of C in organic matter ranges from 48 to 58% by weight. Assuming 58% C, the van Bemmelen factor of $f = 1.724$ has been used often. More realistic values for f are 1.9 for surface soil samples and 2.5 for subsoil samples.[39] A conversion factor of 2.0 is satisfactory for most general applications. It is more appropriate to analyze and report the %OC in a soil than to report an estimate of the %OM. While the %OC can be determined with reasonable accuracy, the estimated %OM value is only as accurate as the conversion factor, which is highly variable from soil to soil, as well as from horizon to horizon within a soil pedon.

There are numerous methods for the quantitative analysis of soil organic carbon (SOC). In the Walkley–Black method, the organic matter is oxidized with potassium dichromate (0.167 M):

$$2Cr_2O_7^{2-} + 3C^{\circ} + 16H^+ \rightleftharpoons 4Cr^{3+} + 3CO_2 + 8H_2O \tag{3-4}.$$

The amount of dichromate reduced is then measured by titration with $FeSO_4$.

Modern analytical methods for quantifying SOC typically use autoanalyzers that totally combust the organic matter and analyze the evolved CO_2 gas with an infrared (IR) sensor. Commonly used instruments are made by Carbo-Erba Instruments (Milan, Italy), Leco Instruments (St. Joseph, MI, USA) and Perkin-Elmer (Norwalk, CT, USA). These instruments analyze total C, which includes both organic and inorganic C fractions. If inorganic C is present, then it needs to be analyzed separately and subtracted from the total C results to obtain the organic C content. The inorganic C analysis typically involves acid dissolution followed by purging of the CO_2 product and analyzing the evolved CO_2 gas. These autoanalyzers have significantly changed the layout of research labs everywhere. The accuracy and reliability of autoanalyzers have replaced the messy and labor intensive digesters.

Quantitative analysis of SOM typically involves oxidation of the OM with H_2O_2 or ignition of OM at high temperatures (400°C for 16 h). It is calculated gravimetrically:

$$\%OM = \frac{\text{weight change of soil sample}}{\text{initial weight of soil sample}} \times 100 \tag{3-5}.$$

The oxidation of organic matter with H_2O_2 is not always complete, and the extent of oxidation varies from soil to soil. The results of the loss-on-ignition method (LOI) may be affected by the decomposition of inorganic carbonates (but this needs temperatures over 750°C) or the dehydroxylation of phyllosilicates (but this is minimal below 450°C) or gibbsite (which loses structural water at 300 to 350°C).[40] The higher the organic matter content, the lower the potential impact of these constituents on the results obtained. The ease of performing the LOI analysis of SOM makes this an attractive procedure to perform. Nevertheless, the conversion of SOM values to SOC values is not reliable (or reasons for the conversion factor chosen are poorly documented) and, hence, this analysis is not highly recommended. Because of the inherent possible problems in the procedures used, the confidence in the conversion factor (Equation [3-3]) for SOM to SOC is lower than for a conversion of SOC to SOM. The LOI method does, however, offer a reasonable rough estimate of the SOM content, which is sometimes all that is desired for a particular discussion or research project.

Quantitative analysis of dissolved organic samples (such as from nearby streams) can also be done based on UV absorbance, typically at 254 nm. Spectroscopic methods of analysis (such as NMR, IR, and EPR) are generally reserved for qualitative research that seeks to identify the nature of the organic functional groups present.

[3.10] Acid–Base Behavior of SOM

Soil organic matter has a strong buffering capacity and, accordingly, tends to stabilize the pH conditions of the soil environment. Figure 3-9 shows a typical acid/base titration for SOM samples. In absence of SOM, the pH would rise rapidly. Instead, however, the SOM acidic and phenolic groups deprotonate as the pH is raised and results in the buffering regions, which are particularly apparent around pH 5 in the sample shown in Figure 3-9. Typical quantities of acidic groups (–COOH groups) in SOM range from 1.5 to 5.7 mmol g^{-1} of organic material (Table 3-5). The amount of acidic groups in a given SOM sample can be inferred from the size of the buffered region in the titration curves. The acidity constants (pK_a values) can also be inferred from the pH values of the center of the buffered plateau.

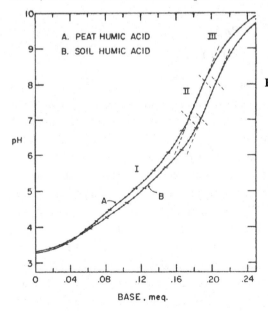

Figure 3-9: Titration curves of a soil and peat humic acid. The small wavy lines on the curves indicate endpoints of ionization of weak-acid groups having different, but overlapping, ionization constants. *From Stevenson (1994)[3], Humus chemistry: Genesis, composition, reactions. 2nd ed., copyright © 1994 by John Wiley & Sons. This material is used by permission of John Wiley & Sons, Inc.*

Modeling the organic matter's titration pattern can be done based on simple assumptions about the protonation of the functional groups present. However, difficulties often arise because of the large variety of functional groups that may be present in a humus sample. That is, a humus sample may have numerous acidic functional groups (–COOH groups), each one being slightly different from the next, and each one having slightly different but overlapping acidity constants (referred to as "ionization constants" in Figure 3-9). Modeling the degree of protonation of the organic material is a required first step toward modeling the affinity of SOM for other ions. By treating the proton (H^+) as the complexing metal, similar modeling procedures applied to complexing ions (discussed below) can be applied toward modeling the titration data.

[3.11] Measuring SOM–Metal Stability Constants

Natural soil organic matter has a strong affinity for binding with metals. The recommended procedure to obtain humic acid that is free of inorganic impurities is as follows: redissolve the HA at pH 7, add 0.1 M KCl, and wash several times with acid (0.3 M HF in 0.1 M HCl) to obtain a precipitate of "ash-free" humic acid.

Much research has been focused on the binding of metal cations by organic soils. In organic soils, the ion with the highest selectivity for OM will be the ion that is most rapidly adsorbed. The rates of reaction typically follow first-order kinetics.

The affinity of a metal toward an organic compound can be described as an equilibrium reaction:

$$a\text{M} + b\text{L} \rightleftharpoons \text{M}_a\text{L}_b \qquad [3\text{-}6],$$

where M = free metal, L = free ligand, and M_aL_b = metal–ligand complex. Equation [3-6] can be modeled mathematically based on the stability constant (K) and the total metals concentration (M_T):

$$K = \frac{(\text{M}_a\text{L}_b)}{(\text{M})^a(\text{L})^b} \qquad [3\text{-}7],$$

$$\text{M}_T = (\text{M}) + (\text{M}_a\text{L}_b) \qquad [3\text{-}8].$$

It is important here to avoid introducing contaminant artifacts in the analytical procedure. One often needs to work in dust-free, "clean" rooms. The total metals concentration (M_T) can be directly determined by speciation modeling (see Section 2.17) and analytical methods such as atomic absorption spectroscopy (AAS) or inductively coupled plasma spectroscopy (ICPS). The free metal concentration (M) is somewhat more difficult to measure than the total metals concentration. Analytical methods include: dialysis, ion exchange, gel permeation chromatography (GPC), bioassay, UV and fluorescence spectroscopy, ion selective electrode (ISE), and anodic stripping voltametry (ASV). Total ligand (L_T), free ligand (L), and complexed ligand (L_C) determinations are less reliable and are frequently, therefore, done indirectly. All these analytical problems lead to large uncertainties in the estimated K values.

Modeling the reaction of metals with organic macromolecules is much more complicated than with simple organic ligands. With macromolecules, the various metal-to-ligand bonding affinities can be very diverse within the same molecule, with similar but not identical overlap in the behavior of each functional group to the next functional group. The three general types of bonding reactions are[38]:

1. Mono–mono binding complexes, where $a = b = 1$ in the M_aL_b complex.
2. Mononuclear complexes with two or more binding substrates. Here, the central molecule of the complex is either the ligand (M_aL, where $a \geq 2$) or the metal (ML_b, where $b \geq 2$). The ligand as the central molecule is very common in SOM.
3. Polynuclear binding with two or more binding substrates and also two or more metal ions, where $a > 1$ and $b > 1$ in the M_aL_b complex. This includes both homopoly-nuclear (involving a single type of metal) and herteropolynuclear (involving more than one type of metal) binding reactions.

What follows is a modeling procedure that is commonly used on matrices where the macromolecule is the central complexing ligand, forming n complexes with a bound metal (M_a), each exhibiting a unique binding constant (Type 2 above).

A. The various species in the matrix are ML, M_2L, M_3L .... M_nL.

B. Each species is given a unique binding constant: K_1, K_2, K_3, ... K_n where

$$K_n = \frac{(M_nL)}{(M)(M_{n-1}L)}$$ [3-9].

C. The formation function is defined as

$$v = \frac{(M_aL)_T}{(L)_T} = \frac{(ML)+2(M_2L)+3(M_3L)+\cdots+n(M_nL)}{(L)+(ML)+(M_2L)+\cdots+(M_nL)}$$ [3-10].

The formation function quantifies the degree of saturation of the possible binding sites of the organic fraction.

When all the binding sites behave identically and independently at the molecular level, then the numerator of Equation [3-10] simplifies to $(ML)_o \sum n = (ML)n(n+1)/2$, where $(ML)_o$ = overall concentration of metal–ligand complexes. If K_o = overall binding constant, then $(ML)_o = K_o(M)(L)$. Let $j = n(n+1)/2$ = number of complexes formed with M. The denominator of Equation [3-10] simplifies to $(L) + (ML)_o$, or $(L) + K_o(M)(L)$. For these conditions, we finally obtain a simplified form of the formation function known as Adair's Equation:

$$v = \frac{jK_o(M)}{1+K_o(M)}$$ [3-11].

If there are several different populations of binding sites containing i populations of binding classes each with j reactive number of sites, we may use:

$$v = \frac{j_1K_1(M)}{1+K_1(M)} + \frac{j_2K_2(M)}{1+K_2(M)} + \cdots + \frac{j_iK_i(M)}{1+K_i(M)}$$ [3-12].

For example, we may wish to apply Equation [3-9] to the titration curves shown in Figure 3-9, where the pH value of each small wavy line is equal to each of the corresponding pK_n values. Note that we are modeling the hydrogen proton as a complexing metal in this example. We may also wish to group the titration curve into broad regions and describe the titration curve with fewer acidity constants based on Equation [3-12]. The amount of buffering of each titration region would control the j_i values.

The published metal–ligand binding constants vary widely in the literature. The primary reason for this variation is that the natural organic ligands investigated vary greatly with source and method of preparation. Secondly, alternate results will be obtained if the ionic strength, ion composition, and pH of the background solutions are varied. And finally, there are numerous other approaches for modeling the metal–ligand complexation reactions, and each will lead to a different quantitative description of the binding constant (see Stevenson, 1994[3], for a review of various other modeling approaches). Binding constants can be compared if and only if they are based on similar models.

[3.12] Factors Affecting the Formation of SOM

Soil scientists have been associating the origins of black soil, also known as chernozem, with aquatic vegetation, animals, or steppe grasses since the second half of the 18th century. The percentage of SOM in a soil is stabilized by numerous factors. If the soil is wet most of the time, then the aeration of the soil is low and the availability of oxygen is quickly depleted by the aerobic microorganisms. The SOM acts as the food source, or electron donor, for the microorganisms, while the oxygen acts as the electron acceptor. These wet environments tend to decompose SOM very slowly, and the annual accumulation of deposited organic matter will typically exceed the amount of decomposed organic matter. Soils with deep organic horizons often form in these areas.

If the soil is well drained, the aeration of the soil is excellent, and the availability of oxygen for microbial decomposition of the SOM is high. The amount of SOM is typically low in dry environments, with the actual percentage present based on the equilibrium conditions of the ecosystem. Five factors that affect the organic matter content in well-drained soils in the USA were described by Jenny (1941)[41], and their importance was sequenced as follows:

Climate > Vegetation > Topography = Parent Material > Age   [3-13].
Environmental factors affecting the organic N content of soils follow the same general sequence.

The residence time of FA fractions in soils is shorter than the residence time of HA fractions. For example, a ^{14}C-dating of humic fractions of two Canadian soils, an Alfisol and a Mollisol, estimated the FA fractions to be 50 and 600 years old, and the HA fractions from 150 and 1300 years old, respectively.[42] The FA fractions, which are more soluble in water than the HA fractions, are more easily leached from the soil environment.

Poorly drained soils will also display more accumulation of SOM and slower SOM decomposition (or degradation) rates as a result of the reduced oxygen levels in soil water compared to a dry, well aerated soil. Figure 3-10 illustrates the general decomposition rates observed for SOM, but note that the actual rates will vary depending on the climatic conditions present. In general, there is a rapid initial decomposition of the organic matter followed by a slower decomposition rate, and both rates are faster in warm climates compared with cooler climates.

The vegetation will significantly affect the content of HA and FA in soils. For example, forest soils (Alfisols, Spodosols, and Ultisols) are high in FA, while peat and grassland soils (Mollisols) are high in HA.[3,43] Furthermore, the HA of forest soils are typically of the brown HA type (less aromatic in nature and more similar to FA than the HA of grassland soils), while the HA of grassland soils are typically of the gray HA type. Figure 3-11 illustrates the general distribution of humus material from one type of soil to another. It is also worth noting that the HA/FA ratio in humus often decreases with depth. Unquestionably, the variation of humus materials in soils is enormous.

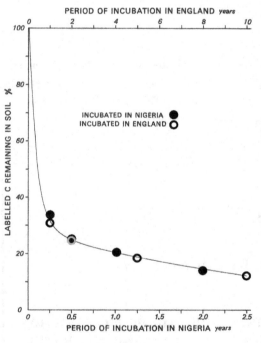

Figure 3-10: Decomposition of uniformly labeled ryegrass in England (Rothamsted Exp. Stn., Harpenden, England) and Nigeria (IITA, Int. Inst. of Tropical Agriculture, Ibadan, Nigeria). *From Jenkinson and Ayanaba (1977)[44], with permission.*

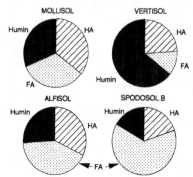

Figure 3-11: Distribution of humus forms in the soils of four soil orders. *From Stevenson (1994)[3], Humus chemistry: Genesis, composition, reactions. 2nd ed., copyright © 1994 by John Wiley & Sons. This material is used by permission of John Wiley & Sons, Inc.*

[3.13] Composition of Plant & Animal Residues

The amount and kind of litter that falls on soils varies seasonably and geographically. As an example, the annual amount of litter in a forest soil of central Europe has been estimated as follows[43]: dying roots (3 to 5000 kg ha^{-1}), leaf fall (3 to 5000 kg ha^{-1}), dying microorganisms (1000 kg ha^{-1}), and bodies of dead animals (several 100s kg ha^{-1}). As is illustrated in Figure 3-12, the degradation of plant and animal residues involves many intermediate products. When a plant falls on the ground, the plant residues decompose within a few years. This is followed by the decay of soil stabilization metabolites and cell wall constituents with half-lives of around 5 to 25 years. The resistant fractions remain for as long as 250 to 2500 years or more.

As a rough estimate, cell walls consist primarily of cellulose (~25%), hemicellulose (~51%), protein (~12%), pectin (~5%), and waxes (~7%). The common materials involved in the formation of humus are very briefly described below.

Lipids:

There are many different kinds of compounds that are lipids. By definition, lipids are any organic molecules in living organisms that are insoluble in water and can be extracted with low polarity solvents, such as ether or chloroform. This catch-all definition includes steroids, terpenes, fats, and many other kinds of compounds.

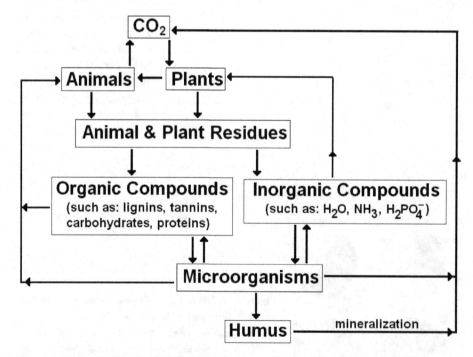

Figure 3-12: General mechanism for the formation of humus. *After Stevenson (1964).*[45]

Fats:

Fats are very abundant in cells, being one of the important food reserves in organisms. Extracted animal and vegetable fats include substances such as corn oil, palm oil, tallow, bacon grease, and butter. Liquid fats are referred to as oils. Fats are carboxylic esters derived from glycerol ($HOCH_2CHOHCH_2OH$), and are known as glycerides or triacylglycerols. They are generally C_3 to C_{18} straight-chain compounds, and can be saturated (with no double bonds present) or unsaturated (with one or more double bonds present) acids.

A glyceride:

$$CH_2-O-\underset{\underset{O}{\|}}{C}-R$$
$$CH_2-O-\underset{\underset{O}{\|}}{C}-R'$$
$$CH_2-O-\underset{\underset{O}{\|}}{C}-R''$$

Waxes:

There are no clear chemical definitions of the word "wax". The word derives from "weax", an Anglo-Saxon word for beeswax. Wax is found as a thin layer of fatty constituents that cover the leaves of plants, or as a surface coating on the skin of animals and insects. Wax contains many lipid components, but the major component tends to be long-chain esters (from long-chain fatty alcohols with long-chain fatty acids): $CH_3-(CH_2)_n-COO-(CH_2)_m-CH_3$, where $n+m$ generally equals 34 to 62. Long chains of fatty acids, alcohols, aldehydes, and ketones will also have around 20 to 35 carbons.

Lignins:

In plant tissues, there are two lignin groups. The first are precursors of other lignins, and these are rapidly involved in humus formation during the early stages of humification of the plant residues. The second are those formed in the lignified tissues (such as xylem vessels and the endodermis), and these are involved in later stages of humus formation. Lignins are the most resistant to microbial decomposition, and this is probably due to the random linkages among the aromatic rings.

Lignins are highly branched and cross-linked polymers that bind to cellulose fibers and, through lignification, form the hard, woody, and rigid material in plant cells, particularly in xylem and sclerenchyma cells. These complex polymers are based on phenolpropanol units. Note that by definition, *fibers*, which include lignins and polysaccharides, are plant materials that are resistant to hydrolysis by the digestive enzymes of humans. Lignins are not carbohydrates.

A lignin polymer:

R and R′ are typically either
–H or –OCH_3 or both.

Tannins:

Tannins are high-molecular-weight phenolic polymers (MW 500 to >20,000 g mol^{-1}) that can precipitate proteins, or that contain sufficient hydroxyls and other functional groups (such as carboxyls) to form strong complexes with proteins. Note that tannins can also complex with starch, cellulose, and minerals. Tannins are used for tanning (waterproofing and preserving) of leather and production of ink. Tannins are natural antioxidants and are commonly found in leaves, unripe fruits, and the bark of trees, and their unpleasant taste may discourage grazing animals. Most tannins are soluble in water and are usually subdivided into two groups: *hydrolyzable tannins* and *condensed tannins*. Hydrolyzable tannins have a polyol (such as D-glucose) at their central core, and can be hydrolyzed by enzymes (e.g., tannase), mild acids or mild bases to yield low-molecular-weight products, such as carbohydrates and phenolic acids. Condensed tannins, more correctly called *proanthocyanidins*, are polymers of flavanoid or flavonoid units linked by carbon–carbon bonds that cannot be cleaved by hydrolysis.

A hydrolyzable tannin: (an ellagitannin)

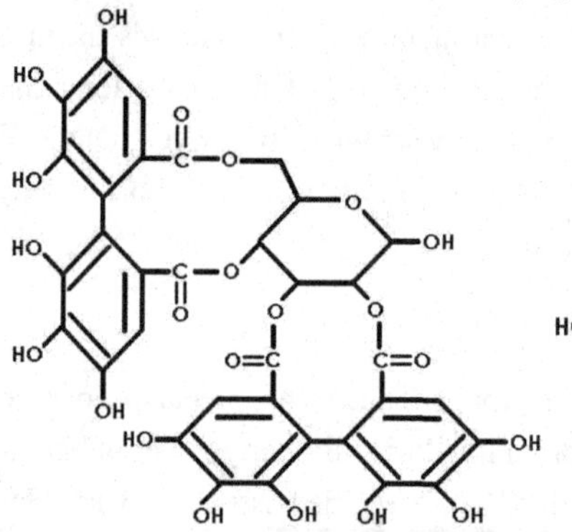

A condensed tannin polymer:

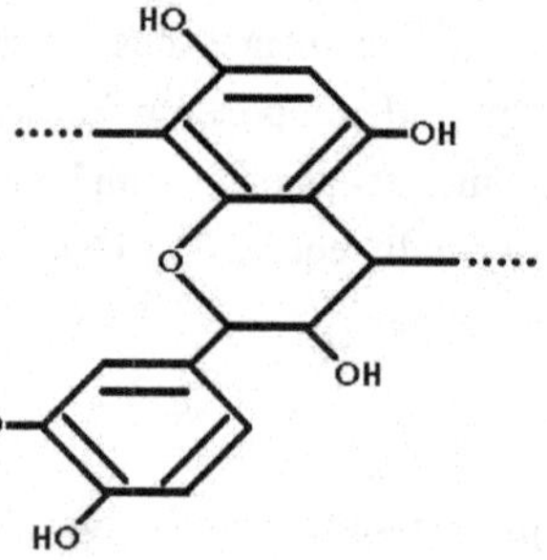

The flavanoid unit:

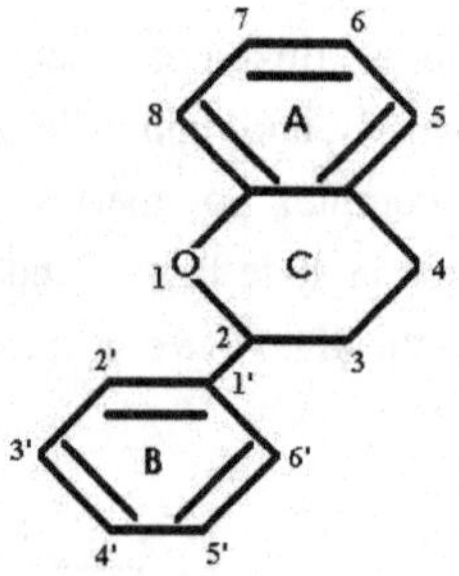

The flavonoid unit:

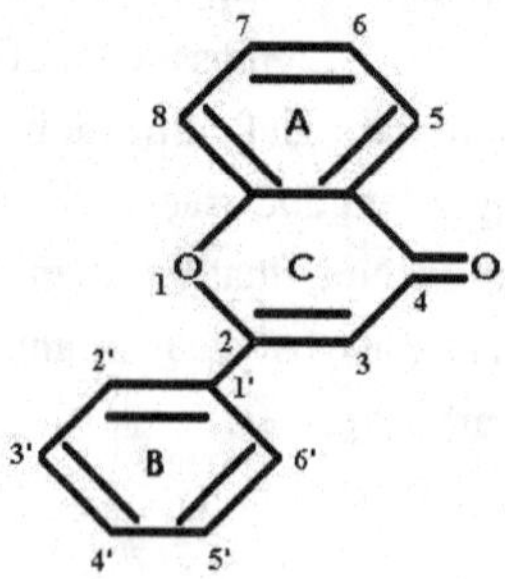

Carbohydrates:

The general formula of carbohydrates is $C_x(H_2O)_y$. They are polyhydroxy aldehydes, polyhydroxy ketones, or compounds that can be hydrolyzed to them. The simplest carbohydrates are the *saccharides*, or sugars. If the carbohydrate, such as cellulose or starch, can be hydrolyzed to many saccharide molecules, then it is called a *polysaccharide*.

Sugars:

These are water-soluble carbohydrates. Notable examples include glucose and sucrose, shown below. Sucrose is common table sugar, which contains glucose and fructose components. The carbon numbering sequence is also show below for glucose.

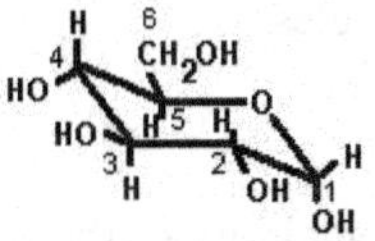

α-D-(+)-glucose 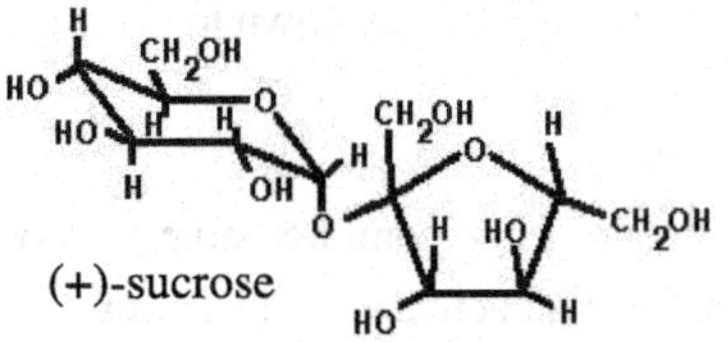   (+)-sucrose

Cellulose:

Cellulose is the main constituent in the cell walls of higher plants, and it is responsible for the rigidity of the cell wall. Cellulose and other carbohydrates are the result of extensive transformations closely associated with metabolic processes occurring in the cells of microorganinsms (such as mold fungi, actinomycetes, and bacteria). These water insoluble cellulose polymers are long unbranched chains of D-(+)-glucose. With molecular weights from 250,000 to 1,000,000, there are at least 1500 glucose units per molecule. Unlike starch, which is all alpha linkages, the links in cellulose are beta linkages. Note that the glucose molecules in adjacent linear polymers will associate through H-bonds, and the cumulative bonding energy of one H-bond per glucose molecule in these long polymers is very large. These H-bonds result in micro fibers, which in turn interact to form fibers.

cellulose:

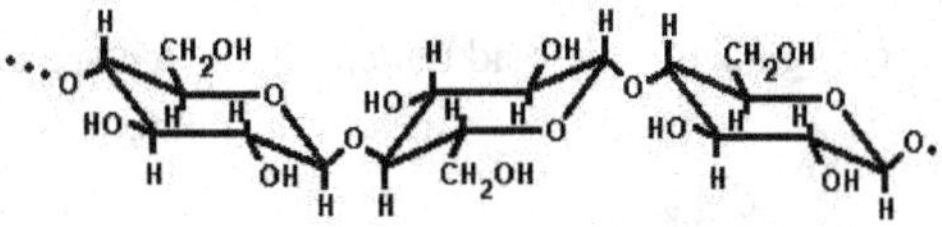

Starch:

Starches in plants are high-molecular-weight carbohydrates consisting of about 20% amylose (a water-soluble fraction), and about 80% amylopectin (a water-insoluble fraction). *Amyloses* are long straight chains of 1000 to 4000 units of D-(+)-glucose with little or no branching of the chain. The attachments of the units are from C-1 in one glucose unit to C-4 in the next, and so forth. *Amylopectins* are similar to amyloses but the structure may contain up to a million glucose units per molecule, consisting of highly branched short chains of 20 to 25 glucose units in each short chain. The chains are joined from C-1 of the end member glucose of one short chain to the C-6 on another glucose of the next short chain. *Glycogen*, the carbohydrate form in animals, is similar to amylopectin, but is more highly branched and has shorter chains (12 to 18 D-glucose units each).

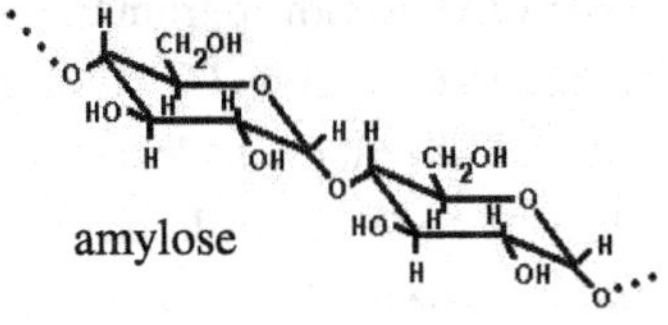

amylose

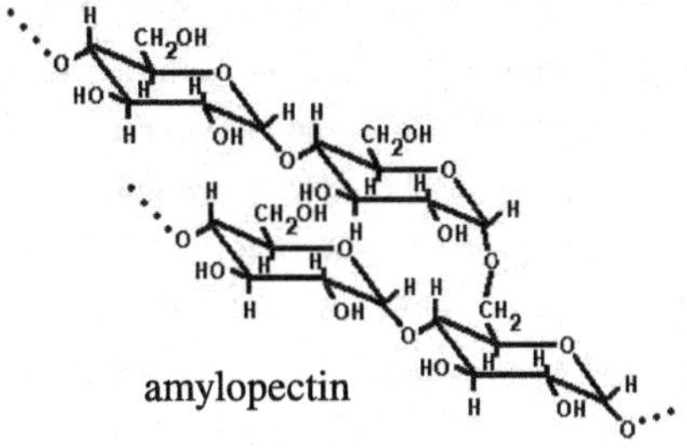

amylopectin

Hemicellulose:

Hemicellulose is a polysaccharide composed of various sugars, such as xylose, arabinose, and mannose. Similar to the amylopectin cross links, the polymer is often branched. These polymers are very hydrophilic, and easily become highly hydrated to form gels. Hemicellulose breaks down to form pectin molecules.

Pectins:

Pectins have complex structures, where the majority of the polymer is partially methylated α-D-galacturonic acid. *Pectin polymer* will contain around 200 galacturonic acid molecules, whereas *pectin acid* will contain around 100 galacturonic acid molecules and fewer methylated groups. Pectin acid is hydrophilic and soluble, while pectin polymer is soluble in hot water. Pectin acids may form Ca^{2+} and Mg^{2+} salt bridges that are insoluble gels. Pectin polymers will also have a substantial number of non-gelling areas of alternating α-(1→2)-L-rhamnosyl and α-(1→4)-D-galacturonosyl sections containing branch points with mostly neutral side chains. These side chains, of 1 to 20 units each, are mainly L-arabinose and D-galactose. Other side chains may be attached to the α-D-galacturonic acid polymer region, and these may contain D-xylose, L-fucose, D-glucuronic acid, D-apiose, 3-deoxy-D-manno-2-octulosonic acid, and 3-deoxy-D-lyxo-2-heptulosonic acid. Pectins do not have exact structures. In solution, the polymer is curved and worm-like with a large amount of flexibility.

α-D-galacturonic acid units: α-D-galacturonosyl and α-L-rhamnosyl units:

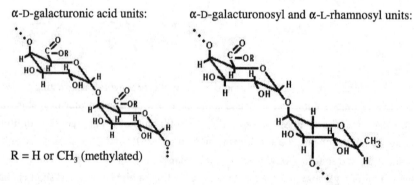

R = H or CH$_3$ (methylated)

Proteins:

Proteins are very long chains of amino acids with molecular weights ranging from 6,000 to several million g mol^{-1}. Amino acids are the building blocks of many essential compounds, such as peptides, proteins, enzymes, and hormones. Amino acids are linked together by peptide bonds, where an amide group (–NH-CO–) is formed from the interaction of the amino group (–NH$_3^+$) of one amino acid with the carboxyl group (–COO$^-$) of another amino acid with the loss of one H$_2$O in the process. Amino acids are soluble in water, but the R-group may be hydrophobic or hydrophilic. Linked amino acids with hydrophobic R-groups do not ionize or participate in H-bonds, and they tend to reside predominantly in the interior of proteins. Linked amino acids with hydrophilic R-groups form H-bonds, interact with water, and are found on the exterior portions of proteins or in reactive centers of enzymes. The acid–base behavior varies, where the general pK_a values for the various groups are as follows: carboxylic acid groups ~2.0, second carboxylic acid groups ~4.0, amino groups from 9 to 11, second amino groups from 10.8 to 12.5 (6.0 for histidine), alcoholic groups ~13.0, sulfur-containing groups ~8.3, and phenolic groups ~10.1. Their chemical formulas are shown in Table 3-6.

Table 3-6: Natural amino acids. Most amino acids are given three-letter codes and single-letter codes, which are used when describing the sequencing of amino acids in larger molecules.

Name (Abbreviation)	Formula	Name (Abbreviation)	Formula
Polar (Hydrophilic) Amino Acids:			
(+)-Arginine (Arg, R)	$H_2NC-NH-CH_2CH_2CH_2CHCOO^-$ (with $\overset{\|}{NH_2^+}$ and NH_3^+)	(−)-Histidine (His, H)	$-CH_2CHCOO^-$, NH_3^+ (imidazole ring)
(−)-Asparagine (Asn, N)	$H_2N-C-CH_2CHCOO^-$ (with O and NH_3^+)	(−)-Hydroxylysine (Hylys)	$^+H_3NCH_2CH\,CH_2CH_2CHCOO^-$ (with OH and NH_3^+)
(+)-Aspartic Acid (Asp, D)	$HO-C-CH_2CHCOO^-$ (with O and NH_3^+)	(−)-Hydroxyproline (Hypro)	HO– ring –COO^-
(−)-Cysteine (Cys, C)	$HS-CH_2CHCOO^-$ (with NH_3^+)	(+)-Lysine (Lys, K)	$^+H_3NCH_2CH_2CH_2CH_2CHCOO^-$ (with NH_3^+)
(−)-Cystine (Cys-Scy)	$^-OOCCHCH_2-S-S-CH_2CHCOO^-$ (with NH_3^+ and NH_3^+)	(−)-Serine (Ser, S)	$HOCH_2CHCOO^-$ (with NH_3^+)
(+)-3,5-Dibromotyrosine	HO– ring(Br, Br) –CH_2CHCOO^- (with NH_3^+)	(−)-Threonine (Thr, T)	$CH_3CH-CHCOO^-$ (with OH and NH_3^+)
(+)-3,5-Diiodotyrosine	HO– ring(I, I) –CH_2CHCOO^- (with NH_3^+)	(+)-Thyroxine	HO– ring –O– ring –CH_2CHCOO^- (with NH_3^+)
(+)-Glutamic Acid (Glu, E)	$HO-C-CH_2CH_2CHCOO^-$ (with O and NH_3^+)	(−)-Tyrosine (Tyr, Y)	HO– ring –CH_2CHCOO^- (with NH_3^+)
(+)-Glutamine (Gln, Q)	$H_2N-C-CH_2CH_2CHCOO^-$ (with O and NH_3^+)		
Nonpolar (Hydrophobic) Amino Acids:			
(+)-Alanine (Ala, A)	CH_3CHCOO^- (with NH_3^+)	(−)-Phenylalanine (Phe, F)	ring–CH_2CHCOO^- (with NH_3^+)
Glycine (Gly, G)	CH_2COO^- (with NH_3^+)	(−)-Proline (Pro, P)	ring–COO^-
(+)-Isoleucine (Ile, I)	$CH_3CH_2CH(CH_3)-CHCOO^-$ (with NH_3^+)	(−)-Tryptophane (Trp, W)	indole ring–CH_2CHCOO^- (with NH_3^+)
(−)-Leucine (Leu, L)	$(CH_3)_2CHCH_2CHCOO^-$ (with NH_3^+)	(+)-Valine (Val, V)	$(CH_3)_2CH-CHCOO^-$ (with NH_3^+)
(−)-Methionine (Met, M)	$CH_3-S-CH_2CH_2CHCOO^-$ (with NH_3^+)		

[3.14] Role of Animals & Microorganisms on SOM Formation

When plant and animal residues are deposited on the soil, various changes can occur. Physical destruction of the material occurs due to natural factors (such as rain and wind) and human activity (such as soil cultivation). Chemical changes occur due to reactions with water, light, and air (such as photochemical reactions, redox reactions, and hydrolysis). The enzymes in the dead cells may still be present and, therefore, may randomly continue to interact with the organic matter in the environment. In dead cells, these reactions are generally of an oxidative character. It is very important to note that none of these changes will cause the humification of the organic matter in the absence of microorganisms and animals. Physical and random chemical changes are not enough to create a rich humus. Humus formation is a complex two-stage process involving (1) decomposition by microorganisms of the original organic matter into simpler chemical compounds or products, and (2) condensation reactions or physicochemical activity without microorganisms where high-molecular-weight humus substances are produced. Accordingly, the formation of humus is based on a non-lignin origin: N-rich plant residues → bacterial plasma → humus products. This theory is better based than the older theory, which assumed a strictly chemical decomposition of lignin: lignin + alkaline oxidation → new "humus" products.

The importance of microorganisms was overlooked for a long time, particularly in the 1800s. In those early years, many misconceptions about humus and its formation process prevailed. Humus was viewed as resulting from chemical decomposition only. Humus was expected to have no N, as did artificial simulations. Humus-N was viewed as a contaminant that needed to be removed by careful purification. Strictly chemical decomposition of different types of organic litter was believed to yield similar by-products.

However, the diversity of microbes in soils does play an important role and, for this reason, the characteristics of the humus are also very diverse. Additionally, changes in the main groups of microorganisms present in a soil will occur. For example, the microbial population in a given soil may shift over time as follows:

mold fungi and non-spore-forming bacteria
> → spore-forming bacteria
>> → cellulose myxobacteria
>>> → actinomycetes   [3-14].

A partial mineralization of the humus occurs with each step, along with the production of CO_2 gas + H_2O. Note also that the rate of decomposition of soil organic matter will be accelerated when organic compounds that are easily utilized by microorganisms are added to the soil (such as carbohydrates, N-rich compounds, and glucose). That is, the "appetites" of microorganisms are significantly improved when certain foods or compounds are also present in the soil.

The activity of animals will greatly accelerate the decomposition of plant residues to humus. Decomposition is enhanced as a result of the mechanical mixing of the soil, particularly by insects, which increases aeration and microbial activity. There is also an intense concentration of digestive chemical activity in the stomachs of animals. There is much humus formed in the intestines of animals and by the animals' oxidizing enzymes. In the absence of animals, humus may form in more than a year; it may form in a few months or days, however, when animals are present. For example, it has been estimated that the Collembola insects can form 180 m^3 of fine-grained humus ha^{-1} yr^{-1} in a deciduous forest with approximately 100,000 Collembola m^{-2}.[43] Collembola (common name "springtails") are wingless insects that live off the fungi that decompose organic matter.

[3.15] Impact of Fires on Soils

Fires may be started by humans (such as to control the growth of mesquite brush [*Prosopis glandulosa*, Torr.] or to quickly clear a forest for farming purposes), or they may occur naturally by lightning strikes. Fires are necessary for the long-term survival of fire-dependent plant species. Trees with serotinous cones die in high-intensity fires but need its heat to melt the cone's wax coating and release the seeds to the fire-bared soil below (that is, onto a bare mineral soil surface with no competing vegetation on it). Examples of such trees are the Table Mountain pine (*Pinus pungens*) and the Pitch pine (*Pinus rigida*) in parts of its range. Some plants will survive a low-intensity fire but need it to promote growth of new vegetation. That is, the same plant continues to grow better after the fire (e.g., blueberries [*Vaccinium* sp.]). Another role of low intensity natural fires is to keep under-stories clear of competing vegetation.

The production of ash from a woody vegetation fire will increase the pH of the surface soil by 0.5 to 1.0 units, and potentially up to 2.5 units.[46] The more organic matter burned, the larger the increase in pH. The soil surface, which burns more thoroughly, will increase in pH after a fire more than the deeper soil horizons. Similarly, the more intense the fire (high-intensity versus low-intensity fires) the higher the increase in soil pH.[47] The increase in the soil pH may peak two months after burning, accompanied by an increase in exchangeable cations, such as Ca and Mg.[48] Following revegetation and renewed soil microbial activity, the soil pH levels return to normal in about 10 years. The high soil pH values begin to drop to normal as a result of the degradation of SOM, which is accompanied by the release of acidic by-products. Respiration processes in the soil produce carbonic acid, which acidifies the soil-water environment: $CO_2 + H_2O \rightarrow H_2CO_3$.

The amount of soil organic carbon (SOC) decreases after a forest fire. This loss will be less if the burn is not complete, as would be expected of areas where the soil moisture content is high. The SOC, however, will normally replenish its supply if the fire is a single or a periodic burn event. Conversely, an annual burning of a field will result in a lack of adequate revegetation and a more permanent loss of the SOC from the burned areas.

The amount of total N in soil usually decreases quickly in the organic layer of the soil after a fire. The amount lost is directly proportional to the amount of forest floor burned.[49] The N is lost to volatilization during burning and to an increase in N trapped in the mineral layers of the soil.[50]

The amount of NH_4–N will increase after a single burn event, but will decrease after multiple burn events. If the soil is not allowed to replenish itself, then the permanent loss of total N will affect the amount of NH_4–N available. The NH_4–N increases in burned soils due to the release of non-exchangeable NH_4^+ from clay interlayers, and from the chemical oxidation of the humic material, which contains organic N.[51] The increase in NH_4–N is directly proportional to the amount of forest floor consumed. If the forest floor is allowed to revegetate, the soils return to their original N levels after one year.[50,52]

Forest fires can also increase the soil hydrophobicity, which is its tendency to repel water. This could lead to greater overland flow of water, soil erosion, transport of nutrient-containing sediments, and difficulty for the forest to revegetate. If the forest fire is of low intensity (<250°C), then a hydrophobic layer on the top of the soil is formed. If the forest fire is of high intensity, then the hydrophobic nature of the soil is decreased. Presumably, a high-intensity fire will break down the hydrophobic organic products of the low-intensity fire.

Fires transform fulvic acids into humic acids or acid-insoluble compounds, and humic acids are further transformed into humin-like alkali-insoluble substances.[53] These changes in solubility are associated with a decrease in the H/C ratio (which indicates an increase in aromatic structures) and a decrease in the O/C ratio (which indicates a higher release of oxygen). The simultaneous dehydration and decarboxylation of humus colloids play an important role in the fire-induced water repellency of soils after burning. Furthermore, as heat is applied to organic soils, the low-molecular-weight compounds should volatilize first, while the larger, less water-soluble compounds tend to remain behind. Generally, the soil's hydrophobic effects are returned to normal within 3 to 10 years.[54]

[3.16] SOM Role in Pedogenic Processes

Only a small part of SOM is present in the free state. Instead, most of it is linked with the mineral part of the soil. As a direct result of this, soil organic matter plays an important role in the weathering of soil minerals and the formation of soil structure.

During the weathering process, SOM initially accumulates on the surface. As this proceeds, the leaching and decomposition of SOM begins to accelerate. Eventually, the percentage of SOM present reaches a stable maximum as the leaching and decomposition rates become equal to the accumulation rates. The SOM can move downward in the soil horizons through the action of small animals or as a result of leaching. Leaching, the downward movement of compounds carried by the percolating water, begins at an early stage of soil development. Burrowing animals can mix soil from deep horizons with the surface. Earthworms, for example, can mix soil to a depth of 2 feet or more.

The migration of SOM into lower soil horizons is often in association with the concurrent downward movement of clay or metal ions. The channels of preferred water flow and ped surfaces quickly become coated with dark-colored mixtures of humus and clay. The organic coatings on the ped surfaces are typically greater than the organic content inside the ped. For various physical reasons, the vertical migration of SOM generally reaches an optimum depth, where a secondary maximum in humus content coincides with the accumulation of clay.

A parent material that is initially devoid of organic matter will change as it begins to react with the organic substances from the biosphere in various ways. First, as the microorganisms consume the SOM, they respire CO_2. This gas quickly forms H_2CO_3 and related species in water, which in turn act as aggressive weathering agents on the mineral components of the soil. Second, soluble SOM will complex with metals and make them mobile. This causes the mineral components of the soil to dissolve more easily in the presence of SOM. Third, the elements taken up by the plants and microorganisms are eventually returned to the soil solution when the SOM mineralizes. These elements can then coat or interact further with the mineral solid phase of the soil. Fourth, SOM also functions as a reducing agent. This allows many microorganisms to catalyze redox reactions that would otherwise proceed at extremely slow rates.

The solubility of minerals is greatly increased in the presence of complexing organic compounds. Table 3-7 lists some examples of the high dissolution of metals in the presence of various organic compounds. In natural environments, an insoluble mineral is attacked by simple organic chelates that are excreted by pioneering microorganisms. This simple organo–metallic complex is generally water soluble and mobile. The metal of this complex can then be traded or sequestered by another organic (or humic) compound to form a more stable organo–metallic complex, which can be viewed as a second generation of organic complexes with the metal. The humic organic compound may continue to sequester metals from the smaller chelators until it is saturated. The more the organic compounds or humus become

Table 3-7: Impact of humic acid and various organic compounds on the solubility of metal ions. Mineral samples were extracted with up to 0.1% (w/v) of extractants at pH 3. Range of values of five HA extractants from five soils of W and NW Tasmania are reported here. *Adapted from Baker (1973)[55], copyright © 1973, reprinted with permission from Elsevier.*

Mineral Name	Formula	Element Extracted	H$_2$O	Humic Acid	Salicylic Acid	Oxalic Acid	Pyrogallol	Alanine
				Metal extracted in 1 h, µg				
Bismuthinite	Bi$_2$S$_3$	Bi	<1	410–1,600	180	4,820	1,640	55
Calcite	CaCO$_3$	Ca	50	730–11,100	11,900	980	2,040	1,400
Bornite	Cu$_5$FeS$_4$	Cu	<1	130–230	260	650	55	15
Chalcocite	Cu$_2$S	Cu	10	1,300–5,600	4,450	9,750	920	1,530
Copper	Cu	Cu	< 1	3,200–6,100	5,500	2,620	1,190	700
Hematite	Fe$_2$O$_3$	Fe	< 1	< 5–490	< 3	80	20	20
Pyrolusite	MnO$_2$	Mn	1	290–1,800	4,200	15,500	5,150	520
Pararammelsbergite	NiAs$_2$	Ni	90	5,400–9,800	10,500	7,620	2,380	1,730
Galena	PbS	Pb	1	80–230	130	95	35	< 5
Lead	Pb	Pb	2	23,000–39,400	41,800	660	1,470	240
Stibnite	Sb$_2$S$_3$	Sb	10	< 5–340	< 5	580	< 5	< 5
Sphalerite	ZnS	Zn	<1	30–90	30	20	8	20

[a] HA: see Figure 3-8 for some proposed structures; Salicylic Acid: ⬡–COOH / OH; Oxalic Acid: HOOC–COOH; Pyrogallol: ⬡(OH)(OH)(OH); Alanine: CH$_3$CHCOO$^-$ / NH$_3^+$.

loaded with metals, the less likely it is for them to remain water soluble. This process of
metal ion translocation is illustrated in Figure 3-13.

Combining the chemical formation of SOM–metal complexes with their physical,
vertical translocation results in significant soil alterations and often sometimes horizonation
of the soil profile. This is particularly well illustrated with the translocation of Fe and Al ions
and the development of Spodosols (Figure 3-14). In this scenario, fulvic acid (FA) or humic
acid (HA) complexes with Fe or Al near the soil surface and proceeds to move down the
profile. As it travels, it continues to form new complexes with Fe and Al ions, essentially
stripping this lower horizon of its original iron color. Finally, when a critical level of

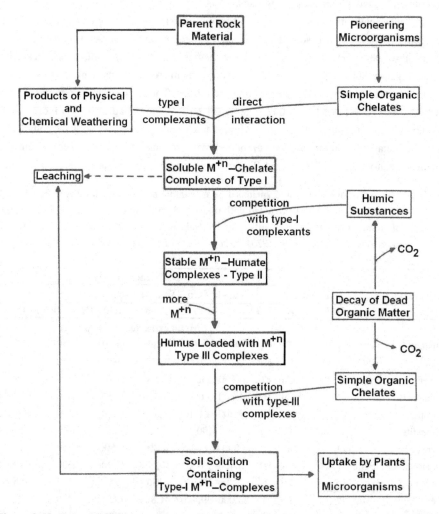

**Figure 3-13: Role of SOM in metal ion translocation. The effective competition by an organic
compound for complexation with metals varies greatly depending on its relative affinity
for the metal ion and its relative concentration in the medium.** *From Stevenson (1994)*[3],
*Humus chemistry: Genesis, composition, reactions. 2ⁿᵈ ed., copyright © 1994 by John Wiley
& Sons. This material is used by permission of John Wiley & Sons, Inc.*

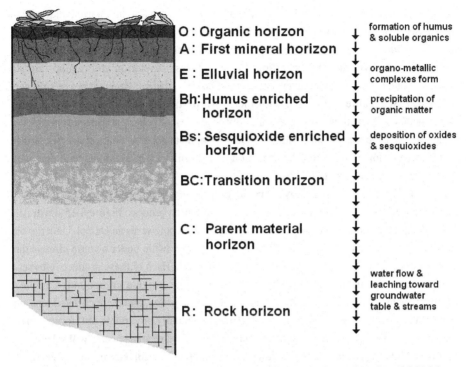

O : Organic horizon

A : First mineral horizon

E : Elluvial horizon

Bh:Humus enriched
     horizon

Bs: Sesquioxide enriched
      horizon

BC:Transition horizon

C : Parent material
     horizon

R : Rock horizon

formation of humus
& soluble organics

organo-metallic
complexes form

precipitation of
organic matter

deposition of oxides
& sesquioxides

water flow &
leaching toward
groundwater
table & streams

Figure 3-14: Illustration of a typical Spodosol profile. The vertical translocation of SOM from the O and A horizons and of organo-metallic complexes from the E horizon results in lower horizons enriched in organic matter and metals. The high humus content of the O and A horizons imparts a dark color to these horizons. The leaching of minerals from the E horizon results in a bleached color in this elluvial horizon. *After Finkl (1979).*[56]

complexation density is reached, the FA–metal or HA–metal complex precipitates, and the B horizon is formed. The B horizons of Spodosols are rich in FA, but it is not known clearly if the FA originated from the forest litter or if it was formed *in situ* (such as, from *in situ* HA degradation to FA). The evapotranspiration of soil water also facilitates the precipitation of humus and inorganic minerals in all horizons. The translocation of some humus and inorganic minerals may be very deep following a strong rainfall event. However, the average depth of translocation of most humus and inorganic minerals will depend on many factors, such as: soil porosity, the frequency and average intensity of the rainfall events for the particular area or pedon, and the average amount of drying that occurs between rainfall events.

The primary chelators of Fe oxides in forest floors are polyphenols, which are extracted from the tree litter. The polyphenol organic compounds are low in MW and are water soluble. The oxidized Fe(III) that is present in the soil minerals is reduced to Fe(II) by the polyphenols, and is translocated down the profile as Fe(II). The organo–Fe complex precipitates in the B horizon for one or more reasons: the organo–Fe complex may have reached its maximum in metal loading and become water insoluble; the pH of the B horizon may be higher than in the surface horizons, causing the metals to precipitate; and, drying of the water that transported the complex creates a supersaturated environment, which induces precipitation and oxidation of Fe(II) to Fe(III). Finally, there can also be a partial mineralization of the organic complexes in the B horizon by microbial oxidation.

Spodosols are common in New England soils as a result of having just the right combination of environmental conditions, such as forest vegetation and regular rainfall patterns (around 10 cm per month, every month). Globally, many different types of soils can be formed (see Table 1-2), each with their own particular combination of chemical SOM–metal complexation or mineral solubility processes and physical transport processes of the water-soluble constituents. Although soil development depends on many factors, which are the same factors as those listed in Equation [3-13], soils rich in SOM and with plenty of rainfall tend to mature most quickly.

The temperature of the environment also plays an important role in the selection of the translocated metals and, therefore, on the soil formation process. Figure 3-15 illustrates this with the formation of two very different soils: a Spodosol versus an Oxisol. Under cool, moist conditions, the organic matter decays incompletely, while under a warm climate the decay to CO_2 and H_2CO_3 products is much more thorough. Each of these products interacts with the elements in the mineral phase differently. The organic compounds complex selectively with Fe and Al ions, a process known as *cheluviation*, and these chelated ions are eluviated downward to a lower horizon where they eventually precipitate out; formation of a Spodosol is in progress. Conversely, the carbonic acid and related species will selectively dissolve Si ions, a process known as *desilication*, and these dissolved ions are leached out of the soil-forming environment, leaving Fe, Al, and Ti to form oxides and other minerals such as kaolinite; formation of an Oxisol is in progress.

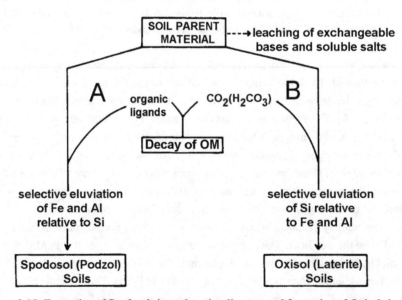

Figure 3-15: Formation of Spodosols in cool, moist climates, and formation of Oxisols in warm climates as postulated by Senstius (1958).[57] *From Stevenson (1994)[3], Humus chemistry: Genesis, composition, reactions. 2nd ed., copyright © 1994 by John Wiley & Sons. This material is used by permission of John Wiley & Sons, Inc.*

[3.17] Clays & the Origin of Life

While we have just discussed some of the impact of SOM on the development and transformation of minerals, as well as the development of various soil profiles, we should realize that the development of SOM and all of life on Earth were also influenced by the minerals present in primitive Earth. Ideas about the origins of life on Earth can be divided into four possibilities: (1) development of a simple organism in a simple environment (*autotrophic origin*), (2) development of a simple organism in a complex environment (*heterotrophic origin*), (3) development of a complex organism in a simple environment (*autotrophic panspermia origin*), or (4) development of a complex organism in a complex environment (*heterotrophic panspermia origin*).[58] The panspermia options require that the first complex organism that developed in a simple or complex Earth come from a prior simple organism that developed on a planet other than Earth. It is, therefore, not easy to evaluate the panspermia options at this time. The first two options involving simple first organisms are in agreement with Darwinian evolution, have been studied extensively, and are further discussed below.

The heterotrophic origin theory was introduced in 1924 by Oparin and 1929 by Haldane[59], and it speculates that a variety of organic compounds (such as sugars and amino acids) were produced in Earth's strongly reducing primitive atmosphere. These simple building blocks accumulated in a pool of water, known as the primordial soup, and combined to synthesize the first simple, self-replicating organisms. This theory is widely accepted as a result of Miller's 1953[60] experiment where he generated two amino acids in a simulated primitive atmosphere consisting of methane, ammonia, water and hydrogen; in his experiment, sparks simulated lightning, and no oxygen was present.

Minerals present in primitive Earth, such as clays and quartz, would adsorb the organic compounds produced in the primordial soup and protect them from destructive transformations.[61, 62] These compounds would then be able to interact and form more complex compounds. There are numerous suggestions that clays were present in primitive Earth, such as: they can be synthesized in reducing environments, they are known to have existed in Precambrian Earth, they are found in meteors, and they've been identified in Martian soil samples. Simple minerals, such as quartz, were definitely present in primitive Earth.

A significant problem with the heterotrophic origin theory is that the atmosphere of primitive Earth might have been slightly oxidizing instead of strongly reducing. It consisted mostly of nitrogen and carbon dioxide. The same organic building blocks can be produced in this environment, but at much lower concentrations where combination into macromolecules is unlikely. The heat needed to concentrate the primordial soup would be enough to further decompose the organic compounds prior to formation into macromolecules. And finally, the compounds present in the primordial soup were very impure, and the chance assembly of highly complex structures such as RNA without the aid of enzymes is unlikely.[63, 64] These problems are resolved with the autotrophic origin theory.

In the autotrophic origin theory, primitive life need not have resembled modern life. More specifically, primitive life need not have originated from the same building blocks found in modern life.[65] The ultimate ancestor need only be a naked gene that can self-repli-cate. In theory, minerals such as silicate clays can function as naked genes. That is, clays and various other minerals were life's ultimate ancestor. Although minerals are simple structures with low intrinsic information, their numerous irregularities can hold large amounts of infor-mation.[66] These irregularities are easily maintained and passed along to new crystal growth.

In the autotrophic origin theory, the clay genes provided the framework for development of adsorbed organic molecules. A close interdependency of organic matter structure with clay structure resulted from the organic–clay complexes formed. Modification of one resulted in a modification of the other. Modifications of the organic molecules and subsequent formation of more complex organic molecules were recorded through concurrent modifications of the irregularities in the inorganic naked gene. Eventually, the complex organic molecules performed these functions better than the clay genes, leaving the latter obsolete. As the scaffolding of the clay gene fell away, it left behind the biochemistry found in modern life. Today, there is still some silica found in plants, animal tissues, and lower life forms such as diatoms and sponges.

The panspermia (literally, "seeds everywhere") theories were fostered by Arrhenius in the early 1900s. The panspermia theories, the spreading of life-bearing spores across the galaxy, has gained some support recently as we now know that some bacteria can survive tens of millions of years and the strong radiations of space. Fossilized evidence of ancient life in meteors also furthers the panspermia theories. Even so, these theories do not eliminate the possibility of a concurrent autotrophic evolution on Earth, nor an autotrophic evolution elsewhere as the ancestor to the space-traveling spores must have developed somehow somewhere.

[3.18] Use of Organic Matter as Soil Modifiers

With the rapid increases in world population, the needs for more food and land have many researchers seeking ways to improve crop production and reduce soil erosion. Seed germination is enhanced by a good crumb structure because this influences the soil's water-holding capacity, water movement, aeration, and heat transfer. Addition of organic matter will improve these soil physical properties as well as increase the fertility of the soil. Soil physical properties improved by organic matter include: (1) increases the stability of crumb structures, (2) stabilizes sandy soils from wind erosion until the crop sufficiently covers the ground to reduce the air speed at ground level, and (3) increases the water-retention capacity of arid soils. In a global scale, organic amendments to soils aid in the establishment of tree belts to stop the encroachment of deserts.

Manure is the most common organic soil amendment, but for economic reasons it is unfortunately generally added to soils that are nearest the point of production, rather than to the soils most in need of the increased fertility. As a result of this mismanagement, we end up with soils testing very high in certain nutrients, particularly phosphates. High leaching and surface erosion of these fields contaminates nearby streams and ponds, resulting in eutrofication and subsequent fish kills in these aquatic resources.

Sewage-sludge compost is another very valuable soil amendment. It is rich in macronutrients and micronutrients essential for plant growth. The primary concern with the use of sewage-sludge compost on soils is quality control. Specifically, no one wishes to have toxic substances accidentally added to their soils, which might occur if a local citizen or industry unknowingly disposed of toxic materials in such a way that it reaches the sewage and composting facility. For this reason, permits for land application of sewage-sludge compost are often highly scrutinized.

Compost from home and yard wastes are excellent soil amendments. Addition of grass clippings adds organic nitrogen to the leaf compost, and is recommended for home composting sites. Carbon/nitrogen ratios of 12:1 in the finished compost are ideal. Constructing long windrows for leaf composting is also a common activity supported by many municipalities. However, these windrows are often made without the addition of grass clippings, usually because of the quality control issue again. That is, grass clippings from home owners, particularly those with lush green lawns, will often have pesticides and herbicides in them that would be incorporated into the compost product if they are added to the municipal windrow. Conversely, home owners rarely ever spray these chemicals on leaves or trees.

Synthetic organic matter, or synthetic polymers, are also used as soil conditioners, but they are often limited to laboratory research or controlled field studies. Microorganisms use polymers to bind themselves to interfaces and to each other. Similarly, synthetic polymers stabilize the soil structure by holding the soil particles together. Polymers attached to soil particles will contain *tails*, *trains*, and *loops* (Figure 3-16). While the train sections are attached to one particle, the tails and loops of the same polymer could potentially attach to other particles, thus resulting in the formation of a floc. Although one of the problems in the use of polymers has been their cost, it is generally accepted that synthetic polymers are ideal soil modifiers and the way of the future for many soil management practices.

If used for growing small-seed crops, polymers can be applied as a surface treatment over the seed drill lines. Limiting the application to this area reduces the cost while significantly aiding the small seeds to germinate and break through to the surface. In the absence of the polymer, some soil surfaces, particularly silts and silty clays with low SOM, can form an impervious cap or crust that interferes with the emergence of the crop, especially a small-seeded crop. By improving emergence, the good start given to the crop is carried through the growing season. A measure of the reduced strength of the crust can be used to quickly assess if a particular polymer would be effective in reducing the formation of the

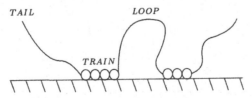

Figure 3-16: An illustration of the loops and tails that may be present in an adsorbed polymer. The train of circles represent surface-bonding sites on the polymer, while the lines connected to these circles represent portions of the same molecule that are not bonding with the surface. *From Schulthess and Tokunaga (1996)[67], with permission.*

cap.[68] The force required to break the crust, the *crust strength*, is measured with a penetrometer before and after application of the polymer. (Following the application of water and polymer to the soil sample, the soil is dried, rewetted, dried again, and then the crust strength is measured.) In general, the soil is better stabilized with polymers of high molecular weight rather than low molecular weight, and with charged polymers rather than uncharged or neutral polymers.

Table 3-8 describes some of the commonly researched, soluble, linear polymers for soil modification. A complete list would probably have thousands of entries. Poly(vinyl

Table 3-8: Description of some common soluble organic polymers used for soil modification.

Poly(vinyl alcohol), PVA, and **Poly(vinyl acetate)**, PVAc: The manufacture of PVA involves the hydrolysis of PVAc, and this hydrolysis is rarely complete. Accordingly, in the structure shown below, PVAc consists only of the group shown on the right, while PVA consists of a mixture of the two groups. For example, the hydrolysis of PVA can be 78, 88, or 98%. The hydrolysis percentage significantly affects the amount of PVA adsorbed onto particles, presumably due to the impact of the acetate groups on the formation of loops.[67] Acetate groups weaken the intramolecular (i.e., within the same molecule) H-bonding in PVA. PVA is linear, and typically has 200 to 2500 repeating units.

PVA: $\left[\text{CH}_2\text{CH} \atop \text{OH} \right]_{\text{hyd.}\%} \left[\text{CH}_2\text{CH} \atop \underset{\text{O}}{\text{OCCH}_3} \right]_{(100-\text{hyd.})\%}$

Polyacrylamide, PAM: There are three types of PAM: anionic with about 100 variations and used extensively in erosion control, cationic with about 1000 variations and used often in water treatment, and non-ionic with rare applications. PAM is a non-toxic, often linear polymer with over 150,000 repeating acrylamide units. Examples of anionic PAM variations include -O⁻Na⁺ substitutions for -NH₂ (such as 1 substitution for every 5 monomers). We can also insert -NH-R groups, resulting in short branching side chains.

An anionic PAM: $\left[\text{CH}_2\text{CH} \atop \text{O=C-O}^- \right]_m \left[\text{CH}_2\text{CH} \atop \text{O=C-NH}_2 \right]_n$

Hydrolyzed polyacrylonitrile, HPAN: Hydrolysis of acrylonitrile yields a carboxylic acid.

HPAN: $\left[\text{CH}_2\text{CH} \atop \text{O=C-O}^- \right]_m \left[\text{CH}_2\text{CH} \atop \text{C≡N} \right]_n$

Poly(vinyl acetate-maleic anhydride), VAMA: $\left[\text{CH}_2\text{CH} \atop \underset{\text{O}}{\text{OCCH}_3} \right]_m \left[\text{CH-CH} \right]_n$

Cellulose xanthate: This is formed when the alcohol groups in cellulose (structure given in Section 3.13) react with carbon disulfide (S=C=S) and aqueous NaOH. This process is an intermediate in rayon production. The product is soluble in the alkali solution, forming a colloidal dispersion known as viscose. The reaction reverses in acidic solutions. When applied to soils, the soil acidity eventually releases the CS₂, leaving behind the insoluble cellulose polymer.

$$\text{RO}^-\text{Na}^+ + \text{S=C=S} \longrightarrow \text{RO-}\overset{\text{S}}{\underset{\|}{\text{C}}}\text{-S}^-\text{Na}^+ \xrightarrow{\text{H}^+} \text{ROH} + \text{CS}_2$$
[a xanthate intermediate]

alcohol) (PVA) works very well on some soils, but poorly on soils that have chlorites (2:1+1 clays) or gibbsite ($Al(OH)_3$). Poly(vinyl acetate) (PVAc) also works well on some soils. Polyacrylamides (PAM) will effectively flocculate clays, are good soil stabilizers, and are the most widely used polymers to prevent irrigation-induced erosion. Hydrolyzed polyacrylonitrile (HPAN) and poly(vinyl acetate-maleic acid) (VAMA) will also improve soil aggregates at low doses. There are various functional groups that are often present in synthetic polymers, such as acrylic acids and acrylates. Polymers consisting of just these functional groups are poly-(acrylic acid) (PAA) for $-\!\!\left[CH_2CH(COOH)\right]_{\overline{n}}$, and polyacrylates for $-\!\!\left[CH_2CH(COOR)\right]_{\overline{n}}$.

After spraying onto the soil, cellulose xanthalate will decompose back to strands of cellulose, which give the soil crumbs mechanical strength. A drawback with using cellulose xanthalate is that its shelf life is less than one week at room temperature. The use of natural polysaccharides, such as cellulose, do not work well because they require very large doses. It also biodegrades rapidly, which leads to rapid denitrification of the soil. Petroleum products and bitumen (a thick, sticky form of crude oil) have also been investigated with little success in soil stabilization.

Another distinct type of polymer marketed for agricultural use are gel-forming polymers. These are highly cross-linked polymers that are insoluble in water. The cross polymerization can occur *in situ*. Cross polymerization can be carried out by a small amount of a divinyl compound or by other means, such as the addition of amine and aldehyde monomers. With *in situ* polymerization, the low-molecular-weight monomers are water soluble and penetrate deep into the soil prior to the condensation process. Use of cross-linked polymers and a deeper, more homogeneous application of the polymers leads to a much higher water-holding capacity (up to 1000 times their weight). Some examples of these gel polymers are: polyacrylonitrile plus starch followed by saponification of the acrylonitrile units, cross-linked polyacrylates, and cross-linked polyacrylamides and cross-linked acrylamide–acrylate copolymers with a high percentage of acrylamide units.

In summary, there are many polymers that can be considered for modification of soil properties. It is also worth emphasizing that any improvement in a soil's structure will also increase the soil's water-holding capacity and its sorptive properties for nutrients. The leaching of nutrients through the soil profile is also reduced. Numerous factors, however, will affect the efficiency of the polymers, such as depth of application, degree of mixing or homogeneity of the application, type of particles present, and particle sizes present. The presence of divalent cations (such as Mg^{2+}, Ca^{2+}, and Fe^{2+}) also seems to decrease the polymer water-holding capacity, presumably by displacement of the trapped water molecules resulting from the strong interactions of these ions with the polymers.

[3.19] Impact of SOM on Agricultural Chemicals

The fate of agricultural chemicals in soils is strongly determined by how the compounds are held by the soil constituents. The agricultural chemicals may bind with the mineral surfaces, soil organic compounds, or a combination of both. Since the SOM is mostly associated with the clay fractions, the reactions of agricultural chemicals in soils is typically toward both the mineral and organic soil components.

There are numerous factors that affect the adsorption and desorption of organic compounds on the soil inorganic minerals.[69] When organic matter is present, a coating of organic material forms on the mineral surfaces, which may affect the distribution of adsorbed chemicals. For example, it has been suggested that s-triazines adsorb onto both the organic and mineral components of soils when the organic matter content is 8% or less, but above 8% OM, the herbicide adsorbs mostly on organic surfaces.[70]

In general, the adsorption of agricultural chemicals follows the Freundlich adsorption isotherm:

$$x/m = KC^{1/n} \qquad\qquad [3\text{-}15],$$

where x/m = quantity of chemical adsorbed per weight of soil (e.g., μmol kg^{-1}), C = equilibrium concentration of the chemical in the aqueous phase, K and $1/n$ = constants. Since it is common for n to be near unity, the constant K is often expressed as the distribution coefficient:

$$K_d = \frac{\text{Amount of chemical adsorbed } (\mu\text{mol kg}^{-1} \text{ of soil})}{\text{Amount of pesticide in solution } (\mu\text{mol L}^{-1})} \qquad [3\text{-}16].$$

A dual-mode sorption mechanism has recently been proposed to describe the retention of organic compounds by SOM.[70] According to this theory, the retention of organic compounds occurs concurrently in two locations: sorption sites on the SOM (referred to as the *dissolution domain*, $S(D)$), and retention by the hole-filling of internal pores or void cavities in the SOM colloid (referred to as the *hole-filling domain*, $S(H)$). The total sorption, S, of an organic compound by the SOM is the sum of the retention by these two domains:

$$S = S(D) + S(H) = K_D C + \sum_{i=1}^{t} \frac{S_i^{\circ} K_i C}{1 + K_i C} \qquad [3\text{-}17],$$

where K_D = lumped coefficient representing all available dissolution regions, C = equilibrium concentration of the chemical in the aqueous phase, S_i° = capacity constant of the hole-filling domain, K_i = affinity constant of the ith hole-filling domain. For multiple hole-filling domains, Equation [3-17] can be approximated by Equation [3-15] with $n \geq 1$. The greater the value of n, the greater the hole-filling contribution to the sorption process.

We can describe SOM as a polymer mesh phase that is dynamic and behaves like a viscous liquid, where partitioning of smaller organic chemicals can occur. Note, however, that the density of these large organic colloids is not constant throughout the colloid structure. The center of the colloid is generally more condensed, referred to as the *"glassy" phase*, while the outer regions of the colloid are generally more expanded with weakly held flexible chains, referred to as the *"rubbery" phase*. The presence of cross-linking cations will tend to increase the condensation of the macromolecule by coordinating with multiple carboxylate groups.

The physical structure of natural SOM colloids probably vary continuously from a highly expanded and flexible structure at the edges, to a highly condensed and rigid structure at the centers.[70] Note, however, that no portion of the SOM structure is ever as rigid as most inorganic minerals. The large SOM colloid will swell as a result of filling the hole-filling domains with organic compounds. Accordingly, calculating $S(H)$ in Equation [3-17] will require some correction factor that attempts to reflect this elasticity of SOM colloids.

Traditional methods for determining surface area and internal porosity of inorganic minerals are not applicable to the string-shaped structures of SOM. This obviously further complicates the optimization of sorption equations because the site densities and pore volumes of the SOM are not known either definitively or easily. As a result of the elasticity of SOM and the lack of reliable evaluation of its surface area and porosity, there is a great deal of loss in the actual physical meaning of the parameters used in sorption equations, such as Equation [3-17] or modified versions of it. This does not mean that these sorption equations are useless. On the contrary, they are extremely useful and insightful. It does mean, however, that you should exercise some caution not to overstate the physicochemical interpretation of the optimized parameters of these sorption equations.

The amount of agricultural chemicals retained by soils varies as greatly as the characteristics of soil organic matter vary from soil to soil. For example, the sorption of triazines and atrazine may be expected to vary according to the following SOM constituents present:

> Humic acids > fulvic acids for *s*-triazines.
> Humic acids and lignins = high affinity for atrazine.
> Proteins = intermediate affinity for atrazine.
> Polysaccharides, fats, waxes and resins = low affinity for atrazine.

Fires will also affect retention of herbicides. Very high retention of herbicides is observed in burned-over fields and those containing wind-blown carbon particles.[3] Do keep in mind, however, that it is risky to establish sorption patterns and sorption generalities for agricultural chemicals based on the limited data sets.

If an agricultural chemical has a high affinity for a soil organic compound, then the environmental fate of the agricultural chemical is closely tied to the mobility and environmental fate of the SOM. For example, the downward movement of the insecticide DDT (dichlorodiphenyltrichloroethane) in soils is due to the downward mobility of water-soluble substances, such as fulvic acids.[71]

In addition to sorption processes, SOM can accelerate the redox reactions in the soil. Consequently, transformations of agricultural chemicals may be catalyzed by soil organic matter.[72] Fulvic acids are particularly involved in this impact on applied chemicals due to their high functional group content. Quinone functional groups, for example, serve as electron acceptors, while the applied agricultural chemical (e.g., *s*-triazine) serves as the electron donor and is transformed to a free radical in the soil environment. The humic substances are also strong reducing agents in soils, and are, accordingly, responsible for numerous transforma-tions in soils.

The fate of agricultural chemicals and their decomposition by-products impacts greatly on the health of all life on the planet. Dr. Jerome Weisner, science counselor to President J.F. Kennedy and the committee chairman to a commission assembled in 1963 to examine the premises of *Silent Spring*, noted that the uncontrolled use of pesticides is more dangerous than atomic fallout.[73] *Silent Spring*, written in 1962 by Rachel Carson, caused a revolutionary awakening about our unchecked faith on our technological progress and the pesticide DDT. It also helped set the stage for the much needed environmental movement.

The combined impact of all toxic agricultural chemicals causes serious damage, such as increases in mentally retarded babies, and sperm count decreases in adult males, to name but a few. Much of these damages are attributed to chlorinated hydrocarbon pesticides. This section of this chapter does not focus heavily on the fate of the agricultural chemicals applied to soils; it seeks, instead, to familiarize you briefly with what they are and how they may interact with soil organic matter. The chemical formula and structure of some commonly used or well known agricultural chemicals are listed in Table 3-9.

Nearly all pesticides are toxic to some degree to all animals that ingest the pesticide, or that eat the plants or animals that had ingested them. Not all pesticides degrade quickly in soil or when ingested and, with some pesticides, their by-products may also be very toxic. Pesticides and their by-products are typically concentrated in the animal's fatty tissues as they work their way up the food chain. This can result in the cumulative ingestion of dangerously high levels of toxic compounds causing high mortality, mutations, or sterility of higher animals. Elucidating the fate of pesticides in soils is an important topic in soil chemistry.

Table 3-9: Chemical formulas of a few pesticides, agricultural chemicals, and other chemicals classed according to use. Most of these are very toxic and dangerous to handle. Many of these also remain in the soil longer than their intended use and eventually pollute our environment or end up in the food chain.

Class: **Acaricide** (Miticide)
Typical Targets: Animal dips; spider mites on fruit & ornamental plants.
Examples:

Dicofol: Dienochlor:

Class: **Algicide**
Typical Target: Algae in swimming pools.
Example: Calcium hypochlorite, $Ca(ClO)_2$. For comparison, clorox is dilute $NaClO$.

Class: **Animal Repellants**
Typical Target: Deer
Examples: garlic oils, ammonium soaps, coyote (*Canis latrans*) urine.

Class: **Avicide** (bird poison)
Typical Targets: Starlings in grain, pigeons, house sparrows.
Example: Avitrol, active ingredient is 4-aminopyridine:

Class: **Bactericide**
Typical Target: Fireblight in fruit.
Example: Streptomycin, an antibiotic that is also produced by soil bacteria of the genus *Streptomyces*. Humans also use this antibiotic, such as to treat for tuberculosis (TB).

Table 3-9: Continued.

Class: **Defoliant** (Also classed as an herbicide.)

Typical Target: Cotton leaves.

Examples: Magnesium chlorate, $Mg(ClO_3)_2$. Agent Orange, a 1:1 mixture of 2,4-dichlorophenoxyacetic acid (2,4-D) and 2,4,5-trichlorophenoxyacetic acid (2,4,5-T), was used extensively in the Vietnam war. 2,4-D and 2,4,5-T are also used domestically for weed and brush control.

R = -H for 2,4-D

R = -Cl for 2,4,5-T

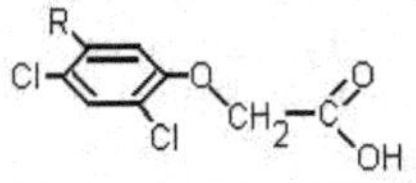

Class: **Desiccant**

Typical Target: Potato vines.

Example: Prometryn. This *s*-triazine compound is also used as an herbicide for control of annual broadleaf and grass weeds.

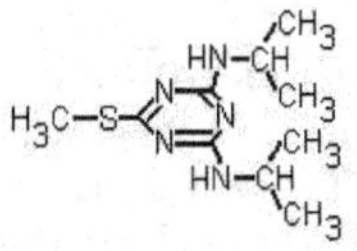

Class: **Fish Poison** (May reach agricultural soils, tidal marsh soils, or other fields by various means, such as through the irrigation system, flooding, or tidal action.)

Typical Target: Sea lamprey.

Example: Niclosamide (or clonitrilide) is also classed as a molluscicide.

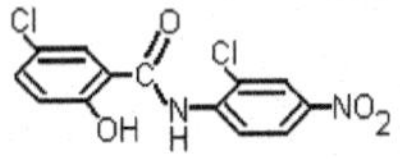

Class: **Fungicide**

Typical Targets: Brown rot on fruit; cereal seed treatment, control of turf diseases.

Examples: Dicloran: Phenylmercuric acetate (PMA):

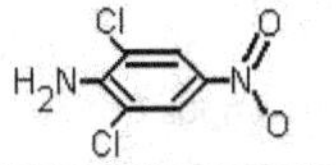

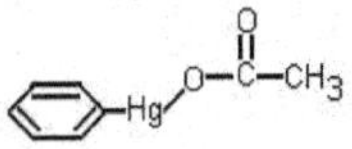

Class: **Growth Regulator**

Typical Target: Olives (thinning).

Example: Naphthalene acetic acid:

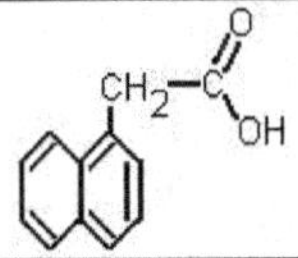

Class: **Herbicide**

Typical Target: Brush on pasture; weed control.

Examples: Paraquat: *s*-Triazines: 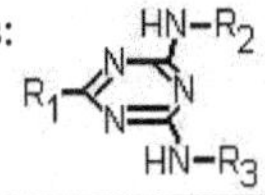   For Atrazine,

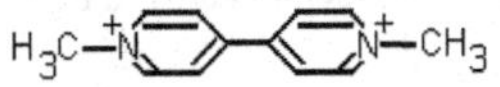

R_1 = -Cl, R_2 = -CH(CH$_3$)$_2$, and R_3 = -CH$_2$CH$_3$.

Class: **Insecticide**

Typical Targets: Earworm on corn; kill mosquitos (which carry malaria) and lice (which carry typhus).

Examples: Carbaryl: Dichlorodiphenyltrichloroethane (DDT):

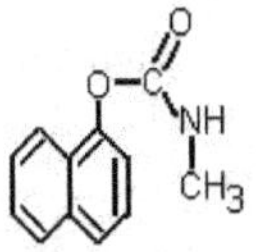

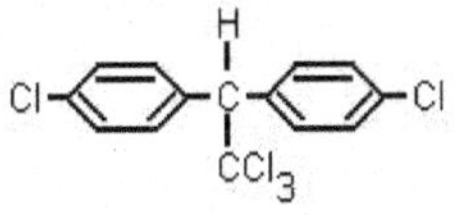

Class: **Molluscicide**

Typical Targets: Snails on flowers; control of slugs and snails.

Example: Metaldehyde:

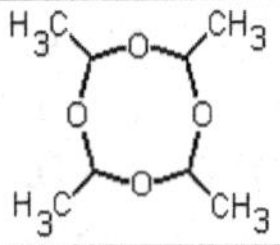

Table 3-9: Continued.

Class: **Nematicide** Typical Targets: Nematodes in vegetables and fruits (e.g., bananas); control of burrowing rootworms or nematodes. Example: Nemagon: 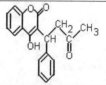
Class: **Rodenticide** Typical Target: Rats Example: Warfarin. (This is an anticoagulant. Also used in humans with heart disease.)
Class: **Slimicide** (Also classed as a bactericide. Slimicides are used to control slime-forming bacteria, fungi, and yeast.) Typical Targets: swimming pools; surfactants in cosmetics. Example: Didecyldimethylammonium chloride:

References Cited

[1] Black, A.P., and R.F. Christman. 1963. Characteristics of colored surface waters. Am. Water Works Assoc. 6:753–770.

[2] Page, H.J. 1930. Studies on the carbon and nitrogen cycles in the soil: I. Introduction. J. Agric. Sci. 20:455–459.

[3] Stevenson, F.J. 1994. Humus chemistry: Genesis, composition, reactions. 2nd ed. John Wiley & Sons, New York, NY.

[4] Schulthess, C.P., J.S. Quigley, M.J. Dapkus, and C.P. Huang. 1997. Are our soil organic extractions too harsh? p. 193–198. *In* J. Drozd, S.S. Gonet, N. Senesi, and J. Weber (ed.) The role of humic substances in the ecosystem and in environmental protection. Polish Soc. of Humic Substances, Wroclaw, Poland.

[5] Schulthess, C.P. 1996. Unpublished data: Laboratory notes from the University of Connecticut, Storrs, CT.

[6] Iota, V., and C.-S. Yoo. 2001. Phase diagram of carbon dioxide: Evidence for a new associated phase. Phys. Rev. Lett. 86:5922–5925.

[7] Lide, D.R. (Ed.-in-Chief). 2002. CRC Handbook of chemistry and physics. 83rd ed. CRC Press, Boca Raton, FL.

[8] Span, R., and W. Wagner. 1996. A new equation of state for carbon dioxide covering the fluid region from the triple-point temperature to 1100 K at pressures up to 800 MPa. J. Phys. Chem. Ref. Data 25:1509–1596.

[9] McHugh, M., and V. Krukonis. 1994. Supercritical fluid extraction. 2nd ed. Butterworth-Heinemann, Boston, MA.

[10] Andrews, A.T., R.C. Ahlert, and D.S. Kosson. 1990. Supercritical fluid extraction of aromatic contaminants from a sandy loam soil. Environ. Prog. 9:204–210.

[11] Schneider, G.M. 1978. Physicochemical principles of extraction with supercritical gases. Angew. Chem. Int. Ed. Engl. 17:716–727.

[12] Peng, D.-Y., and D.B. Robinson. 1976. A new two-constant equation of state. Ind. Eng. Chem. Fundam. 15:59–64.

[13] Akgerman, A., C. Erkey, and S.M. Ghoreishi. 1992. Supercritical extraction of hexachlorobenzene from soil. Ind. Eng. Chem. Res. 31:333–339.

[14] Brady, B.O., C.-P.C. Kao, K.M. Dooley, F.C. Knopf, and R.P. Gambrell. 1987. Supercritical extraction of toxic organics from soils. Ind. Eng. Chem. Res. 26:261–268.

[15] Yazdi, A., and E.J. Beckman. 1994. Highly CO_2-soluble chelating agents for supercritical extraction and recovery of heavy metals. p. 283–288. *In* G. Bunner and M. Perrut (ed.) Proc. Int. Symp. Supercritical Fluids. Vol. 2. 17–19 Oct. 1994. Strasbourg, France. Institut National Polytechnique de Lorraine.

[16] Low, G.K.C., G.J. Duffy, S.D. Sharma, M.D. Chensee, S.W. Weir, and A.R. Tibbett. 1994. Transportable supercritical fluid extractor unit for treating of contaminated soils. p. 275–280. *In* G. Bunner and M. Perrut (ed.) Proc.Int. Symp. Supercritical Fluids. Vol. 2. 17–19 Oct. 1994. Strasbourg, France. Institut National Polytechnique de Lorraine.

[17] Champagne, A.T., and P.R. Bienkowski. 1995. The supercritical fluid extraction of anthracene and pyrene from a model soil: An equilibrium study. Separation Sci. & Technol. 30:1289–1307.

[18] Camel, V., A. Tambuté, and M. Caude. 1993. Analytical-scale supercritical fluid extraction: A promising technique for the determination of pollutants in environmental matrices. J. Chromatogr. 642:263–281.

[19] Janda, V., K.D. Bartle, and A.A. Clifford. 1993. Supercritical fluid extraction in environmental analysis. J. Chromatogr. 642:283–299.

[20] Hawthorne, S.B., D.J. Miller, M.D. Burford, J.J. Langenfeld, S. Eckert-Tilotta, and P.K. Louie. 1993. Factors controlling quantitative supercritical fluid extraction of environmental samples. J. Chromatogr. 642:301–317.

[21] Page, S.H., S.R. Sumpter, and M.L. Lee. 1992. Fluid phase equilibria in supercritical fluid chromatography with CO_2-based mixed mobile phases: A review. J. Microcolumn Sep. 4:91–122.

[22] Shapiro, J. 1961. Freezing-out, a safe technique for concentration of dilute solutions. Science 133:2063–2064.

[23] Town, R.M., and H.K.J. Powell. 1993. Limitations of XAD resins for the isolation of the non-colloidal humic fraction in soil extracts and aquatic samples. Anal. Chim. Acta 271:195–202.

[24] Lobartini, J.C., K.H. Tan, L.E. Asmussen, R.A. Leonard, D. Himmelsbach, and A.R. Gingle. 1989. Humic matter isolation from soils and water by the XAD-8 resin and conventional NaOH methods. Commun. Soil Sci. Plant Anal. 20:1453–1477.

[25] Code of Federal Regulations (CFR). 1984. Volume 40: Protection of environment, Chapter1: Environmental Protection Agency, Part 413.3: National secondary drinking water regulations, Revised July 1, 1984.

[26] Narkis, N., and M. Rebhun. 1977. Stoichiometric relationship between humic and fulvic acids and flocculants. Am. Water Works Assoc. 69:325–328.

[27] Schnitzer, M., and S.U. Khan. 1972. Humic substances in the environment. Marcel Dekker, New York, NY.

[28] Schnitzer, M. 1977. Recent findings on the characterization of humic substances extracted from soils from widely differing climatic zones. p. 117–132. *In* Proc. Symp. Soil Organic Matter Studies, Braunschweig, Germany. 6–10 Sept. 1976. Vol. 2. Int. Atomic Energy Agency, Vienna, Austria.

[29] Schnitzer, M. 1985. Nature of nitrogen in humic substances. p. 303–325. *In* G.R. Aiken, D.M. McKnight, R.L. Wershaw, and P. MacCarthy (ed.) Humic substances in soil, sediment, and

water. John Wiley & Sons, New York, NY.

[30] Stevenson, F.J. 1985. Geochemistry of soil humic substances. p. 13–52. *In* G.R. Aiken, D.M. McKnight, R.L. Wershaw, and P. MacCarthy (ed.) Humic substances in soil, sediment, and water. John Wiley & Sons, New York, NY.

[31] Hatcher, P.G., I.A. Breger, G.E. Maciel, and N.M. Szeverenyi. 1985. Geochemistry of humin. p. 275–302. *In* G.R. Aiken, D.M. McKnight, R.L. Wershaw, and P. MacCarthy (ed.) Humic substances in soil, sediment, and water. John Wiley & Sons, New York, NY.

[32] McKeague, J.A., M.V. Cheshire, F. Andreux, and J. Berthelin. 1986. Organo–mineral complexes in relation to pedogenesis. p. 549–592. *In* P.M. Huang and M. Schnitzer (ed.) Interactions of soil minerals with natural organics and microbes. SSSA Spec. Publ. 17. Soil Sci. Soc. Am., Madison, WI.

[33] Dragunov, S.S., N.N. Zhelokhovtseva, and E.I. Strelkova. 1948. A comparative study of humic acids in soil and peat. Pochvovedenie 7:409–420 (in Russian).

[34] Flaig, W. 1960. Chemie der Humusstoffe. Suomen Kemistilehti A33:229–251.

[35] Fuchs, W. 1931. Die Chemie der Kohle. Springer, Berlin, Germany.

[36] Schnitzer, M. 1978. Humic substances: Chemistry and reactions. p. 1–64. *In* M. Schnitzer and S.U. Khan (ed.) Soil organic matter. Dev. Soil Sci. 8. Elsevier, New York, NY.

[37] Buffle, J.A.E. 1977. Les substances humiques et leurs interactions avec les ions mineraux. p. 3–10. Conference Proceedings de la Commission d'Hydrologie Appliquee de l'A.G.H.T.M., L'University d'Orsay.

[38] Wolt, J.D. 1994. Soil solution chemistry: Applications to environmental science and agriculture. John Wiley & Sons, New York, NY.

[39] Broadbent, F.E. 1953. The soil organic fraction. Adv. Agron. 5:153–183.

[40] Nelson, D.W., and L.E. Sommers. 1996. Total carbon, organic carbon, and organic matter. p. 961–1010. *In* D.L. Sparks (ed.) Methods of soil analysis. Part 3: Chemical methods. SSSA Book Series no. 5. Soil Sci. Soc. Am., Madison, WI.

[41] Jenny, H. 1941. Factors of soil formation. McGraw-Hill, New York, NY.

[42] Campbell, C.A., E.A. Paul, D.A. Rennie, and K.J. McCallum. 1967. Application of the carbon-dating method of analysis to soil humus studies. Soil Sci. 104:217–224.

[43] Kononova, M.M. 1966. Soil organic matter, 2nd ed. Transl. from Russian by T.Z. Nowakowski and A.C.D. Newman. Pergamon Press, New York, NY.

[44] Jenkinson, D.S., and A. Ayanaba. 1977. Decomposition of carbon-14 labeled plant material under tropical conditions. Soil Sci. Soc. Am. J. 41:912–915.

[45] Stevenson, I.L. 1964. Biochemistry of soil. p. 242–291. *In* F.E. Bear (ed.) Chemistry of the soil. 2nd ed. Am. Chem. Soc. Monogr. Ser. no. 160. Reinhold Publ. Corp., New York, NY.

[46] Lynham, T.J., G.M. Wickware, and J.A. Mason. 1998. Soil chemical changes and plant succession following experimental burning in immature jack pine. Canadian J. Soil Sci. 78:93–104.

[47] Jariel, D.M., R.J. Ansley, B.A. Kramp, and D.L. Jones. 2003. Soil chemical and nutrient responses to fire seasonality and frequency in a temperate mixed-grass savanna. Personal Communication.

[48] Ellis, R.C., and A.M. Graley. 1983. Gains and losses in soil nutrients associated with harvesting and burning eucalypt rainforest. Plant Soil 74:437–450.

[49] Little, S.N., and J.L. Ohmann. 1998. Estimating nitrogen lost from forest floor during prescribed fires in Douglas-fir/western hemlock clearcuts. Forest Sci. 34:152–164.

[50] Covington, W.W., and S.S. Sackett. 1992. Soil mineral nitrogen changes following prescribed burning in ponderosa pine. Forest Ecol. Manage. 54:175–191.

[51] Vance, E.D., and G.S. Henderson. 1984. Soil nitrogen availability following long-term burning in an oak–hickory forest. Soil Sci. Soc. Am. J. 48:184–190.

[52] Schoch, P., and D. Binkley. 1986. Prescribed burning increased nitrogen availability in a mature loblolly pine stand. Forest Ecol. Manage. 14:13–22.

[53] Almendros, G., F.J. Gonzalez-Vila, and F. Martin. 1990. Fire-induced transformation of soil organic matter from an oak forest: An experimental approach to the effects of fire on humic substances. Soil Sci. 149:158–168.

[54] Doerr, S.H., R.A. Shakesby, and R.P.D. Walsh. 1998. Spatial variability of soil hydrophobicity in fire-prone eucalyptus and pine forests, Portugal. Soil Sci. 163:313–324.

[55] Baker, W.E. 1973. The role of humic acids from Tasmanian podzolic soils in mineral degradation and metal mobilization. Geochim. Cosmochim. Acta 37:269–281.

[56] Finkl, C.W., Jr. 1979. Leaching. p. 260–265. *In* R.W. Fairbridge and C.W. Finkl, Jr. (ed.) The encyclopedia of soil science, Part I. Dowden, Hutchinson & Ross, Stroudsburg, PA.

[57] Senstius, M.W. 1958. Climax forms of rock-weathering. Amer. Scientist 46:355–367.

[58] Hartman, H. 1975. Speculations on the evolution of the genetic code. Part I, Origins of life 6:423–427; Part II, 9:133–136 (1978); Part III, 14:643–648 (1984).

[59] Bernal, J.D. 1967. The origin of life. Wiedenfeld & Nicolson, London, England.

[60] Miller, S.L. 1953. A production of amino acids under possible primitive Earth conditions. Science 117:528–529.

[61] Bernal, J.D. 1951. The physical basis of life. Routledge & Kegan Paul, London, England.

[62] Bernal, J.D. 1960. The problem of stages in biopoesis. p. 30–45. *In* M. Florkin (ed.) Aspects of the origin of life. Pergamon Press, New York, NY.

[63] Cairns-Smith, A.G. 1982. Genetic takeover. Cambridge Univ. Press, Cambridge, England.

[64] Cairns-Smith, A.G. 1989. The first organisms. Scientific American 252:90–100.

[65] Pirie, N.W. 1960. Chemical diversity and the origins of life. p. 55–62. *In* M. Florkin (ed.) Aspects of the origin of life. Pergamon Press, New York, NY.

[66] Cairns-Smith, A.G. 1971. The life puzzle. On crystals and organisms and on the possibility of a crystal as an ancestor. Univ. of Toronto Press, Toronto, Canada.

[67] Schulthess, C.P., and S. Tokunaga. 1996. Adsorption isotherms of poly(vinyl alcohol) on silicon oxide. Soil Sci. Soc. Am. J. 60:86–91.

[68] Page, E.R. 1979. The effect of poly(vinyl alcohol) on the crust strength of silty soils. J. Soil Sci. 30:643–651.

[69] Bailey, G.W., and J.L. White. 1970. Factors influencing the adsorption, desorption, and movement of pesticides in soil. Residue Rev. 32:29–92.

[70] Pignatello, J.J. 1998. Soil organic matter as a nanoporous sorbent of organic pollutants. Adv. Colloid Interface Sci. 76–77:445–467.

[71] Ballard, T.M. 1971. Role of humic carrier substances in DDT movement through forest soil. Soil Sci. Soc. Am. Proc. 35:145–147.

[72] Crosby, D.G. 1970. The nonbiological degradation of pesticides in soils. p. 86–94. Proc. Int. Symp. on Pesticides in the Soil: Ecology, Degradation & Movement. 25–27 Feb. 1970. Michigan State Univ., East Lansing, MI.

[73] The New York Times. 1964. Rachel Carson dies of cancer; 'Silent Spring' author was 56. The New York Times. 15 Apr. 1964. Obituary Section.

Questions

1.   Why are soil organic matter (SOM) fraction names based on a mechanistic extraction scheme rather than, say, one based on the actual functional groups present in the organic component? List advantages and disadvantages to this nomenclature approach.

2.   Why must you be cautious of humic acid and fulvic acid extracts that involve a strongly alkaline solution?

3.   List several benefits of SOM to the soil physical and chemical properties.

4.   SOM content in soils of a given climatic zone tends to be higher in fine-textured soils than in coarse-textured soils. Present a hypothesis or two about what may be the cause of this correlation.

5.   Based on the titration analysis presentation of Section 2.18, attempt to fit the titration data shown in Figure 3-9. Assume that the SOM titrated has only three distinctly different ionization constants.

6.   There are presently no clear definitions for distinguishing "low intensity" from "high intensity" fires. Present a few of your ideas on what measurable parameter may be used to distinguish these terms. Would there be any benefits in a definition based on the amount and/or rate of SOM consumed?

7.   Why is it that, for a particular soil environment, the amount of SOM present in the soil reaches an optimum concentration that balances the rate of SOM accumulation with the rate of SOM degradation? What may be done to the soil to increase its SOM content?

8.   Why is it that a small percentage of SOM in the soil can have a very significant influence on soil properties?

9.   Why are adsorbed polymers on soil particles held very strongly?
        Answer: There are many adsorption sites for each polymer molecule. Desorption will not occur unless all of these adsorbing links desorb at the same time. If one or more adsorption links remain, the other adsorption sites on the chain will not fully escape into the bulk water, which causes them to readsorb easily all over again.

10.   Why is it that the presence of SOM influences the recommended rates for pesticide applications?

Chapter 4:

Soil Mineralogy

Don't treat our soils like dirt.

Typically, the most studied portion of a soil is its solid phase. The solids present will strongly influence the physical characteristics and chemical reactivity of the soil. We can describe the solids in soils as either *amorphous* or *crystalline*. A solid is crystalline if its elements are arranged in a predictable, repeating fashion — that is, if they have long-range order. Conversely, an amorphous solid has a random arrangement of its elements. The term "mineral" is often used to describe a mined, natural resource (such as coal, petroleum, and natural gas), but strictly speaking *minerals* are solid, crystalline structures with characteristic chemical compositions. *Rocks* are composed of many minerals that are held tightly together.

There are many factors that influence the formation, translocation, and stability of minerals. Accordingly, there is great diversity in the types of minerals found in soils around the world. We cannot fully understand the chemical reactivity of any of these soil minerals without first understanding their chemical composition and physical structure, which is the primary focus of this chapter.

[4.1] Coordination Number (CN)

The coordination number of an element refers to the number of ions, or neighbors, that are packed around a given ion. In minerals, ions tend to pack as close together as possible, which puts them in a stable low-energy state. The geometry of how the atoms are arranged is based on the relative radii of the elements involved in the bonds. Accordingly, the radii of the elements that are part of a mineral greatly controls the structure of the mineral. An element that has a small radius can fit in a comparatively small cavity of a given mineral. The radii of various elements are listed in Table 4-1. Note that as an element loses electrons and acquires a positive charge, the radius of that element decreases significantly. Conversely, if the element gains electrons and acquires a negative charge, the radius of that element increases significantly. The presence or absence of filled electron orbitals clearly affects the size of the element; the size of just the nucleus is extremely small by comparison (on the order of 1/10,000th the diameter of the atom).

Table 4-1: Selected covalent, metallic, and ionic radii for atoms or ions with various coordination numbers (CN). *From A.H. Brownlow (1979)*[1]*, Geochemistry. Copyright © 1979. Reprinted by permission of Pearson Education, Inc., Upper Saddle River, NJ.*

Atom or ion	CN	Radius,[a] Å	Atom or ion	CN	Radius,[a] Å	Atom or ion	CN	Radius,[a] Å
Ag^+	6	1.23	Ga^{3+}	4	0.55	S	4	1.04 (c)
Al	12	1.43 (m)	Gd^{3+}	6	1.02	S^{2-}	6	1.72
Al	4	1.26 (c)	Ge^{4+}	4	0.48	S^{6+}	4	0.20
Al^{3+}	4	0.47	Hf^{4+}	8	0.91	Sb^{5+}	6	0.69
Al^{3+}	6	0.61	Hg^{2+}	6	1.10	Sc^{3+}	6	0.83
As^{5+}	6	0.58	Ho^{3+}	6	0.98	Se^{2-}	4	1.88
Au^{3+}	4	0.78	In^{3+}	6	0.88	Se^{6+}	4	0.37
B^{3+}	4	0.20	K^+	8	1.59	Si^{4+}	4	0.34
Ba^{2+}	8	1.50	K^+	12	1.68	Si^{4+}	6	0.48
Be^{2+}	4	0.35	La^{3+}	6	1.13	Sm^{3+}	6	1.04
Bi^{3+}	6	1.10	Li^+	6	0.82	Sn^{4+}	6	0.77
C	4	0.77 (c)	Lu^{3+}	6	0.94	Sr^{2+}	8	1.33
Ca^{2+}	6	1.08	Mg^{2+}	6	0.80	Ta^{5+}	6	0.72
Ca^{2+}	8	1.20	Mn^{2+}	6	0.75 (LS)	Tb^{3+}	6	1.00
Cd^{2+}	6	1.03	Mn^{2+}	6	0.91 (HS)	Th^{4+}	8	1.12
Ce^{3+}	6	1.09	Mn^{3+}	6	0.66 (LS)	Ti^{3+}	6	0.75
Cl	4	0.99 (c)	Mn^{3+}	6	0.73 (HS)	Ti^{4+}	6	0.69
Cl^-	6	1.72	Mn^{4+}	6	0.62	Tl^{3+}	6	0.97
Co^{2+}	6	0.73 (LS)	Mo^{6+}	6	0.68	Tm^{3+}	6	0.96
Co^{2+}	6	0.83 (HS)	Na^+	6	1.10	U^{4+}	8	1.08
Cr^{3+}	6	0.70	Na^+	8	1.24	U^{6+}	6	0.81
Cs^+	12	1.96	Nb^{5+}	6	0.72	V^{2+}	6	0.87
Cu	12	1.28 (m)	Nd^{3+}	6	1.06	V^{3+}	6	0.72
Cu^+	2	0.54	Ni^{2+}	6	0.77	V^{4+}	6	0.67
Cu^{2+}	6	0.81	O	4	0.73 (c)	V^{5+}	6	0.62
Dy^{3+}	6	0.99	O^{2-}	4	1.30	W^{4+}	6	0.73
Er^{3+}	6	0.97	O^{2-}	6	1.32	W^{6+}	6	0.68
Eu^{3+}	6	1.03	O^{2-}	8	1.34	Y^{3+}	8	1.10
F^-	4	1.23	P^{5+}	4	0.25	Yb^{3+}	6	0.95
Fe	12	1.26 (m)	Pb	12	1.75 (m)	Zn	12	1.39 (m)
Fe^{2+}	4	0.71 (HS)	Pb^{2+}	8	1.37	Zn	4	1.31 (c)
Fe^{2+}	6	0.69 (LS)	Pm^{3+}	6	1.04	Zn^{2+}	4	0.68
Fe^{2+}	6	0.86 (HS)	Pr^{3+}	6	1.08	Zn^{2+}	6	0.83
Fe^{3+}	4	0.57 (HS)	Rb^+	8	1.68	Zr^{4+}	8	0.92
Fe^{3+}	6	0.63 (LS)	Rb^+	12	1.81			
Fe^{3+}	6	0.73 (HS)	Re^{4+}	6	0.71			

[a] These are ionic bonds for ions. Else, c = covalent bond for atoms, and m = metallic bond for atoms. Also, HS = high-spin state (*d* electrons unpaired), and LS = low-spin state (*d* electrons paired) for ionic bonds.

Figure 4-1 illustrates the relative ionic sizes of various cations and anions. Using the periodic table, the size of the ions increases as one moves vertically downward in a given column. This is because there are additional electron shells present in the heavier elements, and each of these shells also partially shields the attraction by the positively charged nucleus. The size of the ions decreases as one moves horizontally from left to right in a given row.

This is because the nuclear charge is increasing as one moves from left to right on the periodic table, but the number of electron shells has remained the same; that is, there is a stronger nuclear attraction on the electrons of a given shell as one moves toward the right in a given row.

         The size of cations and anions are affected by the number of electrons present and by the pull of these electrons by the nucleus. The radius of a cation will increase as more anions (such as oxygen and hydroxides) are surrounding it. Note, for example, the Al^{3+} cation in Table 4-1. As the number of anions surrounding it increases from 4 to 6 (refer to the coordination number column), the radius of the cation increases from 0.47 to 0.61 Å. This is because the surrounding anions, which have a high electronegativity, have the ability to pull on the outer electrons of the central cation. This is not observed much with anions, as

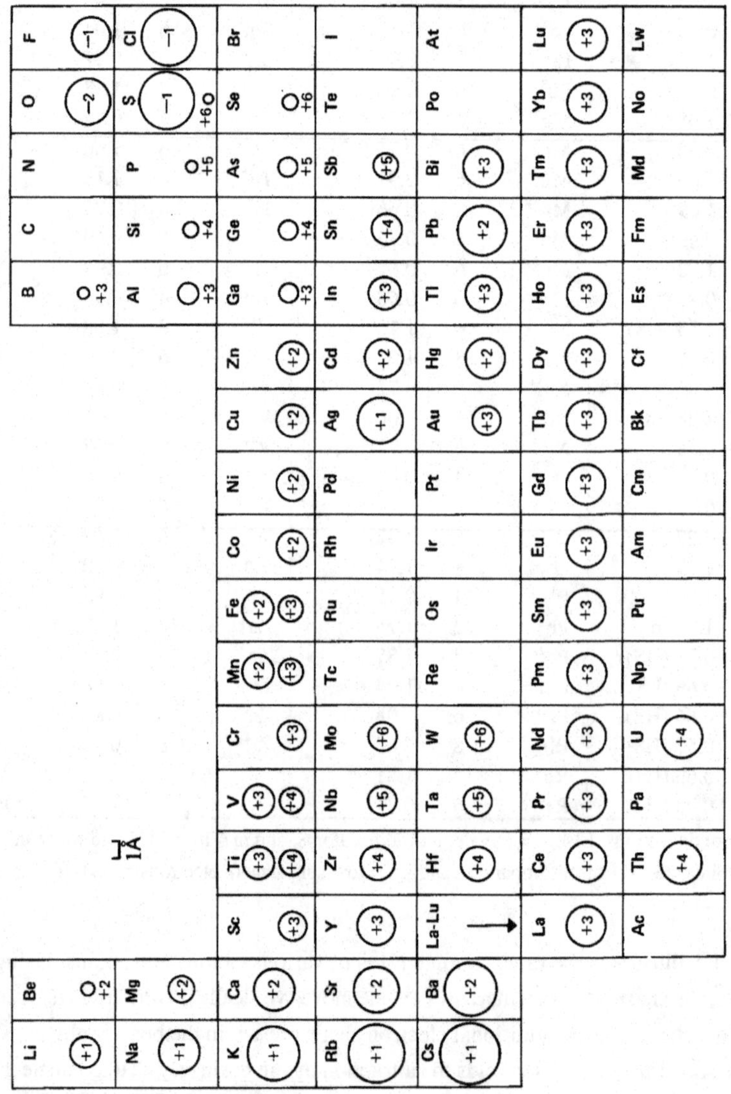

Figure 4-1: Relative ionic sizes of most of the elements for common valences and coordination numbers. Transition metals are in high-spin state. *From A.H. Brownlow (1979)[1], Geochemistry. Copyright © 1979. Reprinted by permission of Pearson Education, Inc., Upper Saddle River, NJ.*

can be seen in Table 4-1 for the O^{2-} anion when in the 4-, 6- and 8-fold coordinations.

The *atomic radii*, which are the sizes of the atoms or ions, are determined by numerous methods, one of them being based on the diffraction of x-rays by the atoms in the crystals. The atomic radii can refer to the covalent radii, metallic radii, or ionic radii of the elements. If the crystal studied consists of elements that are held by covalent bonds, then the resultant calculations will yield the *covalent radii* of the elements. If the elements are held by metallic bonds, such as Cu, then the radii are referred to as *metallic radii*. Estimates of elements held by van der Waals bonds will result in distances that are longer than the sum of the two covalent radii. If hybrid bonds are present (which are formed by a number of elements), then the estimated distances may or may not be different from their normal covalent radii. The size of *ionic radii* is most difficult to determine because there is no purely ionic compound that can consist of only one element. Rough preliminary estimates of the ionic radii of elements have been carried out by first assuming a value for the radius of O^{2-} (1.32 Å, CN = 6), which is the conjugate anion to the cation in many crystals. Some caution is warranted here because the assumed fixed ionic radius for oxygen, which is a reliable estimate for purely ionic minerals, may not be reliable for other minerals (such as silicates).

Crystal structures can be described from either the anion's or the cation's perspective. They both are essentially surrounding each other. Nevertheless, by convention, the cation is chosen as the central element in the description for the following reasons. First, the anions, which are larger in size, arrange themselves in such a way that they are very close or even touching each other in a regular grouping. Second, the variety of anions present in most minerals are not as large as the variety of cations. The anions are typically O^{2-} and OH^-, with both having nearly the same radius. Consequently, the resultant variety in anion radii is not so widespread. For these reasons, it is convenient to place the anions at the corners of the coordination polyhedra, rather than at its center. The cations, however, do not touch each other, and have a much larger range of possible sizes.

The coordination number of cations can be predicted based on the relative size of the cation to the surrounding anions, which is known as the radius ratio (r_{cation}/r_{anion}). As the radius ratios shown in Table 4-2 indicate, a cation that is too small for the interstitial space provided by a given polyhedron will "rattle" inside the excess space and be unstable. The small cation would be much more stable in a polyhedron corresponding to a lower coordination number. In the smaller polyhedron, the small cation will never have excess space between it and any of its surrounding anions; it will not "rattle" in the center of the polyhedral structure.

Using Euclidean geometry, Table 4-3 explains how the lower limit of each radius ratio shown in Table 4-2 can be obtained mathematically based on a rigid spherical model where all the anions are touching each other as tightly as possible, and where the cation is touching all of the anions simultaneously at all times. Values above this lower limit correspond to the spherical anions no longer touching each other; the anions are somewhat loosely held at the corners of the polyhedron, while the cation continues to be touching all of the anions of the polyhedron. As the size of the cation increases, the anions of the polyhedron may become so loose that another anion can be inserted for a more stable arrangement. This occurs at the upper limit of each radius ratio shown in Table 4-2; it is also the lower limit of the next radius ratio entry, which means that no excess interstitial space is created when the additional anion(s) is added. To insert an anion prior to reaching the right cation size would result in an unstable polyhedron with a cation that is held too loosely at its center; it would "rattle" at its center and be unstable. Note that the radii of ions listed in Table 4-1 and the radius ratios listed in Table 4-2 may change slightly from mineral to mineral due to the various configurations possible; consider these values only as averages. Do not confuse these numbers with the hydration radii of solvated ions, which were discussed in Section 2.10.

A particularly interesting cation is the hydrogen cation. Without its sole orbiting electron, this cation is but a proton, an almost volumeless element in space. The hydrogen atom is generally found in association with an oxygen atom, existing typically in a hydroxyl (OH^-) species in most minerals. The oxygen ion and the hydroxide ion have essentially the same ionic radii. It is also important to note that because of its small size, the hydrogen cation can travel through almost any barrier. It is the only element that can travel through the thin

Table 4-2: Prediction of the cation coordination number (CN) based on the cation-to-anion radius ratio. The polyhedron, or geometric figure, that forms is also shown based on imaginary lines that connect the nuclei of the anions. These polyhedra can share one or more anions, resulting in shared corners, edges or faces. Note that the cations (not drawn) are located in the center of the polyhedra and they are very close to each other when the polyhedra are sharing a face, are farther apart when the polyhedra are sharing an edge, and are farthest apart when they share only one corner or none at all. As CN increases, the larger the cation that can fit in the cavity formed at the center of the polyhedron.

Radius Ratio (r_{cation} / r_{anion})	Coordination Number (CN) Predicted	Coordination Polyhedron	Examples of polyhedra sharing corners, edges, and faces.
0.155–0.225	3	Triangle	
0.225–0.414	4	Tetrahedron	
0.414–0.732	6	Octahedron	
0.732–1.000	8	Cube	
1.000	12	Cubo-octahedron	

Table 4-3: Use of Euclidean geometry for calculating the lower limit of the radius ratio, where r_c = cation radius, and r_a = anion radius. Let r_a = 1.000 so that the value of r_c calculated below will also equal the radius ratio's lower limit.

Geometry of Polyhedra	Calculations
Triangle:	$$\cos 30^o = \frac{r_a}{r_a + r_c} = \frac{1}{1 + r_c}$$ $$1 + r_c = \frac{1}{\cos 30^o} = \frac{1}{0.8660} = 1.155$$ $$r_c = 0.155$$
Tetrahedron: D = center of upper anion E = center of cation F = center of base triangle Base Triangle Vertical Triangles In base triangle: $$BF = \frac{r_a}{\cos 30^o} = \frac{1}{\cos 30^o} = \frac{2}{\sqrt{3}}$$ In vertical right triangle: $$DF = \sqrt{BD^2 - BF^2} = \sqrt{(2r_a)^2 - BF^2}$$ $$DF = \sqrt{2^2 - \left(\frac{2}{\sqrt{3}}\right)^2} = \sqrt{4 - \frac{4}{3}} = \sqrt{\frac{8}{3}}$$	Apply two cosine principles to vertical triangles: $$\cos \alpha = \frac{DF}{DB}$$ and $\cos \alpha = \dfrac{DB^2 + DE^2 - BE^2}{2\,DB\cdot DE}$. Combine noting DE = BE, and DB = $2r_a$ = 2 : $$\frac{DF}{DB} = \frac{r_a}{DE} = \frac{1}{DE} .$$ Therefore, $$DE = \frac{DB}{DF} = \frac{2\sqrt{3}}{\sqrt{8}} = 1.225,$$ and $$r_c = DE - r_a = 1.225 - 1 = 0.225 .$$
Octahedron:	$$AD = \sqrt{AC^2 + CD^2}$$ $$2r_a + 2r_c = \sqrt{(2r_a)^2 + (2r_a)^2} = \sqrt{8}$$ $$r_c = \frac{\sqrt{8} - 2}{2} = 0.414$$
Cube: Base of Cube:	At base of cube: $$CD = \sqrt{EC^2 + ED^2} = \sqrt{(2r_a)^2 + (2r_a)^2} = \sqrt{8}$$ Across center of cube: $$AD = \sqrt{AC^2 + CD^2}$$ $$AD = \sqrt{(2r_a)^2 + (\sqrt{8})^2} = \sqrt{12}$$ $$r_c = \frac{AD}{2} - r_a = \frac{\sqrt{12}}{2} - 1 = 0.732$$
Cubo-octahedron:	Angle α = 360° / 6 = 60°; AB = $2r_a$ = 2. BG sin 30° = AB/2 BG(0.5) = 1 BG = 2 r_c = BG - r_a = 2 - 1 = 1.000

glass barrier of a glass pH-electrode; it can travel quickly through your skin and affect your blood's pH balance; and it can travel through a solid mineral and easily interact with the various other elements in the mineral.

Note also that ions that have degenerate orbitals (that is, two molecular orbitals with the same energy levels, such as Mn^{3+}) will result in a distortion of the resulting structure that splits the energy levels of the orbitals. This is known as the Jahn–Teller effect. If the "normal" structure is octahedral, then the Jahn–Teller effect typically tends toward a structure with two opposite ligands further apart, while the remaining four ligands form a nearly square center.

[4.2]   Pauling's Rules

Nearly all soil minerals exhibit a significant amount of ionic character. Ionically bonded solids, such as clays, show regular features that were summarized by Pauling in 1928 and are now known as Pauling's five rules. Pauling's rules are elaborations of rules proposed between 1923 and 1926 by the crystallographer Victor Goldschmidt. Pauling's rules are listed below. The first four rules focus on maximizing the cation–anion attractions and on minimizing the anion–anion and cation–cation attractions.

Rule 1: Each cation has a coordination polyhedron of anions, the cation-to-anion distance is equal to the sum of their radii, and the coordination number (CN) is a function of the radius ratio. This was discussed in Section 4.1. Because of this rule the Si^{4+} cation is always found in a 4-fold coordination as a tetrahedral SiO_4 polyhedron. The Al^{3+} cation is usually found in 6-fold coordination as an octahedral AlO_6 polyhedron, but it can be found in a 4-fold coordination if the temperature of crystallization of the mineral was very high. In clays and numerous soil minerals, the corners of the polyhedra are typically O^{2-} or OH^-.

Rule 2: The Electrostatic Valency Principle: The valence of each anion is equal to the sum of the electrostatic bonds connecting it to adjacent cations. That is, the crystal is electrostatically neutral. Dividing an ion's valence charge (z) by its coordination number (CN) is a measure of the electrostatic valency (e.v.), which is the strength of each bond reaching the coordinating ion. For example, in the tetrahedra SiO_4 polyhedron, the e.v. for each Si—O bond is 1 (= +4/4). For tetrahedral AlO_4, e.v. = 3/4; for octahedral AlO_6, e.v. = 1/2. The higher the absolute e.v. value, the stronger the bond strength.

Rule 3: The presence of shared edges and especially shared faces decreases a crystal's stability. This may also cause some distortion on the polyhedra. That is, two cations tend to be as far apart as possible. Several point-to-point connections between two polyhedra are illustrated in Table 4-2. Note that at the point of connection there is only one anion present (say O^{2-}) that is shared by the two cations in the center of the two connecting polyhedra. Table 4-2 also illustrates several edge-to-edge and face-to-face connections between the polyhedra. With these, the corners at the ends of each edge or face have only one anion that is shared by the two cations in the center of the two connecting polyhedra.

For a given set of two identical polyhedra, it can be seen in Table 4-2 that the two cations present in them will be farthest apart from each other when the polyhedra are sharing a point, closer when they share an edge, and closest when they share a face. According to the rule, sharing a point is the most stable arrangement (e.g., Si in the tetrahedral layer of clays), sharing an edge is a less stable arrangement (e.g., Al in the octahedral layer of clays), and sharing a face is the least stable (this is generally not present in clays). Also note that if two polyhedra share edges or faces, the mutual repulsion of the positively charged cations will cause the shape of the polyhedra to be distorted.

*Rule 4: In a crystal containing different cations, those with large valence and small coordination numbers tend **not** to share polyhedron corners, edges, or faces with each other.* That is, the high-charge cations stay as far from each other as possible, putting more negative charge between the cations. This also lessens their contribution to the crystal's Coulomb energy.

Rule 5: The Principle of Parsimony: The number of essentially different kinds of constituents (i.e., crystallographic configurations, such as tetrahedra and octahedra) in a crystal tends to be small. That is, the structural environment of a given cation tends to be the same throughout a crystal. In this way, the number of distortions and strains, which reduce the stability of a crystal, are minimized. As a result of this, all substances tend to obtain a structure of low energy and greater stability. Note, however, that oxygen anions in silicate minerals are generally present in different structural environments (that is, both in tetrahedral and octahedral environments).

[4.3]   Solid Solutions

It is very important in the study of mineralogy to understand how the elements can mix with each other. Implied in Pauling's rule number 4 (see Section 4.2) and in Lowenstein's rule (discussed in Section 4.10.2), all like cations generally prefer to spread themselves out as much as possible in a crystal. Note, for comparison, that in liquid solutions the dissolved components are evenly spread throughout the solvent medium. By analogy, cations in minerals behave as solid solutions.

The prevalence of impurities and the ability of the elements to mix and form a solid solution results in minerals that nearly always deviate somewhat from that corresponding to the pure substance. The mixed, solid solution that can exist depends on various factors similar to those discussed for liquid solutions in Section 2.12, but an important aspect in the formation of solid solutions is the nature of the solvent. More specifically, the dielectric constant, temperature, and pressure of the solvent will control the extent and kind of solid solution formed.

The presence of substitutions in all minerals is more often the rule rather than the exception. Nearly all soil minerals are salts, where the crystals are held by positively and negatively charged ions. Accordingly, solid solutions form through the substitution of ions of similar size and charge. For example, Ca^{2+} or Fe^{2+} ions may replace Mn^{2+} in rhodochrosite ($MnCO_3$), and MoO_4^{2-} may replace CrO_4^{2-} in other minerals. The stability of the resultant solid solution can be anticipated based on the ionic radius of the substituting ions (refer to Table 4-1). Accordingly, Fe^{2+} (average ionic radius = 0.075 nm) can easily exchange with

Mn^{2+} (average ionic radius = 0.083 nm), while Ba^{2+} (ionic radius = 0.150 nm) will not exchange with either of these.

The stability of solid solutions is typically greater at elevated temperatures. The effective ionic radius of an ion will increase with temperature due to the increased thermal motion. Since this thermal effect is greater on the lighter ions, a hot mixture consists of ions that are more nearly the same in ionic radius than a cooler mixture. As the temperature changes, the lighter ions will expand and contract their ionic radius more extensively than the heavier ions. This is illustrated easily with NaCl and KCl, which are mutually soluble in all proportions above 500°C but exhibit only marginal replacement of K^+ (average ionic radius = 0.164 nm) for Na^+ (average ionic radius = 0.117 nm) at 25°C. *Exsolution*, which generally occurs on cooling, is the process whereby an initially homogeneous solid solution separates into two (or more) distinct crystalline minerals.

Do not confuse solid solution crystalline solids with *isotype* minerals (two minerals that have exactly the same structures), *isostructural* minerals (minerals with similar structures, but not a one-to-one relationship between their atoms), and *isomorphic* minerals (a general term for similarity of mineral structures). An isotype example is the halite–galena pair (NaCl–PbS). An isostructural example is coesite (SiO_2 or Si_4O_8), whose structure is similar to albite ($NaAlSi_3O_8$); coesite has no counterpart for albite's Na atom, which can be described as being stuffed into spaces of a coesite-like structure.

There are three types of solid solutions: substitutional, interstitial, and omission. *Substitutional solid solutions* involve the exchange of anions or cations in the mineral. Two examples are: Br^- for Cl^- in KCl, and Rb^+ for K^+ in KCl or biotite. The substitution of one element by another over the total compositional range is referred to as a complete binary solid solution series. An example is Mg^{2+} for Fe^{2+} in olivine $(Mg,Fe)_2SiO_4$, where the end members are known as forsterite (Mg_2SiO_4) and fayalite (Fe_2SiO_4). Another example is $(Mn,Fe)CO_3$ with end members rhodochrosite ($MnCO_3$) and siderite ($FeCO_3$). It is also possible to have coupled substitution, where one ion is substituted by two or more ions to maintain electrical neutrality. Examples include: Fe^{2+} and Ti^{4+} for $2Al^{3+}$ in corundum (Al_2O_3); Na^+ and Si^{4+} for Ca^{2+} and Al^{3+} in the plagioclase feldspar series with end members albite ($NaAlSi_3O_8$) and anorthite ($CaAl_2Si_2O_8$). Limited solid solution series, rather than a complete solid solution series, also exist, particularly where the coordination polyhedra are not easily accommodated in the atomic sites present. An example is the limited cosubstitution of Ca^{2+} and Mg^{2+} in diopside ($CaMgSi_2O_6$) for Na^+ and Al^{3+} forming jadeite ($NaAlSi_2O_6$).

Interstitial substitution refers to the exchange of ions or molecules (such as H_2O and CO_2) in the interstitial solid solution or regions of the mineral. These regions consist of structural voids such as the channel-like cavities of zeolites or cyclosilicates, or the interlayer region of clays. The exchanges occur in a manner that is consistent with the net charge balance requirements of the mineral.

Omission solid solution refers to the charge-balanced exchange of two or more cations by one ion with a higher charge, resulting in the creation of a lattice vacancy. An example is the substitution of $2K^+$ by Pb^{2+} in microcline feldspar ($KAlSi_3O_8$), resulting in a lattice vacancy and a blue-green color center; this feldspar is known as amazonite.

[4.4]   Prevalence of Si Minerals

The quantity of each element in the Earth's crust is generally expressed in terms of percentage by weight. The atomic percentage (i.e., number of atoms per 100 atoms present) and volumetric percentage (which is estimated based on the atomic percentage and ionic radius of the element) can also be useful. The percent by weight of the most abundant elements in the Earth's crust are O (46.60%), Si (27.72%), Al (8.13%), Fe (5.00%), Ca (3.63%), Na (2.83%), K (2.59%) and Mg (2.09%). These are followed by H, Ti, P, Mn and Ba, which together account for 0.81% of the earth's curst, while all the other elements are present in insignificant amounts. Table 4-4 shows the estimated percentages of common minerals in the Earth's crust. Silicates account for more than 90% of the Earth's crust, with oxides and carbonates present in smaller amounts. Silicates and oxides are also known to exist on the rocky crusts of the moon and the four terrestrial planets of the solar system.

The unit cells of silicates are tetrahedron in shape, and none of these tetrahedra are known to share edges or faces. As summarized in Figure 4-2, the various silicate mineral groups differ from each other primarily based on the number of shared oxygens among the tetrahedra present. If no oxygens are shared, then the silicate units exist as isolated tetrahedra with the cation counter ions evenly distributed around each of the unit cells. These types of minerals are nesosilicates.

If oxygens are shared, then many crystal structures can be constructed, such as sorosilicates, cyclosilicates, and several types of inosilicates. The cation counter ions are evenly distributed around each of the silicate structural units.

The inorganic solid components of soils play a very important role in their chemical behavior. It is essential, therefore, that we understand some of the basic chemical and structural characteristics of soil minerals. Since Si minerals are common on the Earth's crust and contribute significantly to the mineralogy of soils, this chapter has a much greater portion of its discussion focused on these Si minerals. Keep in mind, however, that soils can have a significant or even dominant contribution from other minerals present, such as soils rich in carbonates.

Table 4-4: Estimated volumetric percentages of common minerals in the Earth's crust. All entries contain Si except for the last entry. Clearly, Si is present in most minerals in the Earth's crust, with >50% as feldspars. *Data from Ronov and Yaroshevsky (1969).*[2]

Mineral	Volume (%)	Mineral	Volume (%)
Plagioclase	39	Clays	5
Alkali feldspars	12	Olivines	3
Quartz	12	Other Minerals	5
Pyroxenes	11	(Silicates and Nonsilicates)	
Amphiboles	5	Other Minerals	3
Micas	5	(Carbonates and Oxides)	

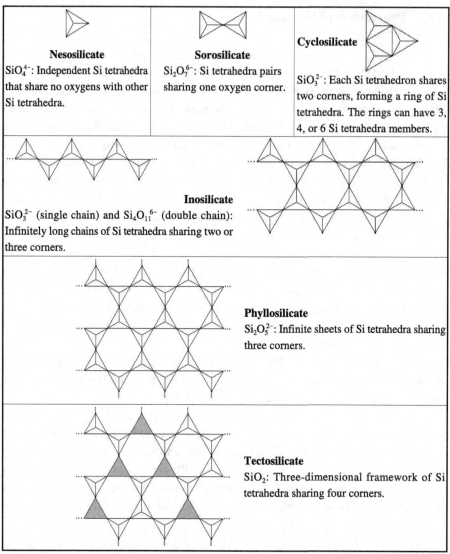

Figure 4-2: Principle classes of silicate crystal structures. Repeat unit formulas and short comments are included with each class illustrated. Note that the counter cations are nearby and maintain the minerals electrically neutral.

[4.5]   Nesosilicates

The structure of nesosilicates, which are sometimes referred to as orthosilicates, is greatly dependent on the size and charge of the interstitial cations that form ionic bonds with the isolated Si tetrahedra. "Neso" is Greek for "island". These isolated SiO_4^{4-} tetrahedra are similar to the isolated tetraheral units of sulfates (SO_4^{2-}) and phosphates (PO_4^{3-}). The atomic packing is generally dense, has a high index of refraction, and pronounced cleavage directions are generally absent. There are numerous gemstones in the nesosilicate class.

The interstitial cations in nesosilicates are typically octahedral polyhedra that surround the Si tetrahedra in numerous complex arrangements. Table 4-5 lists the chemical formula of various nesosilicate minerals. The crystal structure of titanite has the interesting feature of having a 7-coordinated Ca site. Clearly, the list of coordination options in Table 4-2 can be expanded and modified. Another interesting coordination number is found in the crystal structure of andalusite, where half the Al are in octahedral chains and the other half are in 5-coordinated polyhedra. The crystal structure of garnet is illustrated in Figure 4-3.

Table 4-5: Names and formulas of various nesosilicate minerals.

Al$_2$SiO$_5$ Group:		Olivine Group:	
Andalusite	Al$_2$SiO$_5$	Fayalite	Fe$_2$SiO$_4$
Kyanite	Al$_2$SiO$_5$	Fosterite	Mg$_2$SiO$_4$
Sillimanite	Al$_2$SiO$_5$	Olivine	(Mg,Fe)$_2$SiO$_4$
Bredigite	Ca$_7$Mg(SiO$_4$)$_4$	Tephroite	Mn$_2$SiO$_4$
Chloritoid	(Fe,Mg,Mn)Al$_2$SiO$_5$(OH)$_2$	**Natisite**	Na$_2$TiSiO$_5$
Datolite Group:		**Phenacite Group:**	
Datolite	CaB(SiO$_4$)(OH)	Eucryptite	LiAlSiO$_4$
Gadolinite	FeY$_2$Be$_2$Si$_2$O$_{10}$	Phenacite	Be$_2$SiO$_4$
Minasgeraisite	CaY$_2$Be$_2$Si$_2$O$_{10}$	Willemite	Zn$_2$SiO$_4$
Euclase Group:		**Staurolite**	(Fe,Mg,Zn)$_2$Al$_9$(Si,Al)$_4$O$_{22}$(OH)$_2$
Euclase	BeAl(SiO$_4$)(OH)	**Sverigeite**	NaMnMgSnBe$_2$(SiO$_4$)$_3$(OH)
Hodgkinsonite	MnZn$_2$SiO$_4$(OH)$_2$	**Titanite**	CaTiSiO$_5$
Eulytite	Bi$_4$(SiO$_4$)$_3$	**Topaz**	Al$_2$SiO$_4$(F,OH)$_2$
Garnet Group:	**[A_3B_2(SiO$_4$)$_3$]**	**Tornebohmite**	(Ce,La)$_2$Al(SiO$_4$)$_2$(OH)
Almandine	Fe$_3^{2+}$Al$_2$(SiO$_4$)$_3$	**Trimerite Group:**	
Andradite	Ca$_3$Fe$_2^{3+}$(SiO$_4$)$_3$	Larsenite	PbZnSiO$_4$
Grossular	Ca$_3$Al$_2$(SiO$_4$)$_3$	Trimerite	CaMn$_2$(BeSiO$_4$)$_3$
Pyrope	Mg$_3$Al$_2$(SiO$_4$)$_3$	**Uranophane Group:**	
Spessartine	Mn$_3$Al$_2$(SiO$_4$)$_3$	Uranophane	Ca(UO$_2$)$_2$[SiO$_3$(OH)]$_2$·5H$_2$O
Uvarovite	Ca$_3$Cr$_2$(SiO$_4$)$_3$	**Vanadomalayaite**	CaVSiO$_5$
Howlite	Ca$_2$B$_5$SiO$_9$(OH)$_5$	**Wadalite**	(Ca,Mg)$_6$(Al,Fe)$_5$Si$_2$O$_{16}$Cl$_3$
Humite Group:		**Zircon Group:**	
Chondrodite	[(Mg,Fe^{2+})$_2$SiO$_4$]$_2$·Mg(F,OH)$_2$	Coffinite	U(SiO$_4$)$_{1-x}$(OH)$_{4x}$
Humite	[(Mg,Fe^{2+})$_2$SiO$_4$]$_3$·Mg(F,OH)$_2$	Thorite	ThSiO$_4$
Clinohumite	(Mg$_2$SiO$_4$)$_4$·Mg(F,OH)$_2$	Zircon	ZrSiO$_4$

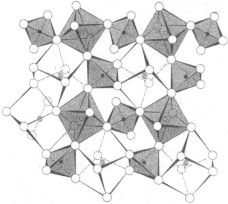

Figure 4-3: The general crystal structure of garnet, a nesosilicate mineral. The structure is projected down z showing, in the shaded portions, the alternating Si tetrahedra and B cations octahedra, and, in the unshaded portions, the A cations triangular dodecahedra (drawn as distorted cubes, CN = 8). Note that none of the Si tetrahedra share oxygen corners with each other. *From Novak and Gibbs (1971)[3], with permission.*

[4.6]   Sorosilicates

Most of the more than 70 known sorosilicate minerals are rare in nature. These "bow tie" Si structures are sometimes also called diorthosilicates, and Table 4-6 lists the chemical formulas of several representative minerals. "Soro" comes from Greek meaning, "group". As with the nesosilicates discussed in the previous section, it should be clear that sorosilicates can have a vast variety of possible structures. Sorosilicates are divided into four subgroups: (1) Si_2O_7 groups with no additional anions (such as fresnoite) , (2) Si_2O_7 groups with additional anions, such as OH, F, and H_2O (such as bertrandite) , (3) insular Si_3O_{10} and larger noncyclic groups, and (4) insular mixture of single and larger tetragonal groups. Examples of the third subgroup include zunyite (Si_5O_{16}, cross-shaped pentamer), ruizite (Si_4O_{11}, linear structure), aminoffite, harstigite, and kinoite (all Si_3O_{10} structures). Examples of the fourth subgroup include epidote, pumpellyite and vesuvianite (all mixtures of SiO_4 and Si_2O_7 structures). Hemimorphite is illustrated in Figure 4-4, epidote in Figure 4-5, and zunyite in Figure 4-6.

Table 4-6: Names and formulas of some sorosilicate minerals.

Bertrandite	$Be_4Si_2O_7(OH)_2$	**Fresnoite**	$Ba_2Ti(Si_2O_7)O$
Danburite	$CaB_2(Si_2O_7)O$	**Hemimorphite**	$Zn_4Si_2O_7(OH)_2{\cdot}H_2O$
Epidote Group:	$[A_2B_3O(SiO_4)(Si_2O_7)(OH)]$	**Ilvaite**	$CaFe_3O(Si_2O_7)OH$
Allanite	$A_2B_3 = (Ca,Ce)_2(Fe^{2+},Fe^{3+})Al_2$	**Kinoite**	$Ca_2Cu_2Si_3O_8OH$
Clinozoisite	$A_2B_3 = Ca_2Al_3$	**Leucophanite**	$(Na,Ca)_2BeSi_2(O,OH,F)_7$
Epidote	$A_2B_3 = Ca_2(Al,Fe)Al_2$	**Suolunite**	$Ca_2Si_2O_5(OH)_2{\cdot}H_2O$
Hancockite	$A_2B_3 = CaPb(Al,Fe^{3+})_3$	**Yttrialite**	$(Y,Th)_2Si_2O_7$
Piemontite	$A_2B_3 = Ca_2(Mn,Fe)Al_2$	**Zunyite**	$Al_{13}(Si_5O_{16})(OH,F)_{16}O_4F_2Cl$

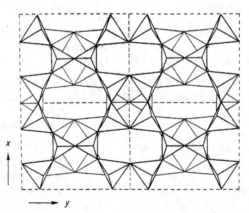

Figure 4-4: A [001] projection of the crystal structure of hemimorphite, a sorosilicate mineral. Four unit cells are shown. The "bow tie" Si_2O_7 structure is easily seen in the center of each unit cell, with one oxygen shared between two Si atoms and each of the other three oxygens are shared with two Zn atoms. Notice the large channels present. The H_2O molecule occupies the center of the cavity formed by 20 oxygen atoms of the tetrahedral framework. The water can be removed on heating without disruption of the structure. *From McDonald and Cruickshank (1967)[4], with permission.*

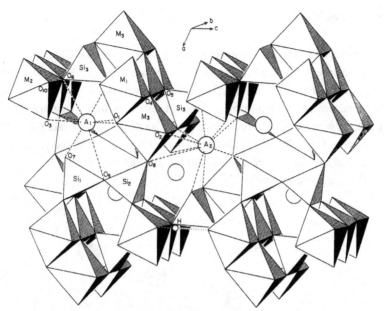

Figure 4-5: The crystal structure of the basic epidote structure. There are two edge-sharing octahedra: a single chain (labeled M2) and a zig-zag chain consisting of a central M1 chain with peripheral M3 octahedra. The SiO_4 and Si_2O_7 crosslink these octahedral chains. The cations are housed in the A1 and A2 cavities, with the A2 site being larger and with a higher coordination number even if they house the same cation (such as Ca): CN(A1) = 9, CN(A2) = 10 (or 11 for allanite). Although these are mixed neso-sorosilicate minerals, they continue to be classed as sorosilicates because of the higher organization of the Si_2O_7 groups. *From Dollase (1971)[5], with permission.*

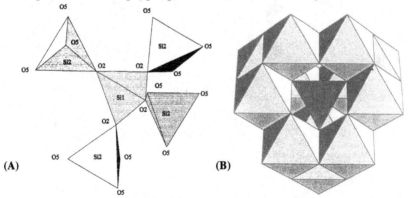

(A)   (B)

Figure 4-6: Two components of the zunyite mineral: (A) the Si_5O_{16} pentamer, and (B) the $Al_{13}O_{16}(OH)_{24}$ Keggin molecule. Each Si_5O_{16} pentamer is surrounded by Al_{13} groups, with a pentamer-to-Al_{13} group ratio of 1:1. The Keggin center is a tetrahedral Al with three octahedral Al on each of its corners. The aluminum polymers $Al_2(OH)_2^{4+}$ and $Al_{13}O_4(OH)_{24}(H_2O)_{12}^{7+}$ are easily formed in aqueous environments by partial neutralization of acidic Al(III) solutions. *Figure (A) from Dirken et al. (1995)[6], with permission; Figure (B) from Furrer et al. (1992)[7], copyright © 1992, reprinted with permission from Elsevier.*

[4.7]   Cyclosilicates

The cyclosilicates contain Si tetrahedral rings of various sizes: three-member rings (Si_3O_9), four-member rings (Si_4O_{12}), and six-member rings (Si_6O_{18}). The four-member rings are rare in silicates. The six-member cyclosilicates are more common. There are also some cyclosilicate minerals with eight-member rings and more complicated ring structures. There can be substitutions present in the rings, such as tetraheral Al substitution in the six-member rings of cordierite. Figure 4-7 illustrates the structure of the cyclosilicate beryl, and Table 4-7 lists the chemical formulas of various cyclosilicates. The [0001] projection designation corresponds to the [$a_1a_2a_3c$] coordinates.

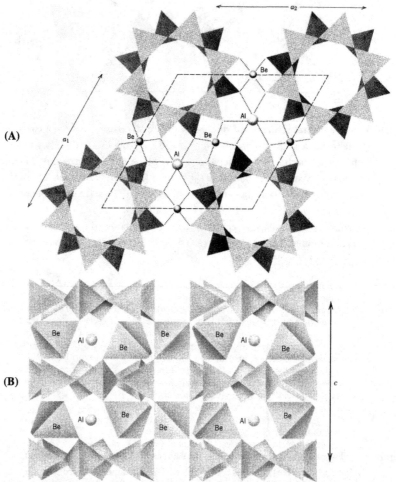

Figure 4-7: (A) Crystal structure of beryl projected onto [0001]. Dashed lines outline unit cell. (B) View of beryl structure with c axis vertical. Note that the overlaid 6-member Si rings form prominent channels parallel to the c axis. These channels can hold a wide variety of ions, atoms, and molecules, such as H_2O and CO_2. *From Klein and Hurlbut (1993)*[8]*, Manual of mineralogy. 21st ed., copyright © 1993 by John Wiley & Sons. This material is used by permission of John Wiley & Sons, Inc.*

Table 4-7: Names and formulas of a few cyclosilicate minerals.

Mineral	Formula	Size of Si rings in unit structure
Bentoite	$BaTi(SiO_3)_3$	3
Axinite	$Ca_2(Mn,Fe,Mg)Al_2(BO_3OH)(SiO_3)_4$	4
Beryl	$Be_3Al_2(SiO_3)_6$	6
Cordierite	$(Mg,Fe)_2Al_4Si_5O_{18} \cdot nH_2O$	6
Eifelite	$KNa_3Mg_4Si_{12}O_{30}$	6
Ferrokentbrooksite	$Na_{15}Ca_6(Fe,Mn)_3Zr_3Nb(Si_{25}O_{73})(O,OH,H_2O)_3(Cl,F,OH)_2$	9 & 3
Megacyclite	$Na_{16}K_2Si_{18}O_{36}(OH)_{18} \cdot 38H_2O$	18

[4.8]   Inosilicates

Inosilicates are characterized by the formation of tetrahedral Si chains. There are two important groups of inosilicate minerals: the single chain *pyroxene* minerals and the double chain *amphibole* minerals, which are illustrated in Figure 4-2. A common physical characteristic of some inosilicates is the presence of sharp needles or fibrous/asbestoslike features. Except for the presence of OH groups in amphiboles, both of these mineral groups have similar chemical compositions.

There are two other inosilicate minerals worth mentioning here. *Pyroxenoid* minerals are similar to pyroxenes but have longer repeat distances along the axis of the chain due to differences in the geometry of the chain (Figure 4-8). There are also *biopyribole*

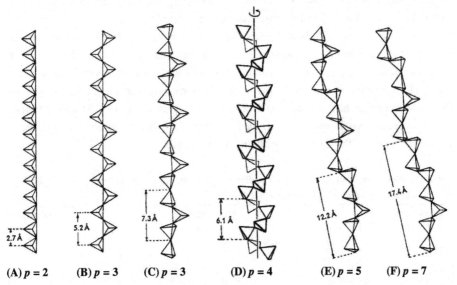

(A) $p = 2$ (B) $p = 3$ (C) $p = 3$ (D) $p = 4$ (E) $p = 5$ (F) $p = 7$

Figure 4-8: Si tetrahedral chains in pyroxenes (B) and pyroxenoids (C, E, and F) of various periodicity (p). The periodicity length increases incrementally from left to right. Chain structures with higher periodicity lengths are possible. Mineral examples with Si chains are: diopside (B), wollastonite (C), rhodonite (E), and pyroxmangite (F). Non-Si-based tetrahedral chains can also exhibit these shapes, such as: K_2CuCl_3 (A), $MgGeO_3$ (B), $(NaPO_3)_x$ (C), and $(AgPO_3)_x$ (D). *Reproduced from Liebau (1959)[9], Acta Crystallogr. 12: 177–181, copyright © 1959, with permission by International Union of Crystallography.*

minerals, which can be described as combinations of single Si chains (as in pyroxenes, and forming the "pyr" part of the name), double Si chains (as in amphiboles, and forming the "ibole" part of the name), and triple Si chains (as in biotite clays or phyllosilicates, and forming the "bio" part of the name).[10] The biopyribole minerals appear to be in intermediate stages of structural development between the anhydrous, high-temperature pyroxenes and the hydrous, lower temperature phyllosilicates.

The general chemical formula of pyroxenes is $XY(Si,Al)_2O_6$, where typically $X = Na^+$, Ca^{2+}, Fe^{2+}, or Mg^{2+} (and more rarely Zn^{2+}, Mn^{2+}, Li^+), and $Y = Mn^{2+}$, Fe^{2+}, Mg^{2+}, Fe^{3+}, Al^{3+}, Cr^{3+}, Sc^{3+}, V^{3+} or Ti^{4+}. The usual method for describing the location of elements within a mineral is with a crystallographic location designation. Using capital letters, "M" is for describing the octahedral (or larger) locations, "T" is for describing the tetrahedral locations, and "A" is for the cavities located between the various polyhedra. These are further subdivided with numbers (such as M1, M2, ..., or A1, A2, ...), where the larger numbers refer to the larger sites. With pyroxenes, the X cations are generally larger in ionic radius than the Y cations. Accordingly, the X cations are found in the larger M2 sites, while the Y cations are found in the smaller M1 sites. Unlike various other silicate minerals, Al tetrahedra do not often substitute for Si tetrahedra in pyroxenes.

The general chemical formula of amphibole is $XY_2Z_5(Si,Al,Ti)_8O_{22}(OH,F)_2$, where typically $X = Na^+$, K^+, or nothing (vacant), $Y = Na^+$, Ca^{2+}, Fe^{2+}, Li^+, Mn^{2+}, Al^{3+}, or Mg^{2+} (and more rarely Zn^{2+}, Ni^{2+}, or Co^{2+}), and $Z = Fe^{3+}$, Mn^{3+}, Cr^{3+}, Al^{3+}, Ti^{4+}, Fe^{2+}, Li^+ or Mn^{2+}. The X cations are in the A site, the Y cations are in the M4 site, and the Z cations are in the M1, M2, and M3 sites. The $(Si,Al,Ti)_8$ cations are in the tetrahedral sites.

Table 4-8 lists the chemical formulas of various inosilicates. The most common pyroxene mineral is augite. Amphiboles are very common minerals and are present in most metamorphic and igneous rocks. Amphiboles form at lower temperatures in the presence of water in the rock-forming melt, while pyroxenes only form when water is scarce. Amphiboles can be used to gauge the water content of the rock-forming melt. Figures 4-9 to 4-11 illustrate a few inosilicate crystal structures. Note the formation of tetrahedral–octahedral–tetrahedral (t–o–t) strips, where those of the amphiboles are twice as wide.

Table 4-8: Names and formulas of some inosilicate minerals. An explanation of the periodicity of the pyroxenoids is given in Figure 4-8.

Pyroxenes:		Pyroxenoids:		periodicity:
Aegirine	$NaFe^{3+}Si_2O_6$	Bustamite	$CaMnSi_2O_6$	3
Augite	$(Ca,Na)(Mg,Fe,Al)(Si,Al)_2O_6$	Ferrosilite III	$FeSiO_3$	9
Diopside	$CaMgSi_2O_6$	Pectolite	$Ca_2NaH(SiO_3)_3$	3
Hedenbergite	$CaFeSi_2O_6$	Pyroxmangite	$MnSiO_3$	7
Jadeite	$NaAlSi_2O_6$	Rhodonite	$MnSiO_3$	5
Pigeonite	$Ca_{0.25}(Mg,Fe)_{1.75}Si_2O_6$	Wollastonite	$CaSiO_3$	3
Spodumene	$LiAlSi_2O_6$	**Amphiboles:**		
Enstatite	$Mg(Mg,Fe)Si_2O_6$	Anthophyllite	$(Mg,Fe)_7Si_8O_{22}(OH)_2$	
Ferrosilite	$(Fe,Mg)_2Si_2O_6$	Cummingtonite	$Fe_2Mg_5Si_8O_{22}(OH)_2$	
		Glaucophane	$Na_2Mg_3Al_2Si_8O_{22}(OH)_2$	
		Hornblende	$Ca_2(Mg,Fe,Al)_5Si_7AlO_{22}(OH)_2$	
		Tremolite	$Ca_2Mg_5Si_8O_{22}(OH)_2$	

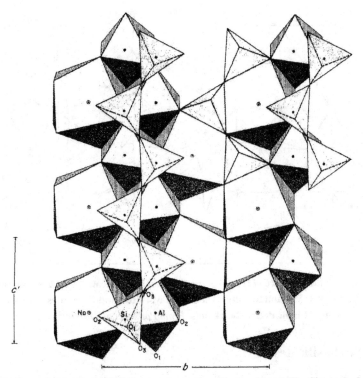

Figure 4-9: Crystal structure of jadeite, a pyroxene inosilicate mineral. The Si tetrahedral chains alternate between pointing up and down. In this drawing the center chain is pointing up, while those on left and right sides are pointing down. The Al^{3+} cations are in the M1 sites (CN = 6), while the Na^+ cations are in the larger M2 sites (CN = 8). *From Prewitt and Burnham (1966)[11], with permission.*

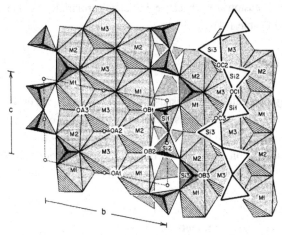

Figure 4-10: Crystal structure of wollastonite, a pyroxenoid inosilicate mineral. The Si chains with a periodicity of 3 are separated a distance three octahedra wide. *Reprinted from Papike (1987)[12], "Chemistry of the rock-forming silicates: Ortho, ring and single-chain structures", Rev. Geophys. 25:1483–1526, copyright © 1987 American Geophysical Union. Reproduced by permission of American Geophysical Union.*

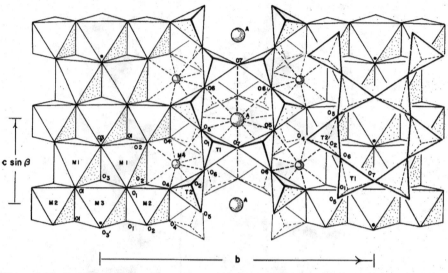

Figure 4-11: Crystal structure of tremolite, an amphibole inosilicate mineral. Magnesium (Mg) fills the M1, M2, and M3 sites; Ca fills the M4 sites; and the A sites are vacant. The H positions are indicated by black dots. *From Papike et al. (1969)[13], with permission.*

[4.9]   Phyllosilicates

The phyllosilicate minerals consist of sheets of Si tetrahedra arranged in six-fold rings. The negative charge of the Si tetrahedra sheet is neutralized by an adjacent octahedral sheet of cations (typically Mg or Al). Phyllosilicates are typically referred to as clays. Although they are usually small, this does not necessarily mean that these minerals are always small in size. This confusion arises because the "clay" term is also used to describe particles <2 µm in diameter (Table 1-4).

[4.9.1] Taxonomy of Clays

Most phyllosilicate clays consist of various combinations of Si tetrahedral sheets adjacent to Al or Mg octahedral sheets. Clays are classified as one of three types: 1:1, 2:1, or 2:1+1. These ratios refer to the number of tetrahedral sheets for every octahedral sheet present. Much substitution can occur in the tetrahedral and octahedral sheets, and this isomorphic substitution process greatly affects how these minerals are classified, particularly with the 2:1 and 2:1+1 clays.

For a quick and easy drawing of the clays, common picture symbols are used (Figure 4-12). A rectangle represents an octahedral sheet, and a trapezoid represents a tetrahedral sheet. These symbols are chosen because a side view of the clay minerals will reveal similar edge angles as those formed by combining rectangles and trapezoids. The terms plane, sheet, and layer are not interchangeable. Each *layer* of a clay mineral is made up of tetrahedral and octahedral sheets. Each *sheet* is, in turn, made up of several *planes* of

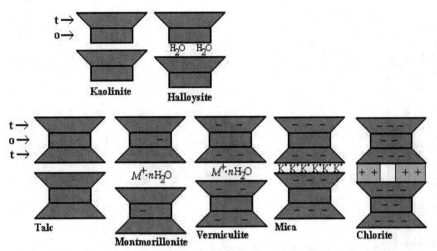

Figure 4-12: Commonly used schematic illustrations of 1:1, 2:1, and 2:1+1 clays. The tetrahedral sheet (t) is represented by a trapezoid, while the octahedral sheet (o) is represented by a rectangle. Isomorphic substitution in the tetrahedral and/or octahedral sheets will result in a negative charge imbalance, which is neutralized by the cations in the interlayer. Expansion may occur when water is present in the interlayer, and this is the primary cause of swelling of some clays.

atoms. For example, a tetrahedral sheet typically has a plane of Si atoms between two planes of O atoms.

The taxonomic description of various clays includes: type (based on 1:1, 2:1 or 2:1+1 layering of the polyhedra), group name (based on the amount and placement of isomorphic substitution), subgroup name (based on the presence of a di- or trioctahedral sheet), and species name. The octahedral sheet is classified as either dioctahedral or trioctahedral. *Dioctahedral* clays have only two octahedra sharing oxygens along the same direction in the mineral (the *b*-axis) in the octahedral sheet. Conversely, *trioctahedral* clays have three octahedra sharing oxygens along the same direction in the mineral per half-chemical formula unit. Note that the trioctahedral clays actually have a solidly filled octahedral sheet. The term trioctahedral merely highlights that there are no empty slots in any sequence of three octahedral places that can exist adjacent to every set of two (for 1:1 clays) or four (for 2:1 clays) tetrahedral elements.

The full-formula units (Table 4-9) are often not used with clays because that would double the number of atoms described in the octahedral sheet and mask the easy connection to the di- and trioctahedral terms for that sheet. The crystal drawings of the clays do not typically require that all the atoms be drawn to communicate the placement of the rest of the mineral's elements. That is, the placement of the second half of a clay's elements nearly always follow the same general pattern as those of the first half, but with a slight offset either forward or backward.

In the half-chemical formulas, the interlayer elements are often not identified, but are instead described generically as *M* for "metal" (as in Table 4-9). This is because they do not typically affect the mineral's structure. They are, however, needed by the mineral to

Table 4-9: Classification, morphology, and formulas of sheet silicate clay minerals. Traces of numerous other elements are typically also present. Layer charge (X) is in equivalents of charge per unit half-cell. Note that the interlayer charge (M_x^+) always neutralizes the layer charge. *After Sparks (2003).* [14]

Type (X = charge)	Group	Subgroup	Example Species	Morphology	Interlayer	Tetrahedral	Octahedral	Anions
1:1 (X ≈ 0)	Kaolin-serpentine	Kaolin (dioctahedral)	Kaolinite	hexagonal plate	—	Si_2	Al_2	$O_5(OH)_4$
			Dickite	hexagonal plate	—	Si_2	Al_2	$O_5(OH)_4$
			Nacrite	hexagonal plate	—	Si_2	Al_2	$O_5(OH)_4$
			Halloysite	tubular / sphere	$2H_2O$	Si_2	Al_2	$O_5(OH)_4$
		Serpentine (trioctahedral)	Antigorite	wavy plate	—	Si_2	Mg_3	$O_5(OH)_4$
			Chrysotile	tubular	—	Si_2	Mg_3	$O_5(OH)_4$
			Lizardite	platy	—	Si_2	Mg_3	$O_5(OH)_4$
2:1 (X ≈ 0)	Pyrophyllite-talc	Pyrophyllite (dioctahedral)	Pyrophyllite	platy	—	Si_4	Al_2	$O_{10}(OH)_2$
		Talc (trioctahedral)	Talc	platy	—	Si_4	Mg_3	$O_{10}(OH)_2$
			Kerolite	platy	H_2O	Si_4	$(Mg,Ni^{2+})_3$	$O_{10}(OH)_2$
(X = 0.2 to 0.6)	Smectite	Dioctahedral-smectites	Montmorillonite	platy	$M_{0.4}^+ \cdot nH_2O$	Si_4	$Al_{1.6}\ (Fe^{2+},Mg)_{0.4}$	$O_{10}(OH)_2$
			Beidellite	platy	$M_{0.4}^+ \cdot nH_2O$	$Si_{3.6}\ Al_{0.4}$	Al_2	$O_{10}(OH)_2$
			Nontronite	platy	$M_{0.4}^+ \cdot nH_2O$	$Si_{3.6}\ Al_{0.4}$	Fe_2^{3+}	$O_{10}(OH)_2$
		Trioctahedral-smectites	Saponite	platy	$M_{0.4}^+ \cdot nH_2O$	$Si_{3.6}\ Al_{0.4}$	Mg_3	$O_{10}(OH)_2$
			Hectorite	platy	$M_{0.4}^+ \cdot nH_2O$	Si_4	$Mg_{2.6}\ Li_{0.4}$	$O_{10}(OH)_2$
(X = 0.6 to 0.9)	Vermiculite	Dioctahedral-vermiculite	Dioctahedral-vermiculite	platy	$M_{0.8}^+ \cdot nH_2O$	$Si_{3.5}\ Al_{0.5}$	$Al_{1.4}\ Mg_{0.3}\ Fe_{0.3}^{3+}$	$O_{10}(OH)_2$
		Trioctahedral-vermiculite	Trioctahedral-vermiculite	platy	$M_{0.8}^+ \cdot nH_2O$	$Si_{3.2}\ Al_{0.8}$	Mg_3	$O_{10}(OH)_2$
(X = 0.6 to 0.9)	Illite	Illite	Illite	platy	$K_{0.8}\ Ca_{0.02}\ Na_{0.01}$	$Si_{3.4}\ Al_{0.6}$	$Al_{1.53}\ Fe_{0.22}^{3+}\ Fe_{0.03}^{2+}\ Mg_{0.22}$	$O_{10}(OH)_2$
(X = 1)	Mica	Dioctahedral-mica	Muscovite	platy	K	$Si_3\ Al$	Al_2	$O_{10}(OH,F)_2$
			Paragonite	platy	Na	$Si_3\ Al$	Al_2	$O_{10}(OH,F)_2$
			Celadonite	platy	K	Si_4	$Mg\ Fe^{3+}$	$O_{10}(OH)_2$
		Trioctahedral-mica	Biotite	platy	K	$Si_3\ Al$	$Mg_{1.5}\ Fe_{1.5}^{2+}$	$O_{10}(OH,F)_2$
			Phlogopite	platy	K	$Si_{2.7}\ Al_{1.3}$	$Mg_{2.7}\ Al_{0.2}\ Fe_{0.1}^{3+}$	$O_{10}(OH,F)_2$
			Lapidolite	platy	K	Si_4	$Li_2\ Al$	$O_{10}(OH,F)_2$

Brittle mica (X ≈ 2)	Dioctahedral-brittle mica	Margarite	platy	Ca	$Si_2 Al_2$	Al_2	$O_{10}(OH)_2$
	Trioctahedral-brittle mica	Clintonite	platy	Ca	$Si Al_3$	$Mg_2 Al$	$O_{10}(OH)_2$
Chlorite 2:1+1 (X = variable)	Di, dioctahedral-chlorite	Dombassite	platy	$Al_{2.27}(OH)_6$	$Si_{3.2} Al_{0.8}$	Al_2	$O_{10}(OH)_2$
	Di, trioctahedral-chlorite	Cookeite	platy	$Li Al_2 (OH)_6$	$Si_3 Al$	Al_2	$O_{10}(OH)_2$
	Tri, trioctahedral-chlorite	Amesite	platy	$Mg Al_2 (OH)_6$	$Si_2 Al_2$	Mg_3	$O_{10}(OH)_2$

maintain electroneutrality. When these clays are in contact with water, their interlayer elements are very easily exchanged with other elements in the liquid phase via a process known as *ion exchange*. The degree of isomorphic substitution in the tetrahedral plus octahedral sheets controls the number of cations needed in the interlayer to maintain electroneutrality. Each group of phyllosilicate minerals has a range of isomorphic substitution (in equivalents of charge per unit half-cell) that can be expected within the group. Clearly, the half-chemical formulas given in Table 4-9 are very narrow examples of what can exist for each group, subgroup, and species listed.

[4.9.2] Aligning Octahedral & Tetrahedral Sheets

It is a challenge in structural geometry to perfectly align the tetrahedral sheet with the octahedral sheet. Based on the ionic radii listed in Table 4-1 and Euclidean geometry similar to that used in Table 4-3, an ideal tetrahedral sheet with no Al^{3+} substitution for Si^{4+} will have the following dimensions: Si—O = 0.164 nm, O—O = 0.268 nm, and Si—Si (distance between Si in adjacent tetrahedra) = 0.309 nm. The ideal b dimension of the Si sheet (equivalent to the end-to-end distance of the six-member Si tetrahedral ring) is 0.928 nm, and this would increase to 0.968 nm if 50% of the Si^{4+} were replaced by Al^{3+} in the tetrahedral sheet.

Similarly, in an ideal Al octahedral sheet we can calculate the following dimensions: Al—O = 0.191 nm, and O—O = 0.270 nm. As can be seen in Figure 4-13, the 0.270-nm spacing between the Al octahedral oxygens is much smaller than the 0.309-nm spacing between the Si tetrahedral oxygens with which it needs to coincide.

We can present the same problem based on x-ray data. X-ray studies measure the b dimension of octahedral-containing minerals to be 0.864 nm for gibbsite ($Al(OH)_3$) and 0.943 nm for brucite ($Mg(OH)_2$). Assuming that these x-ray values of octahedra in the free state approximate the desired dimensions of the octahedra in clay minerals, the b dimension for gibbsite suggests that the dioctahedral sheet of clays is 7% smaller ($-0.07 = 0.864/0.928 - 1$) than the

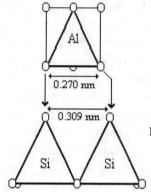

Figure 4-13: Side view (*b* dimension) of two Si tetrahedra misaligned with an Al octahedra assuming ideal structures for both sheets.

b dimension of the ideal tetrahedral sheet, while the *b* dimension for brucite suggests that the trioctahedral sheet of clays is 2% larger ($0.02 = 0.943/0.928 - 1$) than the *b* dimension of the ideal tetrahedral sheet.

The misalignment of the t-o sheets can be resolved by one of several rotating, tilting, and/or thickening distortions of the ideal sheet orientations. A rotation of the Si tetrahedra reduces the distance between the apical oxygens, and these rotations are frequently observed in clays. Figure 4-17 illustrates this with 1:1 clays discussed in the next section. The lateral *a* and *b* dimensions can also be shortened while thickening the *c* dimension. This adjustment involves increasing the angle of the apical oxygen to the center cation to one of the basal oxygens in the tetrahedra, which is ideally 109°28′. The ability of the basal oxygens to compress toward each other increases as the size of the central cation increases.

Adjustments of trioctahedral sheets is basically a pulling of the anions toward the cation plane, resulting in a thinner sheet with larger lateral dimensions. Adjustments on dioctahedral sheets, however, are more complex due to the added effects of the vacant site. Typically, all six anions around the vacant site move away from each other and toward the occupied cation sites. One plane of anions rotates clockwise, while the other rotates counterclockwise. As a result of this attraction of oppositely charged ions (or lack of attraction around the vacant site), the edge length of the vacant-site octahedra can increase from the ideal 0.27 nm to 0.32 – 0.34 nm. Concurrently, as the anions in the upper and lower planes approach each other, the octahedral sheet gets thinner and the lateral dimensions increase. Adjusting these anions also increases the shielding between the cations. The final thickness of the octahedral sheet can be as low as 0.204 to 0.214 nm after these adjustments.

If the distortion of the octahedral and tetrahedral elements do not align the sheets, then the morphology of the mineral cannot be platy. One option is the formation of a tubular or spherical structure with the smaller sheet on the inside of the curve, as is illustrated in Figure 4-14. The resulting tubes or spheres consist of a series of concentric layers or spirals. Note that a simple translation of its unit cell does not produce a macrocrystal. Accordingly, this option does not yield true unit cell dimensions and the crystal is therefore termed *paracrystalline*. A second option is the formation of a wavy morphology resulting from the periodic reversal of the Si tetrahedral sheet, as is illustrated in Figure 4-15.

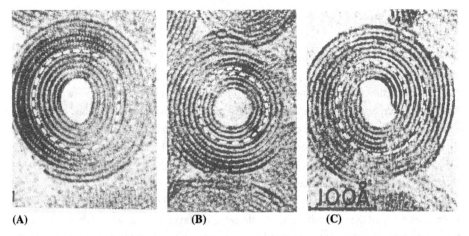

(A)   (B)   (C)

Figure 4-14: Cross section of chrysotile (a 1:1 trioctahedral clay) with high resolution transmission electron microscopy (HRTEM) showing tubular morphology. Highlighted by the dashed lines, some variations are shown: (A) concentric structure, (B) single spiral structure, and (C) multi-spiral structure. The diameter of the hollow of fibers is generally 70–80 Å, but can reach 100 Å. The outside diameters of most fibers are 220–270 Å, but can reach diameters of 650 to 800 Å. Fibers larger than 350 Å will usually have two or more stages of growth that are not concentric with the first stage. *Reproduced from Yada (1971)[15], Acta Crystallogr. A27:659–664, copyright © 1971, with permission by International Union of Crystallography.*

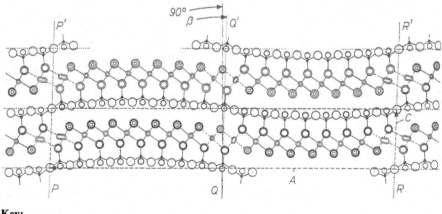

Key:

○ ◯ · ◎ ◦ ◉
O_1 O_2 Si_2 O_2OH Mg_3 $(OH)_3$

Figure 4-15: A [010] projection of antigorite (a 1:1 trioctahedral clay). Key based on $Mg_3Si_2O_5(OH)_4$ formula. The structure reverses polarity at PP', QQ', and RR', resulting in a wavy morphology. *From Kunze (1956)[16], with permission.*

[4.9.3] 1:1 Clays

Figure 4-16 illustrates the structure of kaolinite. Each 1:1 clay layer is held close to the next 1:1 clay layer through hydrogen bonding of the Al-OH groups in one layer with the Si-O groups in the other. Although this type of bond is relatively weak, there are many such bonds in each interlayer. Consequently, this H-bonding deters the expansion and contraction of the interlayer region of these clays. Lacking hydrated interlayer cations, these 1:1 clays do not swell or shrink with exposure to wet and dry cycles. It is also important to note that 1:1 clays do not need to have interlayer ions because the octahedral sheet perfectly balances the charge of the tetrahedral sheet.

Dickite and nacrite are similar to kaolinite. Looking through the Si tetrahedral ring of one layer toward the octahedral sheet of the adjacent layer, we see three H-bonds between the layers with the octahedral Al placed in the center of the Si tetrahedral ring in kaolinite and dickite, or three H-bonds with the Al offset in nacrite (Figure 4-17). Note that the c axis is not the same as true vertical in Cartesian coordinates because, by definition, the direction of the a, b, c axes are based on the planes by which the crystal unit cell is bounded, and these are not necessarily equivalent to the x, y, z axes used in Cartesian geometry. With kaolinite, the angles between axes are as follows: between the b and c axes, $\alpha = 91°48'$; between the a and c axes, $\beta = 104°30'$; and between the a and b axes, $\gamma = 90°$. Looking vertically along the c axis, kaolinite's Al in one unit cell is directly above the Al in the next unit cell. That is,

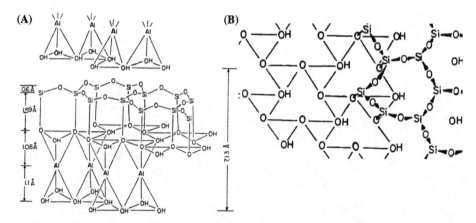

Figure 4-16: Ideal crystal structure of kaolinite: (A) b-axis side view, (B) top view. For side view, the vertical c-axis scale is stretched, and the upper and lower faces of the octahedral polyhedra are drawn to allow easier perception of depth in this two-dimensional drawing. Notice that the full-chemical formula for kaolinite has four Al in the octahedral sheet, with two of these offset from the other two in the b direction. The vacant site following each set of 2 Al is surrounded by six Al atoms. A top view of 1:1 minerals shows OH present in the center of the "nest" formed by the six-member Si ring. For an ideal structure, the Si ring diameter is 9.28 Å measured from centers of end oxygen to opposite-end oxygen. The thickness of each layer is approximately 7.1 to 7.2 Å, and it does not stretch or shrink as a result of the H-bonding between the layers.

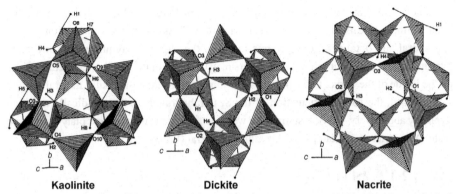

Kaolinite   Dickite   Nacrite

Figure 4-17: Projections of the structure of kaolinite, dickite, and nacrite showing the octahedral sheet of one layer and the tetrahedral sheet of the layer lying above. Line of sight is true vertical (i.e., z axis). The location of the hydroxyl hydrogen positions are also shown. *From Giese (1988)[17], with permission.*

the Al is in relatively the same position in all layers. The Al position, however, shifts in dickite and nacrite. With dickite, the Al position alternates by $\pm b/6$ in the b dimension, such that the Al positions coincide in the 1st, 3rd, 5th, ... layers and also in the 2nd, 4th, 6th, ... layers. With nacrite, the Al position is progressively rotationally translated 60°, such that the Al position repeats every second layer (normally we would predict a repeat every sixth layer, but this is divided by 3 because of the symmetry of the three void spaces around every Al in the dioctahedral sheet).

Halloysite is similar to kaolinite but with two water molecules per unit cell between its layers when fully hydrated. The d_{001}-spacing (that is, the thickness of each layer along the z axis) increases from 7 to 10 Å when halloysite is fully hydrated, but it very easily dehydrates in unsaturated environments and collapses back toward 7 Å as the interlayer water is lost. Dehydrated halloysite is often called metahalloysite or "halloysite (7 Å)." Fully hydrated halloysite is sometimes called "halloysite (10 Å)", and intermediate hydration states of halloysite will identify the d_{001} to the tenth of an angstrom, such as "halloysite (7.9 Å)." Halloysite minerals exhibit various morphologies, such as tubes (similar to Figure 4-14), spheres, plates, and various other irregular shapes. Most tubes roll up along the b axis, and dehydration leaves the rolls unchanged save for some shrinkage along the c axis. These various shapes are probably the result of how the halloysite (10 Å) grew, and not a result of the interlayer water.[17] In effect, halloysite accommodates the misalignment of its tetrahedral and octahedral sheets by curling its layers, with the smaller sheet on the inside of the curve.

The serpentine trioctahedral minerals are similar to the kaolin minerals but there are three Mg^{2+} cations instead of the two Al^{3+} plus one empty site. The net charge of +6 contributed by the cations in this layer is preserved. The alignment of the octahedral sheets with the tetrahedral sheets also needs to be resolved with these trioctahedral minerals, but this time the octahedral sheets are larger than the tetrahedral sheets. The alignment is resolved in one of three basic ways. First, some substitution of ions smaller than Mg^{2+} (such as Zn^{2+} or Fe^{2+}) in the octahedral layer, or substitution of ions larger than Si^{4+} in the tetrahedral layer, will tend to even out these two sheet sizes and result in a platy (i.e., flat) morphology. Lizardite is an example of this kind of mineral. Lizardite will also have some distortions of the tetrahedral and octahedral components to aid in the alignment requirement for a platy morphology. Second, curvature of the layers may occur, with the smaller tetrahedral sheets on the inside of the curve. Chrysotile is an example of this kind of tubular mineral (Figure 4-14). Third, the natural orientation of the Si tetrahedral sheet is periodically reversed (typically about every 40 Å), resulting in a wavy pattern that is more or less platy on a macro scale. Antigorite is an example of this kind of mineral (Figure 4-15).

[4.9.4] 2:1 & 2:1+1 Clays

The 2:1 clay minerals consist of one octahedral sheet sandwiched between two tetrahedral sheets. This sandwich-like structure prevents the curvature of the sheets, and the alignment of the tetrahedra with the octahedra is accomplished by rotating, tilting, and thickening distortions of the ideal sheet orientations. Accordingly, all of the 2:1 and 2:1+1 clay minerals have a platy morphology.

The amount of isomorphic substitution present in the mineral is the primary criterion for determining the type of mineral present. Pyrophyllite-talc has almost no isomorphic substitution, or at least no charge imbalance in its tetrahedral or octahedral sheets. Accordingly, it has no need for interlayer cations ($X \approx 0$ mol$_c$/half-cell).

The smectite group was formerly called the montmorillonite group. It is easy to understand why confusion between the group name and the species name led to the name change. The smectites will have some isomorphic substitution in either the octahedral sheet or the tetrahedral sheet. The resulting negative charge imbalance ($X \approx 0.2$–0.6 mol$_c$/half-cell) is balanced by an equivalent amount of positive charge in the interlayer.

Figure 4-18 illustrates the structure of montmorillonite. The 2:1 clays are not held together by strong hydrogen bonds but by weaker bonds, such as by the electrostatic charge attraction of the hydrated interlayer cations. An important consequence of this very weak "glue" between the 2:1 clay units is that the clay will easily expand and contract in the interlayer region. Exposing the clay to wet and dry cycles will result in extensive swelling and shrinking of the mineral, with some expansions exceeding 100 Å in very dilute, low ionic strength solutions. The layers remain parallel to each other even at these extreme expansions.

As we proceed down the list of clay minerals, the groups with a high charge imbalance ($X > 0.6$ mol$_c$/half-cell) typically have a significant amount of isomorphic substitution in the tetrahedral sheet. The charge imbalance of the vermiculites and illite groups are similar ($X \approx 0.6$–0.9 mol$_c$/half-cell), but they differ in the type of cations present in their interlayers: mostly K^+ for illite while they are nonspecific for vermiculite.

In part because of its very close similarity to mica (such as K^+ cations in the interlayer), illite is often considered a mica. The term illite is often used instead of the group name mica, particularly when referring to the micaceous component of mixed-layered clays, as in illite/smectite. When referring to the species illite, be sure to specify this clearly to avoid confusion with its more general group or with micas. The composition of the species

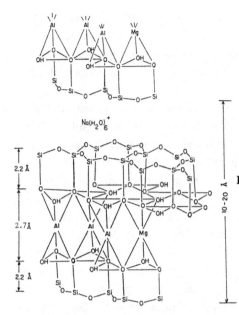

Figure 4-18: A [010] projection (*b*-axis side view) of ideal crystal structure of montmorillonite. Dimensions shown are between oxygen centers assuming no distortions of the ideal sheet orientations. One layer thickness is approximately 9.7 to 10 Å.

illite is difficult to describe. It is a mixture of muscovite and smectite. Furthermore, the existence of the pure species illite is uncertain because all illite minerals seem to have an interstratified component that is typically smectite, but it can be vermiculite or chlorite. The percentage of mixed-layered clays in illite are believed to be too low (up to 5%) to be detected by x-ray methods. Illite will not swell in water or shrink when dry.

The amount of isomorphic substitution in mica and brittle mica is large, resulting in a high charge imbalance ($X \approx 1$ mol$_c$/half-cell for mica, $X \approx 2$ mol$_c$/half-cell for brittle mica). The interlayer cation present in these minerals is an important component of their identity and physical characteristics. Specifically, if the K^+ cation is present in the interlayer and the water molecules are removed, then the 2:1 mineral collapses down upon itself. It turns out that the diameter of the K^+ interlayer cation fits snugly inside the "nest" formed by the six-member tetrahedral Si ring with an OH anion centered at its base. The layers are strongly attracted to each other and are held in place by the electrostatic forces of the bridging K^+ cations.

It is important to reemphasize that the ions in the interlayers are an essential part of all the clay minerals. The isomorphic substitution in the tetrahedral or octahedral sheets typically results in a layer charge imbalance that must be neutralized by nearby counter cations. The minerals must have them to maintain electroneutrality. No ion exchange will occur in any mineral if the electroneutrality criterion is violated. These ions are easily exchangeable in montmorillonites, saponites, and vermiculites, and the exact identity of these interlayer ions will affect the expansibility of the clay. These ions do not, however, significantly affect the general structure or the nomenclature of the clay. By contrast, the interlayer cations of micas and chlorites are not easily exchanged and, hence, have become closely associated with the identity and nomenclature of these particular clay minerals.

In micas, the superposition of one layer over another is controlled by the presence of K^+ cations in the ditrigonal holes. In 1:1 clays, the superposition of one layer over another is controlled by the H-bonding of one layer to another. Hydrated smectites, however, have no such mechanisms in place to align the layers with each other. Because there is no limit to the lateral or rotational movement of each layer, there is no idealized unit cell to describe the stacking symmetry of smectites. Accordingly, smectites are paracrystalline minerals rather than true minerals.

Chlorites are a large group of phyllosilicates, but its members are not well known. The various chlorite minerals are related to each other by complex solid solution substitutions. They are 2:1 clays with a layer of octahedral cations in the interlayer, such as $Al(OH)_x$ (gibbsite-like) or $Mg(OH)_x$ (brucite-like). Hence, chlorites are often referred to as 2:1+1 clays. These interlayer octahedra are also described as either dioctahedral or trioctahedral. For example, cookeite in Table 4-9 is a di,tri chlorite, where the second adjective (di = dioctahedral, tri = trioctahedral) refers to the interlayer octahedral structure. Chlorites are not expansible in water because of the strong hydrogen bonding between the Al-OH groups (or other hydroxyl groups) in the interlayer and the Si-O groups in the tetrahedral sheets adjacent to the interlayer. These minerals will not shrink because the islands of gibbsite or brucite lenses in the interlayer cannot be compressed.

There are some 2:1 clay structures not included in Table 4-9 that have their tetrahedral sheets divided into ribbons by inversion (Figure 4-19). While the tetrahedral sheets are continuous in two dimensions, the octahedral sheets are continuous in only one-dimensional ribbons. A channel exists between the strips, and these zeolite-like interlayer channels can contain water and exchangeable cations.

It is easy to misinterpret the thickness of the unit cell outlined in Figure 4-19. Notice that the tetrahedral inversion results in only one plane of oxygens centered between the tetrahedra. The apical oxygens of these tetrahedra are shared with the adjacent octahedra, and the small Si cations fit snugly in the center of the tetrahedra. Accordingly, the net thickness of each tetrahedral inversion is equal to that of one tetrahedral sheet. Summing it all up, the cell thickness (c) in Figure 4-19 is equal to the thickness of t–o–t–o sheets (where t = tetrahedra, and o = octahedra).

Two examples of clays with tetrahedral inversion are palygorskite and sepiolite. Palygorskite is also known as attapulgite or "mountain leather" in reference to its resemblance to woven cloth. Its ideal formula is $(Mg,Ca)_{0.5}(Mg,Al,Fe^{3+})_2Si_4O_{10}(OH)_2$, with

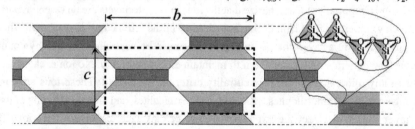

Figure 4-19: Illustration of tetrahedral inversion in 2:1 clays. As with all 2:1 clays, they have a flat, platy morphology. As we move horizontally along these clays, note that the orientation of the Si tetrahedra flip directions resulting in tetrahedral inversion. Insert: the Si tetrahedron that is oriented up shares an oxygen with the Si tetrahedron that is oriented down at each of these tetrahedral inversion points. *After Moore and Reynolds (1989).*[18]

Mg^{2+} as the main cation in the octahedral sheet. It can form large crystals, and its cell dimensions are: $c = 1.27$ nm, $b = 1.806$ nm.

Sepiolite is known as "meerschaum" by pipe smokers because its holes allow draw through the bowl material (hole diameters are about 1 nm). The ideal formula of this trioctahedral clay is $Mg_8Si_{12}O_{30}(OH)_4(OH_2)_4$, with very little isomorphic substitution and with each 2:1 package consisting of 6 Si in each tetrahedral sheet and 4 Mg in the octahedral sheet. Its dimensions are: $c = 1.34$ nm, $b = 2.68$ nm.

[4.9.5] Expansibility of 2:1 Clays

The expansibility of a clay is based on the clay's ability to pick up water in its interlayer. The ability to expand depends on the mechanism holding the layers together, such as H-bonding, which is very difficult to overcome because of the shear number of H bonds involved, or electrostatic attractions to the cations in the interlayer, which are easy to overcome if the cations are hydrated. Kaolinite is an example of a clay whose interlayer does not easily expand as a result of the H-bonding between the layers. Chlorites do not expand as a result the H-bonding of the brucite-like and gibbsite-like islands that form their interlayer regions. Mica is an example of a clay whose interlayer does not easily expand as a result of its preference for the K^+ cation, which is held snugly between the layers to remain unhydrated in the interlayer. Note that an NH_4^+ cation would also fit well in the craters of these six-member rings and collapse the 2:1 clay mineral. Clays that have tetrahedral inversions, such as those illustrated in Figures 4-15 and 4-19, lack the elasticity necessary to allow them to swell.

There are, however, many soils that clearly exhibit swelling and shrinking with wet and dry cycles, which is caused mostly by smectites and vermiculites. These are clays that have interlayer cations that are very easily hydrated. The expansibility of the clay is affected by the charge of the hydrated interlayer cations, with the larger expansions observed when these cations have a low positive charge and/or have a small ionic radius. That is, clay expansion is generally greatest when the interlayer cations have a large hydration radius (see Section 2.10).

The magnitude of the electrostatic attraction to the interlayer cation by the clay layers above and below it is also dependent on the amount and position of the layer charges present in the clays. From Coulomb's law, the force of attraction (F) is a function of the charge of the cation (e^+), the charge of the anion (e^-), and the distance between the charges (d):

$$F = \frac{e^+e^-}{d^2}$$

[4-1].

The smaller the electrostatic attraction to the cation, the larger the expansion of the clay. Using Coulomb's law, the relative expansion of some clays can be anticipated. Montmorillonite and beidellite have a similar amount of charge imbalance in the 2:1 layers due to isomorphic substitution (Table 4-9). Montmorillonite has isomorphic substitution in the octahedral sheet, while beidellite has isomorphic substitution in the tetrahedral sheet. If the charge imbalances are the same, then the beidellite will be less expansible. This is because the attraction for the interlayer cations is stronger when the charge imbalance is in the nearby tetrahedral sheet instead of in the farther away octahedral sheet. Similarly, a saponite mineral will be easier to expand than a trioctahedral vermiculite mineral. Both clays have isomorphic substitution in the tetrahedral sheet, but the amount of charge imbalance is lower in the saponite structure. And finally, the large expansion of montmorillonite versus the much smaller expansion of vermiculite is expected for a given concentration and type of interlayer cation. Montmorillonite's layer charge is smaller and farther away from the interlayer, while vermiculite's layer charge is larger and closer to the interlayer, which is illustrated schematically in Figure 4-12.

[4.9.6] Interstratified Clays

In natural environments, *mixed-layer* clay minerals are common. These are minerals composed of two or more clays stacked together in a repeating sequence. They are also referred to as *interstratified* clay minerals or *interlayered* clay minerals. "Ordered", "regularly stacked", and "regularly interstratified" are interchangeable terms, as are "random", "disordered", and "irregularly stacked." The term *Reichweite* (from German for "reach back"), and designated by R, is used to describe the ordering of mixed-layer clay minerals or the sequence of stacking of the interlayered clay minerals. The Reichweite basically describes the influence of a mineral in one layer on the type of mineral that will be found in the next layer. If $R = 0$, then there is no influence and the mixing of the two minerals in the stacked interlayered clay mineral is completely random. If $R = 1$ (or $R1$ for short), then the sequence is perfectly ordered. An example of $R1$ is I/S/I/S/I/S/I/S... for an illite–smectite mixed-layer clay mineral. If $R = 3$, then there are three of one clay for every one of the other in an ordered sequence, such as: S/S/S/I/S/S/S/I/S/S/S/I/S/S/S/I..., where the sequence here is SSSI. The Reichweite values will vary based on the degree of ordering or randomness of the clay layers in the mineral, but common values for R are 0, 0.5, 1, 1.5, 2, 2.5, and 3.

If the proportions in a mixed-layer clay mineral are evenly distributed in a nearly 50:50 mix, and the clays are regularly stacked ($R = 1$), then unique names can be assigned to them. Table 4-10 lists a few of these. Corrensite is sometimes referred to as a "swelling chlorite" because of the swelling and shrinking characteristics of its smectite component. A high-charge variety of corrensite is observed, but it contains vermiculite instead of smectite, and is sometimes referred to as sangarite (which is not an approved name).

Deweylite is a kerolite–serpentine interlayer mineral with variable proportions of each clay. Tarasovite is a 3:1 mica–smectite interlayer mineral (MMMS), which has also been described as a 2:1 regular interstratification of mica layers and rectorite. Note that rectorite is listed in Table 4-12 as a 1:1 mica–smectite mineral. Hence, a 2:1 mica–rectorite mixture is also MMMS, but this approach of describing interlayer minerals like tarasovite is not typical.

The most common mixed-layer clay minerals in sedimentary deposits are illite–smectites, and they are perhaps more common than either discrete illite or discrete smectite. The second most abundant are chlorite–smectites, which are similar to, and at times difficult to differentiate from, the chlorite–vermiculite mixed-layer minerals. All of the known kaolinite–smectite mixed-layer clays seem to be randomly stratified ($R = 0$). The tri-

Table 4-12: Special names assigned to regularly stacked, 50:50 mixed-layer clay minerals. *After Moore and Reynolds (1989).*[18]

Mineral Name	Interlayered clays present in a 1:1 ratio and $R = 1$.
Aliettite	Talc–trioctahedral smectite
Corrensite	Trioctahedral smectite–trioctahedral chlorite (low-charge variety)
Corrensite	Trioctahedral vermiculite–trioctahedral chlorite (high-charge variety)
Hydrobiotite	Biotite–vermiculite
Kulkeite	Talc–chlorite
Rectorite	Dioctahedral mica–dioctahedral smectite
Tosudite	Dioctahedral chlorite–smectite

octahedral mica–vermiculite minerals are widespread in hydrothermal and soil clay minerals, but are uncommon in sedimentary rocks. Examples of other interlayered clay minerals include dioctahedral illite–vermiculite, celadonite–nontronite, and chlorite–saponite.

[4.9.7] Polytypism

A characteristic that is common to layered structures, but is not restricted to clays or silicates, is their ability to vary on how identical layers are stacked. This type of polymorphism, known as polytypism, is not in any way related to the Reichweite value or the ordering of clay layers discussed above. It refers instead to the actual physical alignment of the atoms in one layer with the atoms in another identical layer that is next to it, or to the set of layers that are next to it. Common structures and corresponding Ramsdell structural symbols used to describe polytypes are hexagonal (H), monoclinic (M), orthorhombic (Or), rhombohedral (R), trigonal (T), and triclinic (Tc). Numbers are also combined with these symbols to denote the number of layers that are involved in a repeating unit (e.g., 2M). If necessary, subscripts are also used to further differentiate among similar polytypes with different stacking arrangements (e.g., $2M_1$ and $2M_2$). Figure 4-17 illustrates several polytypes of kaolin: kaolinite, dickite, and nacrite. In general, polytypes should not receive individual mineral names, but should instead have a single name followed by a structural symbol suffix that describes the layer stacking. Figure 4-20 illustrates various standard polytypes.

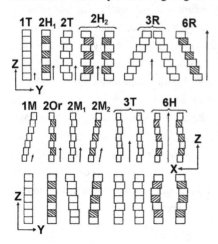

Figure 4-20: Diagrammatic *xz* **and** *yz* **projections of 12 standard polytypes plus 4 enantiomorphic structures (linked by brackets). Blank and ruled rectangles represent layers with octahedral cations in Sets II and I, respectively. Arrows indicate layer periodicity along** *z***.** *From Bailey (1988)*[19]*, with permission.*

[4.10] Tectosilicates

Most of the Earth's rocky crust (approximately 64%) is made up of tectosilicate minerals. These are strongly bonded structures that are often called the "framework silicates." Quartz is the pure form of just Si and O forming the SiO_2 three-dimensional structure. The Al^{3+} ion will easily substitute for the tetrahedral Si^{4+} ion resulting in a very wide range of structures. Additional cations are also incorporated into the structure to balance the charge. Table 4-13 describes the chemical formula of some tectosilicates minerals.

Table 4-13: Names and formulas of some tectosilicate minerals.

Quartz Group:		Feldspathoid Group:	
Cristobalite	SiO_2	Leucite	$KAlSi_2O_6$
Tridymite	SiO_2	Nepheline	$(Na,K)AlSiO_4$
Quartz	SiO_2	**Sodalite Group:**	
Opal	$SiO_2 \cdot nH_2O$	Lazurite	$(Na,Ca)_4(AlSiO_4)_3(SO_4,S,Cl)$
Feldspar Group:		Sodalite	$Na_4(AlSiO_4)_3Cl$
K-Feldspars:		**Scapolite Series:**	
Microcline	$KAlSi_3O_8$	Marialite	$Na_4(AlSi_3O_8)_3(Cl_2,CO_3,SO_4)$
Orthoclase	$KAlSi_3O_8$	Meionite	$Ca_4(Al_2Si_2O_8)_3(Cl_2,CO_3,SO_4)$
Sanidine	$KAlSi_3O_8$	**Zeolite Group:**	
Plagioclase Feldspars:		Chabazite	$Ca[Al_2Si_4O_{12}] \cdot 6H_2O$
Anorthite	$CaAl_2Si_2O_8$	Clinoptilolite	$(Na,K)[AlSi_5O_{12}] \cdot 4H_2O$
Bytownite	$Na_{0.2}Ca_{0.8}Al_{1.8}Si_{2.2}O_8$	Heulandite	$(Ca,Na_2,K_2)_4[Al_8Si_{28}O_{72}] \cdot 24H_2O$
Labradorite	$Na_{0.4}Ca_{0.6}Al_{1.6}Si_{2.4}O_8$	Mordenite	$Na_3KCa_2[Al_8Si_{40}O_{96}] \cdot 28H_2O$
Andesine	$Na_{0.6}Ca_{0.4}Al_{1.4}Si_{2.6}O_8$	Natrolite	$Na_2[Al_2Si_3O_{10}] \cdot 2H_2O$
Oligoclase	$Na_{0.8}Ca_{0.2}Al_{1.2}Si_{2.8}O_8$	Phillipsite	$K(Ca,Na_2)[Al_3Si_5O_{18}] \cdot 6H_2O$
Albite	$NaAlSi_3O_8$	Stilbite	$NaCa_2[Al_5Si_{13}O_{36}] \cdot 14H_2O$

[4.10.1] Feldspar, Feldspathoid, & Scapolite

Feldspars compose the most common minerals on the Earth's crust. The chemical composition of these minerals can be expressed in relative quantities of its three end members: (Or) orthoclase, $KAlSi_3O_8$; (Ab) albite, $NaAlSi_3O_8$; and (An) anorthite, $CaAl_2Si_2O_8$. For example, oligoclase can have the chemical formula $Ab_{75}An_{20}Or_5$. Nearly all feldspars fall into one of two mineral series: *plagioclase feldspars* and *alkali feldspars* (or *K-feldspars*). The various minerals in the Ab–An series (with typically ≤5% Or) are known as plagioclase feldspars. The various minerals in the Ab–Or series (with typically ≤5% An) are known as alkali feldspars or K-feldspars. The solid solutions between Ab–An and Ab–Or are complete, particularly with high-temperature feldspars. High-temperature feldspars crystallized at high temperatures or formed under rapid cooling, while low-temperature feldspars crystallized at low temperatures or formed under slow cooling. The low-temperature feldspars (also referred to as the low-feldspars or low structural state) have more order in the Al–Si and K–Na or Na–Ca distributions, while the high-temperature feldspars (also referred to as high-feldspars or high structural state) have more disorder in the Al–Si and K–Na or Na–Ca distributions. Figure 4-21 illustrates this solid solution and the names given to the various ranges of composition. Figure 4-22 shows the chemical structure of an example feldspar mineral.

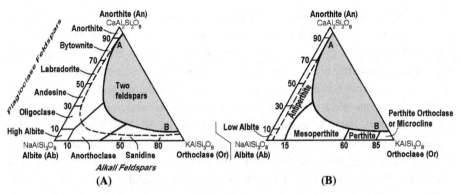

(A)   **(B)**

Figure 4-21: (A) The solid solution and feldspar nomenclature of the Or–Ab and Ab–An series for disordered ternary feldspars (high-temperature series). (B) The ordered ternary feldspars (low-temperature series) in which exsolution results in phase separation and the formation of intergrowths. That is, two feldspars exist in perthite, mesoperthite and antiperthite. Composition is in mole percent. Curve AB = limit of ternary solid solution. *After Deer et al. (1992).*[20]

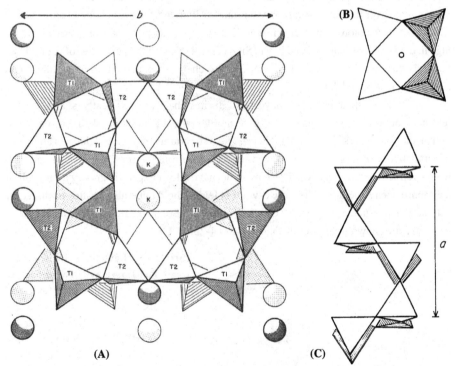

(A)   **(C)**

Figure 4-22: (A) Structure of a high sanidine feldspar, $KAlSi_3O_8$, projected on [$\bar{2}01$]. The Al and Si ions are randomly distributed over the two crystallographically distinct tetrahedral sites (T1 and T2). (B) Ideal four-member tetrahedral ring in the feldspar structure viewed down the *a* axis. (C) Chains parallel to the *a* axis formed by ideal four-member rings. *From Papike and Cameron (1976)*[21]*, "Crystal chemistry of silicate minerals of geophysical interest.", Rev. Geophys. Space Phys. 14:37–80, copyright © 1976 American Geophysical Union. Reproduced by permission of American Geophysical Union.*

The divisions between the K-feldspars are based on the Si–Al ordering and cooling rates of crystal formation. Sanidine crystallizes at high temperatures and the cooling rate is very rapid. Orthoclase crystallizes at intermediate temperatures and the cooling rate is intermediate. Microcline crystallizes at low temperatures and the cooling rate is very slow.

The divisions between the plagioclase feldspars are arbitrary and have no structural significance. The divisions are based on the An percentages, which are as follows: albite, 0–10%; oligoclase, 10–30%; andesine, 30–50%; labradorite, 50–70%; bytownite, 70–90%; and anorthite, 90–100% An. The low-temperature series of Ab–An are structurally complex minerals, while the high-temperature series is almost a complete solid solution. As result of this, the distribution of the Ab component between coexisting solid solutions of K-feldspars and plagioclase feldspars has been used as a geothermometer for estimating the cooling temperatures in which the crystals were formed.

Feldspars consisting of other major cations, such as barium feldspars, are relatively rare. Celsian, abbreviated Cn, is $BaAl_2Si_2O_8$, where the Ba^{2+} and K^+ ions are about the same size and form a solid solution (Cn–Or series). The plagioclase feldspars are widely distributed and more abundant than the K-feldspars. The Or and Ab minerals often predominate over the An minerals in environments with high SiO_2 content in the rocks and with fewer dark-colored minerals present. The An minerals predominate in environments high in dark-colored minerals and low in SiO_2 content. Most of the properties of the feldspar minerals vary uniformly with the change in the chemical composition of the three end members: Or, Ab, and An.

The feldspathoids are anhydrous silicates that are chemically similar to the feldspars, but form from melts that are poor in SiO_2 (about 2/3 that of alkali feldspars) and rich in alkalis (such as Na^+ and K^+). They are also structurally similar to feldspars and silica minerals but have larger structural cavities resulting from the presence of four- and six-membered tetrahedral linkages. These minerals have a lower specific gravity as a result of their more open structures. The large cavities in feldspathoid minerals are often occupied by very large ions, such as Na^+, Cl^-, CO_3^{2-}, SO_4^{2-}, and S^{2-}. Some examples of the feldspathoid structures are shown in Figures 4-23, 4-24, and 4-25.

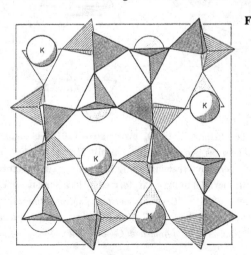

Figure 4-23: Leucite, a feldspathoid mineral, viewed down the c-axis. The tetrahedra form four- and six-member rings, and the 12-coordinated K^+ ions can be substituted for Na^+ ions. *From Papike (1988)[22], "Chemistry of rock-forming silicates: Multiple-chain, sheet, and framework structures", Rev. Geophys. 26:407–444, copyright © 1988 American Geophysical Union. Reproduced by permission of American Geophysical Union.*

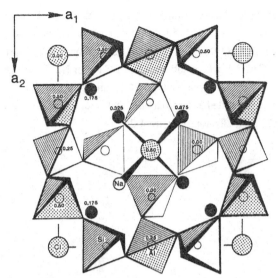

Figure 4-24: **Sodalite, a feldspathoid mineral, with Na$^+$ shown as small solid circles and Cl$^-$ shown as larger shaded circles. Visualize a cube with a four-member ring on each face, and a six-member ring around each corner of the cube.** *From Papike (1988)[22], "Chemistry of rock-forming silicates: Multiple-chain, sheet, and framework structures", Rev. Geophys. 26:407–444, copyright © 1988 American Geophysical Union. Reproduced by permission of American Geophysical Union.*

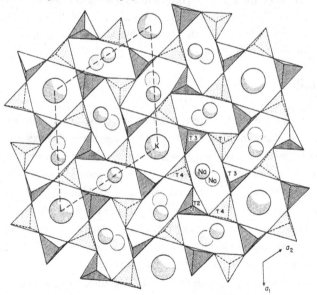

Figure 4-25: **Nepheline, a feldspathoid, projected down the c-axis or [0001]. The K$^+$ and Na$^+$ ions are in the large cavities. It has four crystallographically distinct tetrahedral sites (T1, T2, T3, and T4) in each unit cell outlined by the dashed lines.** *From Papike (1988)[22], "Chemistry of rock-forming silicates: Multiple-chain, sheet, and framework structures", Rev. Geophys. 26:407–444, copyright © 1988 American Geophysical Union. Reproduced by permission of American Geophysical Union.*

The scapolite minerals are similar to feldspars in composition, but the salts NaCl, $CaCO_3$, or $CaSO_4$ are typically added to the formula. A complete solid solution exists between marialite ($3NaAlSi_3O_8 \cdot NaCl$, or Ab_3NaCl) and meionite ($3CaAl_2Si_2O_8 \cdot Ca(SO_4,CO_3)$, or An_3CaSO_4 or An_3CaCO_3), where substitution of Ca^{2+} for Na^+ is charge compensated by substitution of Al^{3+} for Si^{4+}, and where CO_3^{2-}, SO_4^{2-}, and $2Cl^-$ can easily substitute for each other. The structure of scapolite has large cavities that contain both the cationic (Ca, Na) ions and the anionic (CO_3, SO_4, Cl_2) ions (Figure 4-26).

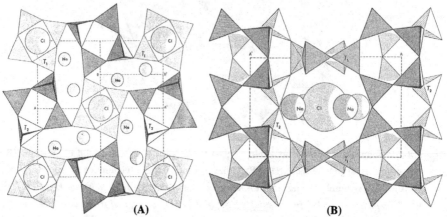

(A)   (B)

Figure 4-26: Structure of marialite scapolite: (A) projected down the *c*-axis or [001], (B) projected down the *a*-axis or [100]. *From Papike and Cameron (1976)*[21]*, "Crystal chemistry of silicate minerals of geophysical interest.", Rev. Geophys. Space Phys. 14:37–80, copyright © 1976 American Geophysical Union. Reproduced by permission of American Geophysical Union.*

[4.10.2] Zeolites

The zeolite minerals, although present in low quantities in soils, have a very high surface area and can, therefore, significantly influence the chemical reactions present in soils. The word "zeolite" comes from the Greek words "zeo", to boil, and "lithos", stone. The first discovery of a zeolite was that of stilbite in 1756 by the Swedish mineralogist Baron Axel F. Cronstedt. These "boiling stone" minerals were so named because they will bubble when heated, as if they were boiling, which is due to the liquid evaporating in the zeolite channels.

Zeolites are tectosilicate minerals with large channels or cavities that can allow water, cations, and organic molecules to enter the structure. Water in the cavities is also called "zeolite water" if it can be removed without destruction of the original structure. The size of the pore and void volumes in the crystal structure, ranging from 0.3 to 1.0 nm, controls the selective passage and exchange of molecules. Zeolites act as *molecular sieves* (term coined by James W. McBain in 1932), or reverse filters, in that they capture small molecules and allow the larger molecules to flow by unimpeded. This size exclusion characteristic of zeolites has led to numerous uses, including catalytic cracking of petroleum products, removal of water from organic liquids (that is, used as drying agents), removal of CO_2 and SO_2 from gaseous effluents, ion exchange of radioactive wastes to remove ^{137}Cs

and ^{90}Sr, and ion removal of NH_4^+ in aquarium waste waters. Some zeolites are hydrophobic, and these can be used to remove hydrocarbon contaminants from water. Zeolites are also amended to soils to reduce heavy metal toxicity, and to control pH or manure odors.

There are over 40 known zeolites. The structure consists of a tetrahedral framework of Si and Al atoms, where the Si content can range from 50% to more than 99.99% pure silica. As the Si-to-Al ratio of zeolites increases, the crystals become increasingly more hydrophobic. The Al makes the zeolites hydrophilic, and prepared materials are normally extensively hydrated. The 50% lower limit of Si content is a result of a generally consistent rule on the chemical structure of zeolites, known as Lowenstein's rule[23], which states that two Al atoms cannot share a tetrahedral oxygen in zeolites. That is, Al–O–Al bridge formations are not allowed. Although rare, non-Lowensteinian distributions are possible. Some zeolite structures show preferences for particular types of ordering (preferred locations or sites) of the Al atoms, but general rules allowing the prediction of order are not currently available.

The zeolite structure consists of tetrahedra sharing all corners, generating a three-dimensional network that has low density as a result of the large pore or void spaces. The structures are described in terms of tetrahedral units known as secondary building units (SBU), as illustrated in Figure 4-27. The use of SBUs was introduced by Meyer (1968), and these are linked together to form any of the zeolite structures.

Figure 4-28 illustrates how SBUs are put together to construct zeolites. In this illustration, the β-cage consists of SBUs 4 and 6. In sodalite, these are fused together via the 4-member rings. The internal free diameter of the β-cage is about 0.6 nm, enough to encapsulate small molecules. In zeolite A, which can be synthesized but is not found in nature, the β-cages are bridging (rather than fusing) via the 4-member rings, resulting in very large voids or supercages of about 0.74 nm diameter. This is large enough for molecules to diffuse in and out of them. In faujasite, a rare mineral, the bridging of the β-cages is via the six-member rings. This is also the structure of zeolite X (which has a low Si/Al ratio) and zeolite Y (which has a high Si/Al ratio). Zeolites X/Y are probably the most important zeolites produced commercially for catalytic systems, such as the cracking of petroleum compounds.

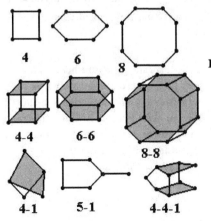

Figure 4-27: Illustration of various secondary building units (SBU) in zeolites. Each corner of the polyhedra represents the center of a MO_4 tetrahedral unit, where M is the central metal cation such as Al or Si. Each line connecting two corners is $M–O–M$, where one oxygen is shared by the two corner tetrahedral units. The MO_4 units are the primary building units.

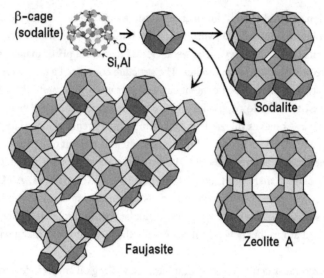

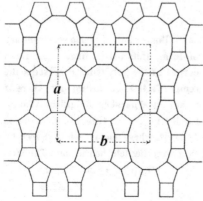

Figure 4-28: Illustration of the structure of the zeolites sodalite, zeolite A, and faujasite. Note that the basic structure here is interchangeably referred to as either a sodalite cage or a β-cage. The latter term is preferred because it reduces confusion with the sodalite mineral. Each β-cage in sodalite and zeolite A is surrounded by 6 cages. Each β-cage in faujasite is surrounded by 4 cages. *After Catlow (1992).* [24]

Figure 4-29 illustrates the structure of mordenite, one of the most widely distributed zeolite minerals in nature. The Si/Al distribution is partially ordered and its ratio is relatively constant at 4.5 to 5.5. The high Si/Al ratio causes the mineral to have a high thermal stability (for example, mordenite loses its water at 80–400°C, but its structure is preserved to 900°C). It has 8- and 12-membered channels parallel to the c axis, and 8-member channels parallel to the b axis. Although the diameter of the 12-member channel is 0.67 nm, all natural mordenite minerals do not sorb molecules greater than 0.42 nm in diameter, while synthetic mordenite obtained at temperatures below 140°C have larger pores. The pore size limitation in the natural mordenites is partially due to a lateral shift of the structural sheets, resulting in an aperture constriction. Another reason is the presence of amorphous materials (such as Si oxides or Fe oxides) occluded at crystallization, resulting in blocked channels. The

Figure 4-29: Mordenite structure built with SBU 5-1. Corners are centers of SiO_4 or AlO_4 tetrahedra, and lines indicate a shared oxygen atom between two tetrahedra in the zeolite structure. The unit cell dimensions are $a = 1.811$ nm, $b = 2.0527$ nm, and $c = 0.752$ nm.

mordenite channels can be unblocked after treatment with HCl (use hot HCl if trying to unblock Na,Ca-mordenites).

The presence of Al^{3+} for Si^{4+} in zeolites results in a net negative charge for the framework. The negative charge is balanced by counter cations (e.g., Na^+, K^+, Ca^{2+}, Sr^{2+}, La^{3+}) in the "extra-framework" sites. These sites are generally well-defined locations in the framework, such as the various six- or four-ring cages present in the framework. Some of the water in zeolites can be ascribed to the hydration of these extra-framework cations; but there is also water in the open cavities of the framework. A typical procedure for dehydrating a zeolite involves heating to 300–400°C for several hours.

The charge imbalance can also be neutralized by protons that are firmly bonded to the lone pairs of the bridging oxygens (Figure 4-30). The catalytic activity of zeolites is closely linked to the presence of these acidic hydroxyl groups, as well as to the presence of other acid groups at defect or surface sites.

It is worth noting that there are a few new synthetic materials that may or may not be considered zeolites depending on how zeolites are defined. Aluminum phosphate zeolite-like crystals (e.g., VPI-5) have been synthesized that are phosphorus rather than silicon based. These Al-P zeolites have uniform intracrystalline channels that are sometimes identical with those of some Si-based zeolites. The channels of the VPI-5 zeolite are larger (1.2 to 1.3 nm in diameter), which gives it a very high commercial potential.

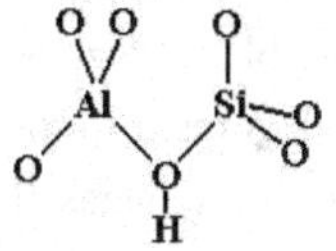

Figure 4-30: The Al–OH–Si bridging hydroxyl group in zeolites.
After Catlow (1992).[24]

[4.10.3] Si Oxides

Silicon oxides are found in all soil fractions: clay, silt, and sand. The general formula of these minerals is SiO_2, and the framework structures are electrically balanced. The three main types of silica minerals are quartz, tridymite, and cristobalite. Each of these can exist in one of two forms: an α form (low-temperature variety) and a β form (high-temperature variety). Note that the adjectives "low" and "high" are sometimes used instead of the Greek letters α and β; hence, we may write "low-quartz" instead of "α-quartz" or "high-quartz" instead of "β-quartz", and similarly for tridymite and cristobalite. The α and β designations here refer only to their temperature of formation and do not imply a particular crystal structure; that is, the β-quartz has a different crystal structure than a β-tridymite. The crystal structures of the α forms are, however, very similar to the crystal structures of the corresponding β forms. The β-to-α conversions are exothermic, and the α-to-β conversions are endothermic. The conversions of each α to β form, or β to α form, occurs at a particular temperature, where these instantaneous and reversible transformations are accompanied by very simple displacive structural changes. The β forms do not "quench in" at low temperatures; or, stated differently, they do not retain their structure when cooled rapidly. Accordingly, if you can hold one of these in your hand without protection, then it must be in an α form.

There are two other very dense silica minerals that form at high pressures (coesite and stishovite) and another that is synthetic (keatite). Thus, there are a total of nine known polymorphs of SiO_2. The stishovite mineral has the Si atoms in octahedral coordination with oxygen, which is a truly rare coordination number to have for Si.

There are also two amorphous Si minerals worth mentioning: lechatelierite (a fused silica or silica glass with variable composition) and opal (a locally ordered structure of silica spheres with a highly variable H_2O content). "Silica glass", "fused quartz", and "quartz glass" are all synonyms of any Si melted at about 1600°C and then cooled. It does not solidify, however, at the original melting-point temperature. At about 1500°C it is too stiff to flow, and is regarded as a supercooled liquid. It is not a crystal because its structure is as random as the liquid.

Of the three common silica polymorphs, quartz has the most compact structure, tridymite has a more open structure, and cristobalite has the most expanded structure. The temperature conversions at 1 atm pressure of the common silica minerals are as follows[25]:

$$\beta\text{-quartz} \xoverset{870\ ^\circ C}{\rightleftharpoons} \beta\text{-tridymite} \xoverset{1470\ ^\circ C}{\rightleftharpoons} \beta\text{-cristobalite}$$

$$\Big\updownarrow 573\ ^\circ C \qquad\qquad \Big\updownarrow 117\text{~}140\ ^\circ C \qquad\qquad \Big\updownarrow 220\text{~}280\ ^\circ C$$

$$\alpha\text{-quartz} \qquad\qquad \alpha\text{-tridymite} \qquad\qquad \alpha\text{-cristobalite} \qquad\qquad [4\text{-}2].$$

The reconstructive transformations between quartz, tridymite, and cristobalite require considerable activation energy. Because of this, the transformations between them are sluggish and the three α forms can coexist for long periods of time, with the α-tridymite and α-cristobalite existing metastably. Figures 4-31 and 4-32 illustrate the structure of various Si oxides.

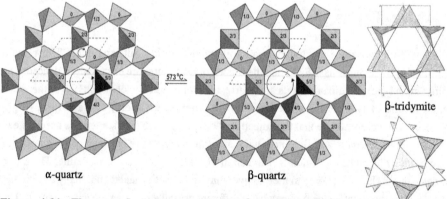

α-quartz   β-quartz

β-tridymite

β-cristobalite

Figure 4-31: The crystal structures of quartz, β-tridymite, and β-cristobalite. The tetrahedral layers of quartz and tridymite are projected down *c*, or [001]. The cristobalite illustration is projected onto (111). The Si tetrahedra in quartz form six- and three-membered helices, with each adjacent tetrahedra offset by 1/3 *c* as indicated. The effect of temperature on structure also impacts their relative densities: 2.65 for α-quartz and 2.53 for β-quartz. *For quartz after Papike and Cameron (1976).[21] For tridymite and cristobalite from Papike and Cameron (1976)[21], "Crystal chemistry of silicate minerals of geophysical interest", Rev. Geophys. Space Phys. 14:37–80, copyright © 1976 American Geophysical Union; reproduced by permission of American Geophysical Union.*

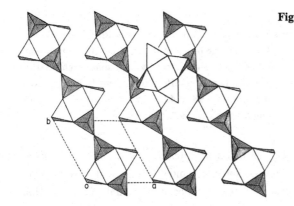

Figure 4-32: The coesite structure with four-membered tetrahedral rings that lie parallel to (001). *From Papike and Cameron (1976)[21], "Crystal chemistry of silicate minerals of geophysical interest.", Rev. Geophys. Space Phys. 14:37–80, copyright © 1976 American Geophysical Union. Reproduced by permission of American Geophysical Union.*

[4.11] Oxides & Hydroxides

Metal oxides (X_mO_n), hydroxides ($X_m(OH)_p$), and hydroxyoxides ($X_mO_n(OH)_p$) are very important soil minerals in terms of quantity and reactivity. They can collectively be referred to as metal oxides because they all contain oxygen in one form of another, and this book will often use the term "oxide" with this broad interpretation; note, however, that this practice is not always acceptable.

These minerals are solid, three-dimensional structures of metal polyhedra, and the bonding present in these structures is generally strongly ionic. There are several types of oxides based on the metal to oxygen (X:O) ratio:

X:O ratio	Group name	Example minerals
X_2O		Cuprite, Cu_2O
XO		Zincite, ZnO
X_2O_3	Hematite	Corundum, α-Al_2O_3; hematite, Fe_2O_3; ilmenite, $FeTiO_3$
XYO_3	Perovskite	Perovskite, $CaTiO_3$
XY_2O_4	Spinels	Spinel, $MgAl_2O_4$
XO_2	Rutile (SiO_2 excluded)	Tectosilicates, SiO_2; Rutile, TiO_2; Pyrolusite, MnO_2

Cuprite and zincite are both tightly packed tetrahedral structures. Although SiO_2 is an oxide with the ratio XO_2, it is grouped as tectosilicate and is not part of the rutile group oxides. Because of their importance to soils, tectosilicates were discussed separately in Section 4.10. Similarly, Al oxides, Fe oxides, and Ti oxides will be further discussed in the Subsections 4.11.1, 4.11.2, and 4.11.3, respectively.

The XYO_3 oxides, referred to as the perovskite group, include a wide range of minerals: $X^+Y^{5+}O_3$ (such as $NaNbO_3$ and $KTaO_3$), $X^{2+}Y^{4+}O_3$ (such as $SrTiO_3$, $BaZrO_3$, and $CaTiO_3$), and $X^{3+}Y^{3+}O_3$ (such as $LaCrO_3$ and $YAlO_3$). The total valence of the two cations must equal +6 to balance the −6 of the three O^{2-}. The Y atoms are in octahedral sites, while the large X atoms are in 12-fold coordination (Figure 4-33). These perovskite oxides are believed to be common under very high-pressure conditions such as the center and deep mantle of Earth.

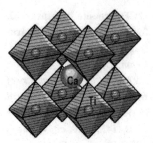

Figure 4-33: The cubic structure of perovskite ($CaTiO_3$), with Ca in center of cube and a Ti octahedron at each corner. Perovskite minerals form under high pressure environments, resulting in extremely high relative densities of 3.98 to 4.84. *After Náray-Szabó (1943)*[26] *and Deer et al. (1992).*[20]

The XY_2O_4 oxides (or $X_8Y_{16}O_{32}$ per unit cell), referred to as the spinel group, are further divided into normal spinel and inverse spinel structures. Examples of these are given in Table 4-14. In the normal spinel structures, the X cations occupy tetrahedral sites, while the Y cations occupy octahedral sites (Figure 4-34). In the inverse spinel structures, there are two tetrahedral sites for every octahedral site, resulting in $Y(YX)O_4$. Natural spinel minerals will typically have cation distributions between the normal and inverse spinel structural types.

The coordination number of the cations in spinels does not follow the radius ratio rule discussed in Section 4.1 or Pauling's rule number 1 discussed in Section 4.2. For example, the larger Mg ion is found in tetrahedral sites and the smaller Al is found in octahedral sites — the opposite of what would be predicted. The explanation for the larger cation occupying the tetrahedral sites rather than the octahedral sites is based on their crystal field stabilization energies rather than the geometric aspects of the ions.

The nesosilicate olivine structure ($X_2^{2+}Y^{4+}O_4$) is similar to the spinel structure ($X^{2+}Y_2^{3+}O_4$), but the olivines are less dense. Under very high pressures, the olivine structure transforms to a spinel structure. This time, however, the Si remains in tetrahedral sites, and Mg and Fe^{2+} remain in octahedral sites in the resultant spinel structure.

Table 4-14: Names and formulas of some spinel end minerals. Solid solutions exist consisting of mixtures of the various end members, including normal spinels with inverse spinels. Except for common spinel, most spinels are rare. Spinels are often further divided into three groups based on the identity of the most abundant cation in the mineral, namely Al-spinels, Cr-spinels, and Fe-spinels.

Normal Spinel Structure, XY_2O_4		Inverse Spinel Structure, $Y(YX)O_4$	
Chromite	$Fe^{2+}Cr_2^{3+}O_4$	Jacobsite	$Fe^{3+}(MnFe^{3+})O_4$
Franklinite	$ZnFe_2^{3+}O_4$	Magnesioferrite	$Fe^{3+}(MgFe^{3+})O_4$
Galaxite	$MnAl_2O_4$	Magnetite	$Fe^{3+}(Fe^{2+}Fe^{3+})O_4$
Hercynite	$Fe^{2+}Al_2O_4$	Ulvöspinel	$Fe^{2+}(Fe^{2+}Ti^{4+})O_4$
Magnesiochromite	$MgAl_2O_4$		
Spinel	$MgAl_2O_4$		
Trevorite	$NiFe_2^{3+}O_4$		

A→
B→
C→
A→

Figure 4-34: The structure of spinel ($MgAl_2O_4$). The alternating layers of octahedral and octahedral–tetrahedral polyhedra are parallel to (111). Right side has top removed to expose center sheet. *After Clark (1972).*[27]

[4.11.1] Al Oxides

Table 4-15 lists many different kinds of natural Al oxides, hydroxides, and hydroxyoxides. Note that bauxite, a common ore of aluminum, is not the name of a mineral but the name of a rock consisting of a mixture of the minerals gibbsite, boehmite, and diaspore, any of which may be dominant. Gibbsite is the most abundant free hydroxide of Al in soils, particularly in highly weathered soils. Boehmite is also formed naturally in soils, but the hydroxide minerals are more common than the hydroxyoxide minerals. Diaspore and corundum are usually found together, but are not commonly present in soils. Bayerite is rarely found in nature. Misra (1986)[28] notes that there is a lack of uniformity in the literature about the Greek letter designations used to describe these minerals. For example, gibbsite is sometimes erroneously designated as an α-Al(OH)$_3$, but the α designation according to general usage in crystallography is for the most densely packed structure, which is bayerite among the various Al(OH)$_3$ polymorphs noted here.

Al(OH)$_3$ Minerals:

Gibbsite (γ-Al(OH)$_3$), which is sometimes referred to as hydrargillite in the European literature, will be discussed here concurrently with brucite (Mg(OH)$_2$). Gibbsite-like and brucite-like components are an important part of clay minerals, particularly in the interlayer regions of 2:1+1 clays. These hydroxide minerals are composed of parallel layers of Al (or Mg) in octahedral coordination with OH$^-$ anions. Figure 4-35A illustrates the brucite structure, which would be referred to as a trioctahedral sheet if present in clays. For dioctahedral clays, a single sheet of the gibbsite structure would be present. Figure 4-35B illustrates the structure of gibbsite. The *c* axis spacing is held by weak H-bonding between the OH groups in adjacent layers. Ideally, the OH groups in each layer are nearly directly opposite to each other in brucite and gibbsite. If the OH layer is designated as either an A or B layer based on its position (Figure 4-35C), then the arrangement of the gibbsite octhahedral sheets results in an AB–BA–AB–BA sequence. The octahedral OH groups are arranged in a slightly polar position in the structure due to the H-bonding along the edges of vacant sites. This results in some distortions of the Al octahedra, where the shared edges are longer than the edges facing the vacant sites. The octahedral layers are somewhat displaced from one another in the direction of the *a* axis.

Bayerite (α-Al(OH)$_3$) will readily form from other Al minerals in the presence of pure water. Some reports indicate that this mineral may revert irreversibly to gibbsite in the presence of alkali metal ions. The bayerite structure is very similar to gibbsite, but the OH layers are stacked in an AB–AB–AB sequence. Relative to gibbsite, the alternating octahedral layer is flipped around so that the lower OH ions of one octahedral layer is in the

Table 4-15: Chemical formulas of various naturally occurring Al minerals.

Bayerite	α-Al(OH)$_3$	Diaspore	α-AlOOH
Nordstrandite	β-Al(OH)$_3$	Boehmite	γ-AlOOH
Gibbsite	γ-Al(OH)$_3$	Corundum	α-Al$_2$O$_3$

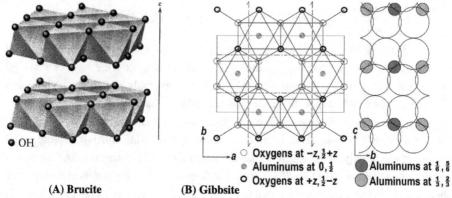

(A) Brucite   **(B) Gibbsite**

Figure 4-35: (A) Structure of brucite ($Mg(OH)_2$) with dimensions $a = 0.314$ nm, and $c = 0.476$ nm; (B) structure of gibbsite (γ-$Al(OH)_3$) with dimensions $a = 0.864$ nm, $b = 0.507$ nm, $c = 0.972$ nm, and $\beta = 94°34'$. Note that the OH groups are stacked directly over each other in these minerals. *For brucite: from Klein and Hurlbut (1993)[8], Manual of mineralogy. 21st ed., copyright © 1993 by John Wiley & Sons. This material is used by permission of John Wiley & Sons, Inc. For gibbsite: adapted from Megaw (1934)[29], with permission.*

same location as the lower OH ions of any other octahedral layer. The OH ions lie in the depressions formed by the three OH ions of the facing octahedral pairs of the adjacent layer, which is consistent with the higher density of bayerite compared with that of gibbsite. Since the orientation of the stacked Al octahedra is not easy to modify, the bayerite structure is not a simple displacive structural change of gibbsite. Crystal dimensions for bayerite are $a = 0.501$ nm and $c = 0.476$ nm. Note that the bayerite unit cell is one octahedral layer deep, while it is two layers deep for gibbsite.

Nordstrandite (β-$Al(OH)_3$) is a mineral with a crystal structure that can be described as a combination of alternating gibbsite and bayerite layering patterns. The stacking of the OH ions follow the sequence AB–AB–BA–BA.

AlOOH Minerals:

Diaspore (α-AlOOH) has each octahedral Al surrounded by three oxygens and three hydroxyls. Double chains of octahedra run continuously up the c axis, with two vacant sites between the chain pairs as viewed from the b axis (Figure 4-36). The two octahedra forming the chain pair are held by shared OH groups. The rows of chain pairs are joined by sharing apical oxygens.

Boehmite (γ-AlOOH) is similar to diaspore but with a somewhat different linkage between the octahedra (Figure 4-37). Each octahedral Al is surrounded by two OH groups and four oxygens. All of the OH groups are located on the edges nearest to the adjacent sheets above and below the corrugated sheet formed by the linked rows. The adjacent sheets are weakly held together by the H-bonding between pairs of oxygens. Boehmite is a lighter mineral than diaspore, as can be expected from this H-bonding and the more open structure resulting from it.

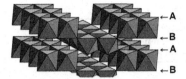

Figure 4-36: Structure of diaspore (α-AlOOH). The double chains run parallel to [001]. *After Clark (1972).*[27]

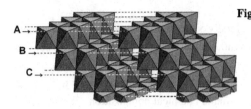

Figure 4-37: Structure of boehmite (γ-AlOOH) with the corrugated sheets parallel to [010]. The octahedral corners are at the O and OH centers. The OH edges in each row are touching the OH edges in the adjacent rows and H-bonding with them. *After Clark (1972).*[27]

Al₂O₃ Minerals — Corundum:

Corundum (α-Al₂O₃) is the most tightly packed of the Al minerals. The mineral consists of vertically stacked sheets of dioctahedral Al, where 2/3 of the octahedra are occupied along each of the axes (Figure 4-38). The octahedra are aligned directly above each other in the c direction.

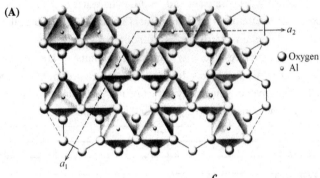

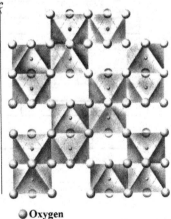

Figure 4-38: (A) Basal sheet of octahedra in corundum, α-Al₂O₃, with one octahedron vacant for every two with Al in the center. (B) Vertical section through the corundum structure showing the locations of filled and empty octahedra. *From Klein and Hurlbut (1993)*[8]*, Manual of mineralogy. 21ˢᵗ ed., copyright © 1993 by John Wiley & Sons. This material is used by permission of John Wiley & Sons, Inc.*

Table 4-16: Chemical formulas of some β-aluminas.

Sodium β-alumina	$Na_2O \cdot 11Al_2O_3$	Calcium β-alumina	$CaO \cdot 6Al_2O_3$
Potassium β-alumina	$K_2O \cdot 11Al_2O_3$	Strontium β-alumina	$SrO \cdot 6Al_2O_3$
Magnesium β-alumina	$MgO \cdot 11Al_2O_3$	Barium β-alumina	$BaO \cdot 6Al_2O_3$

Al_2O_3 Minerals — The β-Alumina Minerals:

In addition to studying the various naturally occurring Al oxides, soil scientists have researched the reactivity of various other synthetic Al minerals. Therefore, this section continues with synthetic products in its brief review of Al minerals.

The β-aluminas, sometimes expressed as $β$-Al_2O_3 minerals, represent a group of minerals having similar spinel structure but variable chemical composition. The chemical formula of various β-aluminas is given in Table 4-16. A detailed review of these minerals was given by Stevens and Binner (1984).[30]

Al_2O_3 Minerals — Activated Al Minerals:

Synthetic Al oxides commonly used by environmental research scientists are activated alumina. Activated alumina minerals are formed when an Al hydroxide or hydroxyoxide is thermally dehydrated at 250–800°C. The water in the mineral is driven off and a highly porous structure of Al oxide remains. Figure 4-39 illustrates the dehydration sequence of various aluminum hydroxides as a function of temperature.

At lower temperatures, various "transition aluminas" are formed. The low-temperature aluminas are formed below dehydration temperatures of 600°C and are known as the γ group. These were once generically designated as γ-aluminas. They have a low order of crystallinity and include the forms ρ-, χ-, η-, and γ-aluminas.

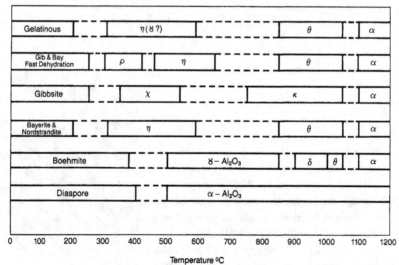

Figure 4-39: Decomposition sequence of Al hydroxides. *Reprinted from Misra (1986)*[28], *copyright © 1986, with permission from the American Chemical Society.*

The high-temperature aluminas are formed at 800 to 1000°C and are known as the δ group. These have a high order of crystallinity with sharper x-ray diffraction patterns and include the forms κ-, δ-, and θ-aluminas. These transition aluminas are differentiated by their x-ray diffraction analysis, and a table of their x-ray diffraction patterns is found in Misra (1986).[28] At 1100°C, the final anhydrous form is corundum (α-Al_2O_3).

The ρ-Al_2O_3 form is amorphous. The χ-Al_2O_3 has a strongly disordered stacking sequence of the layers, and has a highly disordered hexagonal layer structure similar to that of gibbsite. The κ-Al_2O_3 form is nearly anhydrous and also has a hexagonal layer lattice. The κ form does not have any OH ions and is better ordered than the low-temperature χ form, which still contains appreciable amounts of OH ions.

The remaining transition aluminas all have spinel structures. Their chemical formula can be expressed as ranging from $(Al_{0.67}\square_{0.33})Al_2O_4$ to $Al(Al_{1.67}\square_{0.33})O_4$ (refer also to Table 4-14), where the 2.67 Al atoms and 0.33 vacant sites (denoted by $\square$) are distributed on octahedral and tetrahedral positions, and the Al:O ratio of 2/3 is maintained. The greatest disorder is expected for the atoms in the tetrahedral positions.[31] There is no distinct division between γ-aluminas and δ-aluminas because the transformation from the γ to δ phases is gradual and continuous. The γ-Al_2O_3 is a defect spinel with randomly distributed vacancies, while the δ-Al_2O_3 has ordered vacancies on octahedral sites resulting in a spinel superstructure. The η-Al_2O_3 structure is similar to γ-alumina, but with fewer cations in the tetrahedral positions. Accordingly, the η-Al_2O_3 is less dense than the γ-Al_2O_3 phase. The θ-Al_2O_3 structure is a deformed spinel with the Al ions predominantly in tetrahedral positions. Figure 4-40 shows the spinel structure resulting from the dehydration of boehmite, and this spinel structure is similar to that shown in Figure 4-34 except for the orientation of the polyhedra. Again, as mentioned above, not all of the spinel polyhedral positions are filled with Al atoms in these alumina structures.

The formation of corundum (α-Al_2O_3) from these transition aluminas requires a gross reorganization that only becomes kinetically feasible at high temperatures (~1100°C). An exception is the transformation of diaspore (α-AlOOH) to corundum at low temperatures

Figure 4-40: The spinel structure resulting from the dehydration of boehmite (γAlOOH) to an activated alumina (θ-Al_2O_3). The θ-Al_2O_3 mineral is isostructural with β-Ga_2O_3. *After Geller (1960).*[32]

(~500°C), where relatively small rearrangements of oxygen and Al positions are involved and no transition aluminas are formed (compare Figure 4-36 with Figure 4-38).

These activated aluminas, particularly the γ- and δ-Al_2O_3, are studied extensively by soil and environmental scientists. They are attractive to study because they have a very high surface area, they are available commercially, they are generally free of impurities, and they are well characterized. Although these activated aluminas are not found naturally in soils, they will rehydrate into minerals that are more representative of soil constituents. For example, some activated aluminas have been shown to mimic the chemical behavior of the Al sites on kaolinite and smectite clays. The similarity of the chemical reactivity of activated aluminas with those of naturally occurring Al minerals is due in part to their rapid rehydration when in contact with water. Amorphous ρ-alumina will rehydrate rapidly to pseudo-boehmite (particularly when the mixture is hot, with 60% conversion in 4 h at 90°C) and then more slowly to bayerite. Bayerite will not form if the temperature is above 90°C. The γ- and η-aluminas will completely rehydrate to boehmite within 6 h at 250°C, or slowly at room temperature to Al hydroxides. Bayerite forms within three to four weeks from a suspension of γ-Al_2O_3 in pure water at room temperature.

[4.11.2] Fe Oxides

There are about 13 known Fe oxide and hydroxide minerals, and several of these are listed in Table 4-17. When iron rusts, a mixture of FeOOH and maghemite (γ-Fe_2O_3) is formed. Goethite (α-FeOOH) is the most widespread Fe oxide in natural environments.

Solid solutions are easily formed in natural environments because the Fe^{3+} cation is readily replaced by isomorphic substitution by other cations of similar size. Thus, a variety of solid solutions exist, such as pure FeOOH to pure AlOOH. Common partial cation substitutions include Al^{3+}, Mn^{3+}, Cr^{3+}, and many others.

Fe(OH)$_2$ Minerals & Fe(OH)$_3$(am) "Minerals":

Amakinite (Fe(OH)$_2$) forms by precipitation of Fe(II) solutions with alkali under an inert atmosphere. This white precipitate quickly develops a brown color as the iron is oxidized to Fe(III). Although extremely sensitive to oxygen, the Fe(OH)$_2$ structure is very stable and is similar to CdI_2 (which has an AB–AB–AB sequence on the layer stacking) or the brucite (Mg(OH)$_2$) structure illustrated in Figure 4-35A. The structure is maintained even with some oxidation of Fe^{2+} to Fe^{3+}, until the Fe^{3+} content is about 10% of the total iron.[33]

The research literature contains extensive notation of an Fe(III) hydrolysis product.

Table 4-17: Fe oxides, hydroxides, and oxyhydroxides.

Magnetite	Fe_3O_4	Amakinite	Fe(OH)$_2$	Goethite	α-FeOOH
Hematite	α-Fe_2O_3	Ferrihydrite	$Fe_5HO_8 \cdot 4H_2O$	Akaganeite	β-FeOOH
Maghemite	γ-Fe_2O_3	Wüstite	FeO	Lepidocrocite	γ-FeOOH
	β-Fe_2O_3			Feroxyhyte	δ'-FeOOH
	ε-Fe_2O_3				

It is often considered to be amorphous and it is designated by "am" in chemical formulas. The am-$Fe(OH)_3$ formula is typically used as a generic representation of the hydrolysis product of $Fe^{3+} + nH_2O$, such as the product formed from the addition of acids or bases to Fe^{3+} salt solutions. Even when this solid product is not properly characterized with x-ray diffraction, it is generally referred to as $Fe(OH)_3(s)$ or am-$Fe(OH)_3$. As it turns out, the neutral dissolved product of the hydrolysis of Fe^{3+} solutions is $Fe(OH)_3(aq)$, but the precipitated compound will loose its water molecules. The common solids formed are hematite or akaganeite when the method involves acidic solutions. An unstable two-line ferrihydrite forms when the method involves fast alkaline hydrolysis, which later transforms into goethite or hematite, or both.

Limonite ($Fe(OH)_3$) is a generic term for hydrous iron oxides. The precipitated hydrolysis product may be considered amorphous if it shows no x-ray diffraction peak. Keep in mind, however, that very small particles may appear x-ray amorphous but are obviously ordered when viewed through high-resolution electron microscopy. Limonite is no longer considered a soil mineral, and ferrihydrite often may be a more accurate description of what is present in the soil or experimental sample than "limonite" or "am-$Fe(OH)_3$".

FeOOH Minerals:

The α- and γ-FeOOH structures have much H-bonding between the OH–O atoms, and are similar to the corresponding α- and γ-AlOOH structures shown in Figures 4-36 and 4-37. The akaganeite (β-FeOOH) structure is illustrated in Figure 4-41. The 2×1 octahedral tunnels in goethite (α-FeOOH) are only large enough to permit the passage of protons, but the 2×2 akaganeite (β-FeOOH) 0.5×0.5 nm^2 channels can accommodate anions or H_2O molecules. If an anion is in the channels, the electrical neutrality of the unit cell is maintained by the addition of a proton, resulting in $Fe_4O_3(OH)_5 \cdot Cl$, for example.[34]

In addition to viewing the structure of the unit cells, a look at a mineral's superstructure may reveal important physical formations. The presence of channels and pores, for example, will enhance the effective surface area of a mineral and, hence, its relative reactivity with the other elements in the environment. This is illustrated with β-FeOOH in Figure 4-42. Gallagher (1970)[34,35] proposed a superstructure for a single tubular crystal consisting of rings of 16 subcrystals of β-FeOOH, resulting in large square channels

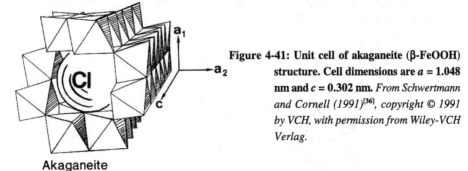

Figure 4-41: Unit cell of akaganeite (β-FeOOH) structure. Cell dimensions are $a = 1.048$ nm and $c = 0.302$ nm. *From Schwertmann and Cornell (1991)[36], copyright © 1991 by VCH, with permission from Wiley-VCH Verlag.*

Akaganeite

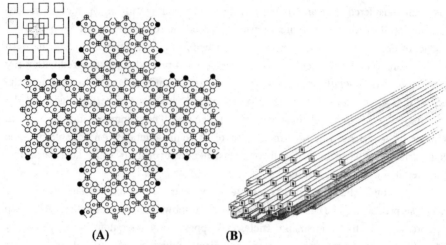

(A)   (B)

Figure 4-42: (A) One unit cell of the idealized superstructure of β-iron(III) oxide hydroxide, $Fe_{128}O_{116}(OH)_{140} \cdot 12(OH)$, in which the numerous small cells shown in Figure 4-41 can be recognized. The inset shows how the replication of the supercell produces the regularly repeating large tunnels within the structure. The A site anions (uncrossed), occupied by OH^- anions, are coordinated to three Fe^{3+} ions in a trigonal pyramid. The B site anions (crossed), occupied by O^{2-} anions, are coordinated to three Fe^{3+} ions in a plane. The surface sites are solid filled. (B) Cigar-shaped crystals of β-FeOOH formed by packing of parallel tubular rods. Each rod cavity is formed by 16 small cells shown in Figure 4-41, or four corners of the larger cell shown in (A). *Figure (A) reproduced from Howe and Gallagher (1975)[35] by permission of The Royal Society of Chemistry. Figure (B) from Gallagher (1970)[34], reprinted with permission from Nature 226:1225–1228, copyright © 1970, Macmillan Publishers Ltd.*

of 3.14 nm on the side. While synthetic β-FeOOH contains large channels, natural akaganeite may lack these large channels and exist in a compressed form.

A synthetic FeOOH that is formed under high pressures and temperatures does not yet have a Greek letter designation, but an incorrect ε designation, as ε-FeOOH, is sometimes used for this mineral.[37] It is formed by hydrothermal conversion of hematite (α-Fe_2O_3) in NaOH at 500°C at 80–90 kbars pressure for one hour.[36] The crystal structure is isostructurally similar to InOOH, which has single chains of octahedra running parallel to the c axis and linked by hydrogen bonds (Figure 4-43). Scant evidence exists for the presence of this mineral in natural environments.[38]

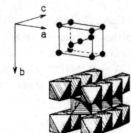

Figure 4-43: Structure of high-pressure FeOOH, which is isostructurally similar to InOOH. *From Pernet et al. (1973)[37], copyright © 1973, reprinted with permission from Elsevier.*

The structure of feroxyhyte (δ'-FeOOH) is not presently well understood. The prime designation (δ' rather than δ) is added to highlight the poorly ordered nature of this mineral. It is believed to be similar to hematite but with a periodicity (or repeating unit) along the c axis of every two octahedra ($c = 0.46$ nm).[36,39]

Fe_2O_3, Fe_3O_4, & FeO Minerals:

Hematite (α-Fe_2O_3) has edge and face-sharing octahedra with a structure that is similar to corundum (α-Al_2O_3) (Figure 4-38). The face sharing and the hexagonal close packing results in a mineral with a high density (5.26 g cm^{-3}). The face sharing also causes some distortion of the octahedra and a regular displacement of the Fe ions.

Another mineral similar to hematite is ilmenite ($FeTiO_3$), which forms a completely solid solution with hematite above 950°C. The Fe and Ti atoms are ordered in alternating layers perpendicular to the c axis in a structure similar to that of hematite and corundum (Figure 4-44). The formula of this group of minerals is more realistically expressed as (Fe, Mg, Mn)TiO_3, where the end member $MgTiO_3$ is geikielite and $MnTiO_3$ is pyrophanite.

Maghemite (γ-Fe_2O_3) has a spinel structure (Figure 4-34), where 1/3 of the interstices are tetrahedrally coordinated with oxygen and 2/3 are octahedrally coordinated. Only 5/6 of the sites are filled with Fe^{3+} ions and 1/6 are vacant ($Fe^{3+}_{2.67}\square_{0.33}O_4$, where $\square$ = vacant sites). The degree of ordering of the vacancies varies widely in maghemite minerals.

The β-Fe_2O_3 mineral will form from the dehydration of akaganeite (β-FeOOH) in high vacuum at 170°C.[36] The dehydration of β-FeOOH to β-Fe_2O_3 results in a 2% contraction of the c axis, but the structure, morphology, and surface area remain the same.[35]

The ε-Fe_2O_3 mineral is formed by reacting alkaline potassium ferricyanide ($K_3Fe(CN)_6$) with sodium hypochlorite (NaOCl).[36] The ε-Fe_2O_3 structure is not known, but it does convert exothermically to hematite (α-Fe_2O_3) at 767°C.[40] These β- and ε-Fe_2O_3 minerals have not been found in nature.

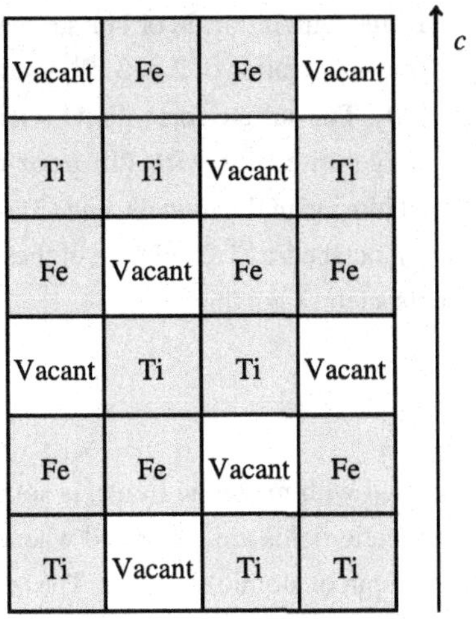

Vacant	Fe	Fe	Vacant
Ti	Ti	Vacant	Ti
Fe	Vacant	Fe	Fe
Vacant	Ti	Ti	Vacant
Fe	Fe	Vacant	Fe
Ti	Vacant	Ti	Ti

c

Figure 4-44: Schematic vertical cross section through the ilmenite structure, which is isostructurally similar to hematite (α-Fe_2O_3) and corundum (α-Al_2O_3). The structure of ilmenite (and of this vertical cross section) is illustrated in Figure 4-38B. *After Klein and Hurlbut (1993).*[8]

Magnetite (Fe_3O_4) has an inverse spinel structure, with the tetrahedral positions completely occupied by Fe^{3+}, while the octahedral positions have equal amounts of Fe^{3+} and Fe^{2+}.

Wüstite (FeO), or feroxite, decomposes to Fe_3O_4 and Fe when the temperature drops below 570°C. It is formed by heating $Fe(OH)_2$ to dryness at 200°C under an N_2 atmosphere. Its partial oxidation results in a defect structure approaching the magnetite (Fe_3O_4) structure.[36] Its chemical formula is better expressed as $Fe^{2+}_{(1-x)}Fe^{3+}_{2x/3}O$.

Ferrihydrite:

The formula of ferrihydrite is not well established, and several formulas have been proposed, namely: $5Fe_2O_3 \cdot 9H_2O$, $Fe_5HO_8 \cdot 4H_2O$, $Fe_6(O_4H_3)_3$, and $Fe_2O_3 \cdot 2FeOOH \cdot 2.6H_2O$. The structure, of course, is also not well understood. Its structure is believed to be similar to hematite but with a periodicity along the c axis every four octahedra ($c = 0.94$ nm).[36,41,42] Ferrihydrites vary in their degree of structural order, and are sometimes named according to the number of broad x-ray peaks that they exhibit (such as two-line and six-line ferrihydrite, where the two-line ferrihydrite is also called protoferrihydrite). Ferrihydrite forms easily with rapid hydrolysis upon raising the pH of Fe(III) solutions. Ferrihydrite will transform to goethite and/or hematite in water at ambient temperatures, but the transformation can be blocked in the presence of adsorbates (such as silicate, phosphate, and organics) or by coprecipitated Al.

Green Rusts:

Blueish-green iron minerals, known as green rusts, contain anions as an essential part of their structural component. They are a family of Fe(II)M(III) hydroxy salts, where M(III) refers to a trivalent metal atom, such Al or Fe(III). The excess positive charge of the trivalent atom is balanced by intercalation of anions between the layers. Depending on the degree of oxidation of the Fe^{2+} atoms to Fe^{3+}, the forms with chloride or sulfate may have Fe^{2+}/Fe^{3+} ratios of up to 4, and the carbonate forms may have ratios of 2 or 3.[36] A suggested formula for the carbonate form is $Fe^{2+}_4 Fe^{3+}_2(OH)_{12}CO_3$. The sulfate form, $Fe_8O_8(OH)_6SO_4$, presumably has a structure that is similar to akaganeite, but with (1) poor crystal development, (2) the presence of sulfate rather than chloride in the tunnels, and (3) some of the sulfate anions sharing oxygens with the $Fe(O,OH)_6$ octahedra.[36] Oxidation of these green rusts will result in the formation of FeOOH products such as goethite.

Magnetism of Fe Minerals:

Magnetism, a group of phenomena associated with magnetic fields, is sometimes involved in the characterization of Fe minerals. Magnetic fields are generated whenever an electric current flows, or by the orbital motion and the spin of atomic electrons. The *magnetic moment* of an atom is the vector sum of the magnetic moments generated by all the orbital motions and spins of the electrons in the atom. If a non-zero net magnetic moment exists, then the individual atoms create magnetic fields around them. The *magnetic susceptibility* of a substance refers to the magnetic field contribution made by the substance when subjected to a magnetic field. There are four main types of magnetic behavior:

A. *Diamagnetism* has magnetization in a direction opposite to that of the applied magnetic field and the susceptibility is negative. All substances are diamagnetic, but this weak form of magnetism is easily masked by other stronger forms of magnetism.

B. *Paramagnetism* has a small positive susceptibility because the net spin magnetic moment of the atoms or molecules are capable of being aligned in the direction of the applied field. It occurs in all atoms and molecules that have unpaired electrons.

C. *Ferromagnetic* substances will remain magnetized after the removal of the applied field. Above a certain temperature, known as the *Curie point* or *Curie temperature*, the ferromagnetic material becomes paramagnetic. Below the Curie point, however, as the external applied magnetic field increases, the magnetization of the substance also increases up to a high value, called the *saturation magnetization*. The saturation magnetization occurs when all the small magnetized regions (about 0.1 to 1 mm across) of the substance, known as domains, become lined up in the direction of the field. If a magnetic field was not applied to the substance, then the ferromagnetic material is not magnetized and the magnetic moments of the domains are not aligned.

D. *Antiferromagnetism* occurs below a certain temperature, known as the Néel temperature, where an ordered array of atomic magnetic moments spontaneously arrange themselves so that alternate moments have opposite directions. There is no magnetic moment observed in the absence of an applied field. Below the Néel temperature, the spontaneous ordering opposes the normal alignment tendencies sought by the applied field. *Ferrimagnetism* is a special form of antiferromagnetism, and occurs when the magnetic moments in one direction are greater than those in the opposite direction. The adjacent ions are antiparallel but of unequal strength. Ferrimagnetic materials of specific magnetic strengths can be synthesized where the antiparallel components are adjusted. An example of ferrimagnetism is exhibited by ferrites, which are ceramic materials that do not conduct electricity and are mixed oxides of $MO \cdot Fe_2O_3$, where M is a metal such as Co, Mn, Ni, or Zn.

In addition to these four main types of magnetic behavior, two more are worth mention here. Occurring under an applied field, *mictomagnetism* refers to the freezing of spin orientations at low temperatures without any long-range spin order, but short-range ordered clusters are present that possess giant moments.[43] Alloys of CuMn or AuFe are known as mictomagnetic alloys or spin glasses. *Sperimagnetic* substances have a randomization of spin arrangements under a longitudinally applied external magnetic field.[43,44] Sperimagnetism should be anticipated in ultrafine particles and amorphous compounds. Table 4-18 describes

Table 4-18: Density and magnetic properties of common iron minerals. *After Schwertmann and Cornell (1991).*[36]

Mineral Name	Formula	Density, g cm^{-3}	Type of Magnetism	Néel Temperature, K
Goethite	α-FeOOH	4.26	antiferromagnetic	400
Akaganeite	β-FeOOH	3.56	antiferromagnetic	295
Lepidocrocite	γ-FeOOH	4.09	antiferromagnetic	77
Feroxyhte	δ'-FeOOH	4.20	ferromagnetic	440
Hematite	α-Fe$_2$O$_3$	5.26	weakly ferromagnetic, or antiferromagnetic	955
Maghemite	γ-Fe$_2$O$_3$	4.87	ferromagnetic	unknown
Magnetite	Fe$_3$O$_4$	5.18	ferromagnetic	850
Ferrihydrite	Fe$_5$HO$_8$·4H$_2$O	3.96	sperimagnetic	25–115

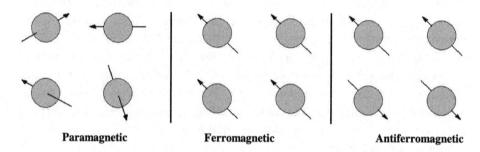

| Paramagnetic | Ferromagnetic | Antiferromagnetic |

Figure 4-45: Illustration of the orientation of atomic moments that result in paramagnetic, ferromagnetic, and antiferromagnetic behavior.

the magnetic behavior of various natural Fe oxides. Figure 4-45 illustrates some of the main magnetic spin structures described above.

[4.11.3] Ti Oxides

Titanium is the ninth most abundant element in the Earth's crust, averaging 0.44%. The amount of Ti in soils is typically less than 1%. However, since these minerals are resistant to weathering, highly weathered soils may be enriched in Ti oxides. The three natural polymorphs of titanium dioxide (TiO$_2$) are rutile, brookite, and anatase. The Ti octahedra vary in the number of edges that are shared: rutile shares two edges, brookite shares three edges, and anatase shares four edges. There is a fair amount of distortion in the polyhedra of these Ti minerals resulting in variable Ti–O bond distances in the polyhedra, such as 0.187 to 0.200 nm for brookite. There can also be a fair amount of substitution in these minerals. As with all minerals, when substitutions occur, electrical neutrality must be maintained. For example, we may have Fe^{2+} or Fe^{3+} substitution for Ti^{4+} with concurrent substitution of Nb^{5+} or Ta^{5+} for Ti^{4+}. Figures 4-46, 4-47, and 4-48 illustrate the structure of these Ti oxide minerals.

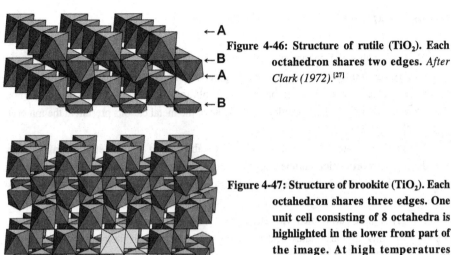

Figure 4-46: Structure of rutile (TiO$_2$). Each octahedron shares two edges. *After Clark (1972).* [27]

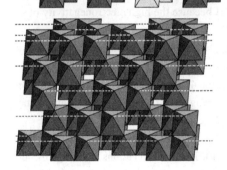

Figure 4-47: Structure of brookite (TiO$_2$). Each octahedron shares three edges. One unit cell consisting of 8 octahedra is highlighted in the lower front part of the image. At high temperatures (about 750°C) brookite will convert to rutile. *After Povarennykh (1972).* [45]

Figure 4-48: Structure of anatase (TiO$_2$). Each octahedron shares four edges. Each row is a sequence of two octahedra followed by two empty spaces, as highlighted by the dashed lines. *After Clark (1972).* [27]

[4.12] Minerals of Important Anions

Several of the minerals discussed below have structures with pronounced *anisodesmic* character, which means that the bonds are of unequal strength. Their opposites are *isodesmic* crystals. Based on the electrostatic valency (e.v.) discussed in Pauling's rule number 2 in Section 4.2, the strength of the bonds within various anionic complexes is higher than the strength of the bond between the anionic complex and the cation. For example, e.v.= 1.33 for the C–O bonds in the CO_3^{2-} anion, which leaves only an e.v.= 0.67 for the bond between the oxygen and cation since the net e.v. of the bonds with oxygen must equal 2.0. The bond with the central carbon atom in the anionic complex is much stronger than the bond between the anionic complex and the cation. This is also true for NO_3^- in nitrates, PO_4^{3-} in phosphates, SO_4^{2-} in sulfates, CrO_4^{2-} in chromates, WO_4^{2-} in tungstates, and AsO_4^{3-} in arsenates to name but a few. Another phenomenon to be aware of is *mesodesmic bonding*, where the anion groups polymerize to form chains, sheets, or networks by sharing oxygen ions. Silicates, discussed above, are the most important example of mesodesmic crystals. Borates are another important group of mesodesmic crystals.

Carbonate Minerals:

Carbonates are common constituents in soils. They are generally found in sedimentary and oxidizing environments. Typically, they are transparent, have average to above average density, and are soluble in acidic solutions. Divergence from these common characteristics is observed as a result of the effects of the metal cations present in the mineral, such as Pb, Cu, Mn, and Fe.

The carbonate (CO_3^{2-}) anion is not very stable, and it tends to strengthen its bonds by reducing its coordination number, such as via $MCO_3 \rightarrow MO + CO_2$. The stable carbonates are those with weak M–O bonds in the polyhedra, such as those with an M element with low valency and low electronegativity, because this allows the C–O bonds to be stronger. Accordingly, simple carbonates with elements of valency three or four are absent in nature. Also rare in soils are carbonates with elements of valency two if their electronegativity is high, such as Cu^{2+}. As a result of this, three-forth of all carbonates will contain additional anions (O^{2-}, OH^-, Cl^-, F^-) or water molecules because that weakens the direct interaction of the cation with the CO_3^{2-} group, and thus helps stabilize the CO_3^{2-} structure. Note also that CO_3^{2-} does not have any replacement of O by OH because the high electronegativity of C would split the OH and form an H bond.

The list of carbonate minerals is very long, but there are three important, structurally different, anhydrous carbonate groups: the calcite group, the aragonite group, and the dolomite group. Examples of these are listed in Table 4-19. The calcite structure is similar to that of the NaCl structure (Figure 1-2), but the triangular CO_3 replaces the Cl anion and the Ca (or other metal in this group) replaces the Na cation (Figure 4-49).

The aragonite group is chemically similar to the calcite group, but structurally the divalent cations are much larger (ionic radii > 0.1 nm). The divalent cations have large metal-to-oxygen distances and form a nine-fold coordination with the nearest oxygens, forming a very distorted cube-octahedron (Figure 4-50). In strontianite, for example, the C–O bond distances are 0.129 nm for two of these bonds and 0.130 nm for the other, while the nine Sr–O bonds average 0.264 nm (one 0.256, two 0.258, two 0.264, two 0.265, and two 0.273 nm).[45] Except for aragonite ($CaCO_3$), the various other aragonite group carbonates are stable at room temperature. The Ca in $CaCO_3$ is stable as a calcite structure at room temperature although it is somewhat large for six-fold coordination, while the Ca in $CaCO_3$ prefers the aragonite structure under high pressure although it is somewhat small for nine-fold coordination. The calcite structure is also preferred over the aragonite structure for $CaCO_3$ as the temperature is increased. Another difference between these two groups is that there is somewhat more solid solution within the calcite group than within the aragonite group.

The dolomite group is structurally similar to calcite but the metal pairs are arranged in alternating layers along the c axis. The large difference in size, say between the Ca and Mg ions, causes cation ordering with the two cations in specific and separate levels in the structure. All the divalent cations are in octahedral coordination, with a mean distance Ca(Mg,Fe)–O_6 of 0.230 nm. Solid solutions between dolomite ($CaMg(CO_3)_2$), calcite ($CaCO_3$), and magnesite ($MgCO_3$) are formed only at elevated temperatures.

Table 4-19: Names and formulas of various carbonate minerals.

Calcite Group:		Aragonite Group:	
Calcite	$CaCO_3$	Aragonite	$CaCO_3$
Gaspeite	$(Ni,Mg,Fe)CO_3$	Cerussite	$PbCO_3$
Magnesite	$MgCO_3$	Strontianite	$SrCO_3$
Otavite	$CdCO_3$	Witherite	$BaCO_3$
Rhodochrosite	$MnCO_3$	**Dolomite Group:**	
Siderite	$FeCO_3$	Ankerite	$Ca(Fe,Mg,Mn)(CO_3)_2$
Smithsonite	$ZnCO_3$	Dolomite	$CaMg(CO_3)_2$
Sphaerocobaltite	$CoCO_3$	Kutnohorite	$Ca(Mn,Mg,Fe)(CO_3)_2$

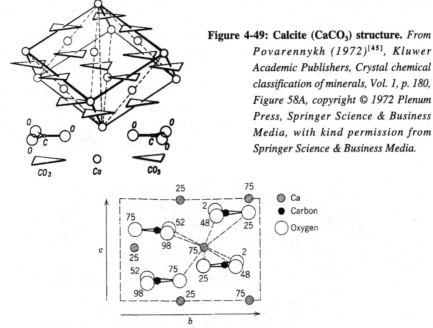

Figure 4-49: Calcite ($CaCO_3$) structure. *From Povarennykh (1972)*[45], *Kluwer Academic Publishers, Crystal chemical classification of minerals, Vol. 1, p. 180, Figure 58A, copyright © 1972 Plenum Press, Springer Science & Business Media, with kind permission from Springer Science & Business Media.*

Figure 4-50: The structure of aragonite, $CaCO_3$, as projected down [100]. Oxygens that would be normally superimposed have been made visible by some displacement. Numbers represent heights of atomic positions above the plane of origin, marked with respect to *a*. The Ca is in nine-fold coordination. *From Klein and Hurlbut (1993)*[8], *Manual of mineralogy. 21st ed., copyright © 1993 by John Wiley & Sons. This material is used by permission of John Wiley & Sons, Inc.*

Nitrates and Borates:

The nitrate minerals are structurally similar to the carbonates. The N–O bonds in the planar triangular NO_3^- groups are held very strongly (e.v. = 1.67 versus e.v. = 1.33 for C–O in CO_3), which is the reason why nitrates are less readily decomposed by acids than carbonates. Nitratite ($NaNO_3$) is isostructural with calcite; niter (KNO_3), also known as saltpeter, is isostructural with aragonite.

Borate mineral units can exist as BO_3 triangles, BO_4 tetrahedra, complex ionic groups, or complex infinite chains of tetrahedra and triangles (such as in colemanite,

$CaB_3O_4(OH)_3 \cdot H_2O)$. Complex ionic groups include $B_3O_3(OH)_5^{2-}$ with one triangle and two tetrahedra, or $B_4O_5(OH)_4^{2-}$ with two triangles and two tetrahedra (such as borax, $Na_2B_4O_5(OH)_4 \cdot 8H_2O$). Boron chemistry is studied extensively for use in glass manufacture. The B–O bond has an e.v. = 1, and the equal sharing of the oxygen atom by the two adjacent boron atoms permits borate triangles to be linked into expanded structural units, such as double triangles, triple rings, sheets, and chains. Most of the common borate minerals are found in arid regions.

Sulfates and Chromates:

The sulfate (SO_4^{2-}) group is tetrahedral and has very strong S–O bonds (e.v. = 1.5) that are covalent; that is, the oxygens are not shared. In barite ($BaSO_4$), each barium ion is coordinated to 12 oxygens belonging to seven different sulfate groups. This structure is also observed in manganates (with tetrahedral MnO_4^{2-}) and chromates (with tetrahedral CrO_4^{2-}) when they also contain large divalent cations. With smaller cations, such as in anhydrite ($CaSO_4$), the cation is in eight-fold coordination. The Ca is also in eight-fold coordination in gypsum ($CaSO_4 \cdot 2H_2O$) (Figure 4-51), which is the most abundant of the hydrous sulfates.

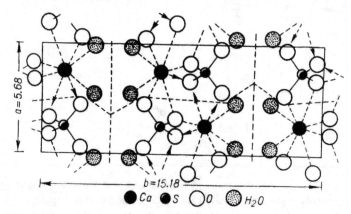

Figure 4-51: The structure of gypsum, $CaSO_4 \cdot 2H_2O$, projected down [001], with layers shown by broken lines. The Ca is in eight-fold coordination: six with oxygens, and two with water molecules. *From Povarennykh (1972)[45], Kluwer Academic Publishers, Crystal chemical classification of minerals, Vol. 2, p. 606, Figure 263, copyright © 1972 Plenum Press, Springer Science & Business Media, with kind permission from Springer Science & Business Media.*

Phosphates, Arsenates, and Vanadates:

Phosphate (PO_4^{3-}), arsenate (AsO_4^{3-}), and vanadate (VO_4^{3-}) are all tetrahedral anionic groups and, because of their similar sizes, can substitute easily with each other. Apatite ($Ca_5(PO_4)_3(OH,F,Cl)$) is the most abundant phosphate. The apatite Ca is in two sites: in an irregular nine-fold coordination and an irregular seven-fold coordination (Figure 4-52). Apatite has extensive solid solution with other anions and cations.

(A)

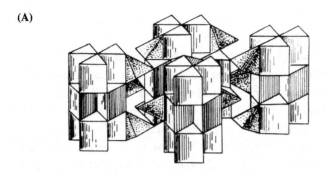

(B)

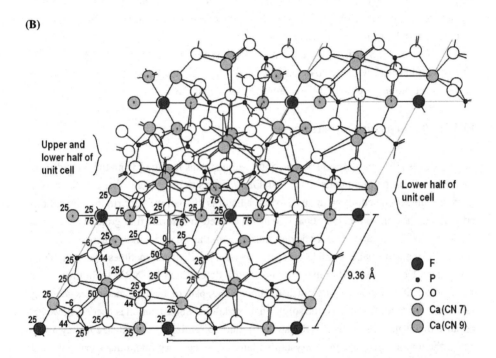

Upper and lower half of unit cell

Lower half of unit cell

9.36 Å

F
P
O
Ca (CN 7)
Ca (CN 9)

Figure 4-52: The structure of fluorapatite, $Ca_5(PO_4)_3F$. (A) In polyhedra, and (B) detail of the structure, where the top left section shows the full unit cell and the remaining three sections only show the lower half of the unit cell. The numbers refer to the location of the atoms in percentages of one c unit; cell dimensions are $a = 9.36$ Å and $c = 6.85$ Å. Notice in (A) the four "towers" at the corners of the structure — these are the four fluoride towers shown in (B). Each F is surrounded by three Ca atoms, and these are rotated by 60° with each $0.5c$ increment. The coordination numbers are 4 for P, and 7 and 9 for Ca. *For (A) from Povarennykh (1972)[45], Kluwer Academic Publishers, Crystal chemical classification of minerals, Vol. 2, p. 542, Figure 243A, copyright © 1972 Plenum Press, Springer Science & Business Media, with kind permission from Springer Science & Business Media. For (B) after Mehmel (1932)[46], Bragg (1937)[47], Chen et al. (2002)[48], and Calderin et al. (2003).[49]*

Sulfides:

The sulfide class of minerals includes the majority of the ore minerals. Using a general formula for the sulfide class as $X_m Z_n$, where X is a metallic element and Z is a nonmetallic element, the sulfide class of minerals include the sulfides (S^{2-}), sulfarsenides (AsS^{2-}), arsenides (As^{3-}), tellurides (Te^{2-}), and several others. Due to the similarity in sizes and charges, all or some of the S in these minerals can be substituted for Se, Te, Sb, As, or Bi. The chemical bonds displayed by these minerals are diverse, where some are ionic and covalent, while others also have metallic bonding characteristics. They also have a broad range of structures that include tetrahedral and octahedral coordination with sulfur or with the nonmetallic element. Most of the recent attention in soil chemistry to sulfides has been on pyrites and their impact in acid mine waste sites.

Galena (PbS) is isostructural with NaCl (Figure 1-2). The pyrite (FeS_2) structure is also similar to NaCl, with the covalently bonded S_2 pairs in the Cl position. Pyrite is easily altered to Fe oxides, typically into the amorphous or hydrous Fe oxides.

[4.13] Amorphous Substances

Crystalline particles display various unique forms resulting from the three dimensional growth of their unit cells. The crystal *habit* refers to the general shape of the crystal (such as cubic, octahedral, prismatic). When a habit is visible, we confirm that the solid is crystalline. The term *microcrystalline* is sometimes used to describe fine-grained aggregates whose crystalline nature can only be determined with the aid of a simple microscope.

For much smaller particles, the term *cryptocrystalline* is sometimes used to describe substances whose crystalline nature can only be determined by XRD techniques. If a substance exhibits a featureless x-ray diffraction (XRD) pattern, then the material is generally said to be amorphous — that is, the solid is not crystalline. Some caution is needed, however, because this is not always true. XRD studies require a minimum crystal size if they are to detect the long-range order in its crystalline lattice. Hence, very fine, micro, crystal forms may be present and observable by other methods, such as high resolution transmission electron microscopy (HRTEM), but remain undetected by XRD. Glasses are examples of true amorphous solids. The term *mineraloids* is also used to describe naturally occurring amorphous substances.

Allophane:

The term *allophane* has been redefined several times since it was first introduced by Stromeyer and Hausmann in 1816 (it was named and analyzed by Stromeyer, but reported by Hausmann)[50] for naturally occurring hydrous aluminosilicates. From the Greek words meaning "to appear" and "other", this term was given in allusion to the frequent changes the material exhibits with loss of water: from a glassy material to an earth appearance. Since then, many materials have been classed with it, and because these materials were amorphous, allophane came to be associated with amorphous materials.

Allophane is x-ray amorphous and electron diffraction amorphous. Allophane has been commonly associated with halloysite. Currently, when this occurs, it is called halloysite-like allophane, with a hypothetical formula of $0.5Al_2O_3 \cdot SiO_2 \cdot 1.4H_2O$ and with the Al in octahedral coordination and the Si in tetrahedral coordination.[25,51] The current consensus is that allophane is a group name for amorphous aluminosilicate gels that are clay size with short-range order and long-range disorder (like glass). It also contains Si, Al, and water, with a Si:Al ratio of around 1:2 to 1:1 ($Al_2O_3 \cdot xSiO_2 \cdot yH_2O$). It also may contain Fe_2O_3 and minor amounts of Mg, Ca, K, and Na. Allophane is one of the more common groups of amorphous materials in soils.[51] Allophane is present in volcanic rock fragments (often referred to as detritus) and in most soils as an intermediate weathering product. (Note: Most dictionaries define detritus as loose material, such as stone fragments and silts, that is typically worn away from rocks; this is the intended meaning in the sentence above. In soil science, however, detritus refers only to the dissolved and particulate dead organic matter.)

Imogolite:

Imogolite is a poorly crystalline mineral with an assumed chemical formula of $Al_2O_3 \cdot SiO_2 \cdot 2.5H_2O$. Using electron microscopy, the mineral appears as threads consisting of assemblies of a tube or hair-like crystal forms. The outside diameter is 3.5 to 5.0 nm, while the hollow internal diameter is approximately 1.0 nm. The structure contains Al only in octahedral coordination, and a defective, gibbsite-like surface is exposed on the external portion of the tubular morphology. It is commonly found in soils derived from volcanic ash, and often in association with allophane. Allophane and imogolite are also found in association with halloysite and other clays of the kaolin group.

References Cited

[1] Brownlow, A.H. 1979. Geochemistry. Prentice-Hall, Englewood Cliffs, NJ.

[2] Ronov, A.B., and A.A. Yaroshevsky. 1969. Chemical composition of the Earth's crust. p. 37–57. *In* P.J. Hart (ed.) The Earth's crust and upper mantle. Am. Geophys. Union Monogr. no. 13. Am. Geophys. Union, Washington, DC.

[3] Novak, G.A., and G.V. Gibbs. 1971. The crystal chemistry of the silicate garnets. Am. Mineral. 56:791–825.

[4] McDonald, W.S., and D.W.J. Cruickshank. 1967. Refinement of the structure of hemimorphite. Z. Kristallogr. 124:180–191.

[5] Dollase, W.A. 1971. Refinement of the crystal structures of epidote, allanite and hancockite. Am. Mineral. 56:447–464.

[6] Dirken, P.J., A.P.M. Kentgens, G.H. Nachtegaal, A.M.J. van der Eerden, and J.B.H. Jansen. 1995. Solid-state MAS NMR study of pentameric aluminosilicate groups with 180° intertetrahedral Al–O–Si angles in zunyite and harkerite. Am. Mineral. 80:39–45.

[7] Furrer, G., C. Ludwig, and P.W. Schindler. 1992. On the chemistry of the Keggin Al_{13} Polymer: I. Acid–base properties. J. Colloid Interface Sci. 149:56–67.

[8] Klein, C., and C.S. Hurlbut, Jr. 1993. Manual of mineralogy. 21st ed. John Wiley & Sons, New York, NY.

[9] Liebau, F. 1959. Über die Kristallstruktur des Pyroxmangits (Mn, Fe, Ca, Mg)SiO$_3$. Acta Crystallogr. 12:177–181.

[10] Veblen, D.R., P.R. Buseck, and C.W. Burnham. 1977. Asbestiform chain silicates: New minerals and structural groups. Science (Washington, DC) 198:359–365.

[11] Prewitt, C.T., and C.W. Burnham. 1966. The crystal structure of jadeite, NaAlSi$_2$O$_6$. Am. Mineral. 51:956–975.

[12] Papike, J.J. 1987. Chemistry of the rock-forming silicates: Ortho, ring, and single-chain structures. Rev. Geophys. 25:1483–1526.

[13] Papike, J.J., M. Ross, and J.R. Clark. 1969. Crystal-chemical characterization of clinoamphiboles based on five new structure refinements. p. 117–136. In J.J. Papike (ed.) Pyroxenes and amphiboles: Crystal chemistry and phase petrology. Mineral. Soc. Am. Spec. Paper no. 2. Mineral. Soc. Am., Washington, DC.

[14] Sparks, D.L. 2003. Environmental soil chemistry. 2nd ed. Academic Press, San Diego, CA.

[15] Yada, K. 1971. Study of microstructure of chrysotile asbestos by high resolution electron microscopy. Acta Crystallogr. A27:659–664.

[16] Kunze, G. 1956. Die gewellte Struktur des Antigorits, I. Z. Kristallogr. 108:82–107.

[17] Giese, R.F. 1988. Kaolin minerals: Structures and stabilities. p. 29–66. In S.W. Bailey (ed.) Hydrous phyllosilicates (exclusive of micas). Rev. Mineral., Vol. 19. Mineral. Soc. Am., Washington, DC.

[18] Moore, D.M., and R.C. Reynolds, Jr. 1989. X-ray diffraction and the identification and analysis of clay minerals. Oxford Univ. Press, New York, NY.

[19] Bailey, S.W. 1988. Polytypism of 1:1 layer silicates. p. 9–27. In S.W. Bailey (ed.) Hydrous phyllosilicates (exclusive of micas). Rev. Mineral., Vol. 19. Mineral. Soc. Am., Washington, DC.

[20] Deer, W.A., R.A. Howie, and J. Zussman. 1992. An introduction to the rock-forming minerals. 2nd ed. Longman Group, Essex, UK.

[21] Papike, J.J., and M. Cameron. 1976. Crystal chemistry of silicate minerals of geophysical interest. Rev. Geophys. Space Phys. 14:37–80.

[22] Papike, J.J. 1988. Chemistry of the rock-forming silicates: Multiple-chain, sheet, and framework structures. Rev. Geophys. 26:407–444.

[23] Loewenstein, W. 1954. The distribution of aluminum in the tetrahedra of silicates and aluminates. Am. Mineral. 39:92–96.

[24] Catlow, C.R.A. 1992. Zeolites: Structure, synthesis and properties — An introduction. p. 1–17. In C.R.A. Catlow (ed.) Modelling of structure and reactivity in zeolites. Academic Press, New York, NY.

[25] Tan, K.H. 1998. Principles of soil chemistry. 3rd Ed. Marcel Dekker, New York, NY.

[26] Náray-Szabó, St.V. 1943. Der Strukturtyp des Perowskits (CaTiO$_3$). Naturwissenschaften 31: 202–203.

[27] Clark, G.M. 1972. The structures of non-molecular solids. John Wiley & Sons, New York, NY.

[28] Misra, C. 1986. Industrial alumina chemicals. ACS Monogr. 184. Am. Chem. Soc., Washington, DC.

[29] Megaw, H.D. 1934. The crystal structure of hydrargillite, Al(OH)$_3$. Z. Kristallogr. 87:185–204.

[30] Stevens, R., and J.G.P. Binner. 1984. Review: Structure, properties and production of β-alumina. J. Material Sci. 19:695–715.

[31] Lippens, B.C., and J.H. de Boer. 1964. Study of phase transformations during calcination of aluminum hydroxides by selected area electron diffraction. Acta Crystallogr. 17:1312–1321.

[32] Geller, S. 1960. Crystal structure of β-Ga_2O_3. J. Chem. Phys. 33:676–684.

[33] Bernal, J.D., D.R. Dasgupta, and A.L. Mackay. 1959. The oxides and hydroxides of iron and their structural inter-relationships. Clay Mineral. Bull. 4:15–30.

[34] Gallagher, K.J. 1970. The atomic structure of tubular subcrystals of β-iron(III) oxide hydroxide. Nature (London) 226:1225–1228.

[35] Howe, A.T., and K.J. Gallagher. 1975. Mössbauer studies in the colloid system β-FeOOH-β-Fe_2O_3: Structures and dehydration mechanism. J. Chem. Soc. Faraday Trans. I, 22–34.

[36] Schwertmann, U., and R.M. Cornell. 1991. Iron oxides in the laboratory: Preparation and characterization. VCH, New York, NY.

[37] Pernet, M., J. Chenavas, J.C. Joubert, C. Meyer, and Y. Gros. 1973. Caracterisation et etude par effet Mössbauer d'une nouvelle variete haute pression de FeOOH. Solid State Commun. 13:1147–1154.

[38] Ostwald, J. 1985. An occurrence of ε-FeOOH in a black shale. Mineral. Mag. 49:139–140.

[39] Drits, V.A., B.A. Sakharov, and A. Manceau. 1993. Structure of feroxyhite as determined by simulation of x-ray diffraction curves. Clay Minerals 28:209–222.

[40] Dézsi, I., and J.M.D. Coey. 1973. Magnetic and thermal properties of ε-Fe_2O_3. Physica Status Solidi, A: Appl. Res. 15:681–685.

[41] Manceau, A., and V.A. Drits. 1993. Local structure of ferrihydrite and feroxyhite by EXAFS spectroscopy. Clay Minerals 28:165–184.

[42] Drits, V.A., B.A. Sakharov, and A. Manceau. 1993. Structural model of ferrihydrite. Clay Minerals 28:185–207.

[43] Coey, J.M.D., and P.W. Readman. 1973. New spin structure in an amorphous ferric gel. Nature (London) 246:476–478.

[44] Murad, E., L.H. Bowen, G.J. Long, and T.G. Quin. 1988. The influence of crystallinity on magnetic ordering in natural ferrihydrites. Clay Minerals 23:161–173.

[45] Povarennykh, A.S. 1972. Crystal chemical classification of minerals. Vol. 1 and 2. Transl. from Russian by J.E.S. Bradley. Plenum Press, New York, NY.

[46] Mehmel, M. 1932. Beziehungen zwischen Kristallstruktur und chemischer Formel des Apatits. Z. Physikalische Chemie 15B:223–241.

[47] Bragg, W.L. 1937. Atomic structure of minerals. Cornell Univ. Press, Ithaca, NY.

[48] Chen, N., Y. Pan, and J.A. Weil. 2002. Electron paramagnetic resonance spectroscopy study of synthetic fluorapatite: Part I. Local structural environment and substitutional mechanism of Gd^{3+} at the Ca2 site. Am. Mineral. 87:37–46.

[49] Calderin, L., M.J. Stott, and A Rubio. 2003. Electronic and crystallographic structure of apatites. Phys. Rev. B 67:134106 (7 pages).

[50] Hausmann, J.F.L., and F. Stromeyer. 1816. Göttingische Gelehrte Anzeigen 2:1251-1253.

[51] Bohn, H.L., B.L. McNeal, and G.A. O'Connor. 1985. Soil chemistry. 2nd ed. John Wiley & Sons, New York, NY.

Questions

1. Section 4.4 lists the percentage by weight of the most common elements in the Earth's crust. Convert these percentages to atom percentage and volume percentage.

    Answer:

	(A)	(B)	(C)	(E)	(F)	(G)	(I)
		Atomic		**Atom % =**	Ionic Radius		**Volume % =**
Element	**Weight, %**	Weight	$A \div B$	$C \div D \times 100$	from Table 4-1	$E \times F$	$G \div H \times 100$
O	46.60	16	2.9125	62.6	1.32	82.632	79.4
Si	27.72	28.1	0.9865	21.2	0.34	7.208	7.9
Al	8.13	27	0.3011	6.5	0.61	3.965	3.8
Fe	5.00	55.8	0.0896	1.9	0.73	1.387	1.3
Ca	3.63	40.1	0.0905	1.9	1.08	2.052	2.0
Na	2.83	23	0.1230	2.6	1.24	3.224	3.1
K	2.59	39.1	0.0662	1.4	1.59	2.226	2.1
Mg	2.09	24.3	0.0860	1.8	0.80	1.440	1.4
Sum = 98.59		Sum (D) = 4.6554			Sum (H) = 104.134		

Note that the volume percentage estimates will vary based on the ionic radius values used in the estimate. Some published volume percentage estimates for oxygen are as high as 94%, suggesting that other properties need to be considered, such as bulk density or pore spaces. Sums of Columns E and I are not exactly 100% due to rounding errors. Column A does not equal 100% because not all the elements in the Earth's crust are represented in the list. The values in Columns E and I will change slightly if all of the elements in the Earth's crust are included in the list, but the change will be minor since the present list already represents most of the crust by weight. The atom percentage of hydrogen atoms and other elements in soils (that is, in the surface and near surface regions of the Earth's crust) can be quite large, resulting in significant changes to the values presented here. Additional studies on the typical distribution of elements in soils are still needed.

2. Why is the hydrogen cation, H^+, not illustrated in Figure 4-1?

3. What is the difference between [100] and (100)?
    Answer: In the Miller indices of minerals, the direction of the axes that are hugging the sides of the crystal unit cell are denoted by brackets. Hence, the a,b,c directions of the unit cell are expressed numerically by $[h,k,\ell]$. This also applies to the Bravais–Miller indices, where the a_1,a_2,a_3,c directions of the unit cell are expressed numerically by $[h,k,i,\ell]$. Use parentheses when referring to the face of a mineral. Hence, a view of the (100) face is a view that is down the [100] direction and the drawing shows the b and c axes.

4. Why do vermiculites "pop" under heat?

5. Compare the morphology and unit cell size of 1:1 and 2:1 clays that have tetrahedral inversion, such as those illustrated in Figure 4-15 and 4-19.

6.   How can the platy habit of gibbsite or the foliation of brucite (which is not an elastic one like that of mica) be explained?

7.   Although it may appear to some readers that this chapter on mineralogy covered all the aspects of mineralogy that are important to soil chemists, it really only scratched the surface. The field of mineralogy is very, very extensive indeed, and this chapter tried to bring only a portion of it to you in a somewhat simplified manner. To understand the validity of this statement, research the literature on a specific mineral or class of minerals and present a one- or two-page summary of your findings. You will quickly see that, regardless of the mineral you choose, there is much to be said about it.

Chapter 5:

Surface Characteristics & Analysis

In many applied environmental science studies, the primary use of mineralogy is to describe the type of minerals present. The emphasis is generally placed only on the internal structure of the minerals studied. For soil reaction studies, some applied environmentalists prefer to describe the soil reactivity on a per-mass unit basis (such as moles per kilogram of soil) because it can be easily converted as needed into other units used in general soil management practices. However, this per-mass unit basis can be misleading because the size of the inert internal structure of the minerals present are not constant in nature. The size of soil particles is affected by various factors, such as temperature, moisture regime, and pH.

Conversely, soil reactivity on a per-surface area basis (such as moles per square meter of particle surface area present), focuses only on the reactivity of the mineral surfaces present. In soils, nearly all of the chemical reactions that involve minerals are controlled by their surface physical and chemical characteristics. Accordingly, basic environmental chemists, geochemists, and soil chemists are primarily concerned with the solid–liquid or solid–air interface properties and reactions of minerals. The first two sections in this chapter discuss the general mineralogical properties of mineral surfaces. Subsequent sections discuss various important measurements and parameters that can be obtained for mineral surfaces.

[5.1] Physical Properties of "Clean" Mineral Surfaces

The elements on the surface of a mineral are not exposed to the same stresses and forces that are exerted on the elements buried inside the mineral. This is an obvious result of the fact that the elements on the surface do not have neighboring elements in the direction away from the mineral, while the elements inside the mineral have neighboring elements in all directions. This very often results in a surface mineral structure that differs from the bulk mineral structure.

If the symmetry of the surface is the same as that of the bulk material, then the surface is said to be *relaxed*. The surface is said to be *reconstructed* when its symmetry is different from that of the bulk material as a result of a surface *relaxation* process. The terms *unreconstructed surface* or *ideal surface* can be used when referring to a surface that can relax but the required relaxation process has not yet occurred, or when the reconstructed structure is not the one being illustrated in the drawing. Reconstruction may include bond elongation, bond angle changes, changes in centering, and rotation of surface angles. A *dangling bond charge density* refers to the charge density remaining at the surface from the bonds that normally were joined with neighbors that were present in the bulk material but are now missing at the surface (the bond is said to be "dangling"). Electrons are sometimes described based on their electronic state (such as ground state or excited state). The dangling bond charge density is localized at the surface in electronic states known as *surface states*.

As summarized by Gibson and LaFemina (1996)[1], there are five principles that can be used to qualitatively understand the relaxations and reconstructions of mineral surfaces:

Principle 1: Stable surfaces are autocompensated. That is, a stable surface is able to remain charge neutral by transferring the electrons to induce neutrality. The electrons are transferred from the cation-derived dangling bonds to the anion-derived dangling bonds, and there is an equal number of each on the surface for this to occur. The bonding electron orbitals (or valence bonds) associated with the anions (which are the most electronegative atoms) are completely filled, while the antibonding orbitals (or conduction bands) associated with the cations (which are the most electropositive atoms) are completely emptied. A surface that is not autocompensated is thermodynamically unstable and will eventually reconstruct in order to become autocompensated.

Principle 2: The dangling bond charge density is rehybridized. This determines the nature of the surface relaxation or reconstruction. The reconstruction of a stable stoichiometry is a function of the energy that is gained by rehybridizing the surface dangling bond charge density in response to the reduced atomic coordination at the surface.

Principle 3: An insulating surface is formed The pairing of the electrons in the dangling bonds opens up a gap between the occupied and unoccupied states, forming an insulating surface. This lowers the net surface energy by stabilizing the occupied surface states while destabilizing the unoccupied surface states.

Principle 4: Near-neighbor bond lengths are conserved. While the formation of local strain fields are energetically unfavorable, bond compression or stretching requires more force than bond bending by as much as one to two orders of magnitude. Conserving the near-neighbor bond lengths in surface relaxations and reconstructions minimizes their elastic energy costs. The elastic energy cost of distorting the local bonding environment must be balanced with the energy lowering from the rehybridization of the dangling bond charge density and the formation of an insulating surface. The topology, or atomic connectivity, of the surface will influence the ease of surface relaxations and reconstructions. Large atomic motions are more likely to occur if the surface topology allows for bond-length-conserving motions of the surface atoms.

Principle 5: Kinetics are important. The surface observed is not necessarily the most stable surface thermodynamically because the activation energy to relaxation or reconstruction may be less than the energy supplied by the cleavage process. A more stable and irreversible surface structure may sometimes require high temperatures to initiate rehybridization of the surface dangling bond charge density. Regardless of the surface observed, it must be autocompensated.

It is fascinating to note that the rehybridization of the dangling bonds (Principle 2), a concept that was originally developed for covalent bonds, is also applicable to ionic substances. As was mentioned in Section 1.3, there is no sharp boundary between covalent and ionic bonding because this classification is neither a complete nor a definitive description of bonds. Traditional views of ionic surfaces have focused on the cleavage process as merely the separation of ions and its properties as being controlled solely by classical electrostatics, resulting in nonpolar surfaces of ionic crystals that are unrelaxed and unreconstructed. Electrostatic considerations successfully predict the surface relaxation when the topology of the surface precludes any large bond-length-conserving motions, but fails when the surface topology allows it to undergo large atomic motion relaxations and redistribution of the dangling bond charge density at the surface.[1] The five principles on generalized autocompensation framework outlined above will, however, successfully describe these two diverse surfaces. In this generalized framework, the bulk atoms seek to completely fill the valence band and completely empty the conduction band, and the surface atoms seek to fill the surface anion dangling bonds and empty the surface cation dangling bonds.

The best way to explain the five principles outlined above is through examples. Hence, Principles 1 to 4 are reviewed below with the minerals zincblende, rutile, corundum, and perovskite. Note that each mineral typically has several different types of surfaces or planes of cleavage, which in turn results in a diverse set of surface termination symmetries.

Zincblende (ZnS):

The zincblende mineral and several of its surfaces are illustrated in Figure 5-1. Inside the mineral, the Zn atoms offer 1/2 electron per bond (note, Zn has two valence electrons for four bonds, and e.v. = 2/4 = 1/2), while the S atoms offer 3/2 electrons per bond (note, S has six valence electrons for four bonds, and e.v. = 6/4 = 3/2). These are stable bonds with 2 electrons each (1/2 + 3/2 = 2). With cleavage of the mineral along the (110) surface, notice that both the cations and anions went from a four-fold to a three-fold coordination, resulting in dangling bonds. Applying Principle 1, however, this surface is autocompensated, charge neutral, and stable by transferring the 1/2 dangling electrons on Zn to the 3/2 dangling electrons on S. The surface is also insulating due to the energy gap that exists between the anion- and cation-derived surface states (Principle 3).

For comparison, Figure 5-1 also illustrates a zincblende mineral with a cleavage along the (111) plane, which results in a surface that is not autocompensated, and remains unstable. Here, notice that one type of surface ion (say, cations for example) went from a four-fold to a two-fold or three-fold coordination, while the other type (say, the anions for example) remained in four-fold coordination. The transfer of dangling electrons is not possible on this surface. The extra electron density on the surface ions form a conduction band character on the surface. From the point of view of a classic electrostatic point ion model, the (111) zincblende surface is charged and is referred to as a *polar surface*. Similarly, the (100) zincblende surface shown in Figure 5-1 is also unstable and not autocompensated.

(110) Surface   **(111) Surface**   **(100) Surface**

Side View:

Top View:

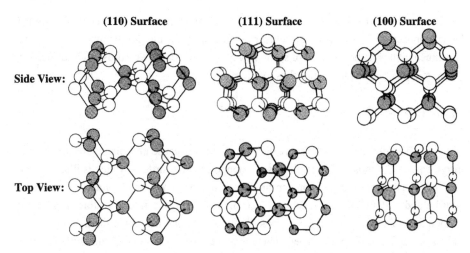

Figure 5-1: The (110), (111), and (100) unreconstructed surface structures of zincblende (ZnS). For the side view of each surface, the surface is drawn facing up toward the top of the page. For the top view of each surface, the surface is drawn facing the reader. *From Duke (1988)[2], with permission.*

The (110) zincblende surface can further lower its energy by rehybridizing the dangling bond charge density (Principle 2). As is illustrated in Figure 5-2, the surface cation bonds rehybridize from sp^3 (tetrahedral conformation) to sp^2 (trigonal planar conformation), and the surface anion bonds rehybridize from sp^3 to p^3 (trigonal pyramidal conformation). This change in local conformation can occur because the resulting change in surface structure does not significantly distort the near-neighbor bond lengths (Principle 4). The surface cations move down toward the bulk crystal and the surface anions move up and out of the surface plane.

In Figure 5-2, a theoretical tilt of about 35° at the surface of the zincblende (110) structure would result in a surface cation that is in a perfect trigonal planar conformation while precisely conserving near-neighbor bond lengths. Experimental observations estimate the tilt to be 29 ± 3°, suggesting that the cation does not move into a perfect trigonal planar conformation and that the near-neighbor bond lengths are not rigorously conserved.[1] An

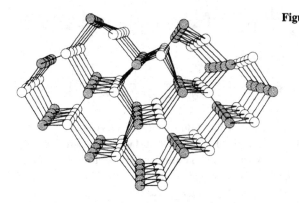

Figure 5-2: Side view of the reconstructed mineral cleavage along the zincblende (110) plane following rehybridization of the surface ions. The (110) surface is drawn facing up toward the top of the page. Anions = open circles, cations = solid circles. *From Duke (1988)[2], with permission. See also Duke and Wang (1988).[3]*

important point to note here is that the final surface structure is predominantly determined by the connectivity of the surface (that is, by its topology), while the chemical nature of the constituent atoms in the lattice is of secondary importance. In other words, each surface has a relaxation that is characteristic of its topology.

Rutile (TiO₂):

The rutile crystal surfaces are difficult to study because this mineral fractures rather than cleaves. The most stable surface is the nonpolar (110) surface, which is illustrated in Figure 5-3. The surface is autocompensated because the nonbonding orbital of the two-fold coordinated surface O atom receives electrons from the five-fold coordinated Ti atom (Principle 1). This surface symmetry does not allow any bond-length-conserving motions for the surface atoms. Only motions perpendicular to the surface are allowed. Furthermore, these surface atoms will not rehybridize because here it would result in too much bond compression (Principle 4). Accordingly, some rumpling of the surface atomic layer and smaller counterrumples of the subsurface layers are predicted.[1]

The rutile (100) surface (Figure 5-4A) is autocompensated when the bridging oxygens are present at the surface (Principle 1). These surface atoms can move in bond-length-conserving motions (Principle 4). The surface Ti atoms are predicted to have an inward relaxation of 0.07 nm but negligibly small relaxations for the O atoms, while the equatorial plane of the surface octahedra with the surface normal is reduced from 45 to 30° in this relaxed structure.[1] Although it is energetically unfavorable for the (100) surface to form (110) facets, the rutile (100) surface can also have a (110) microfaceted surface (Figure 5-4B), and these features are certainly not desired whenever we wish to study the reconstruction mechanism of these surfaces.

(110) Ideal Structure   **(110) Reconstructed Structure**

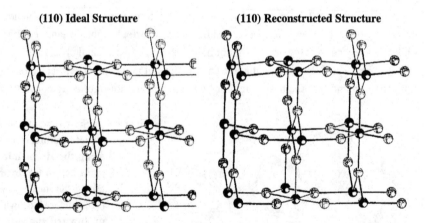

Figure 5-3: Ideal and reconstructed side views of the (110) surface structure of rutile (TiO₂), which is isostructural with cassiterite (SnO₂). The (110) surfaces are drawn facing up toward the top of the page. Dark circles are Ti atoms, light circles are O atoms. *Reprinted figure with permission from Godin and LaFemina (1993)[4], Phys. Rev. B 47:6518–6523. Copyright © 1993 by the American Physical Society.*

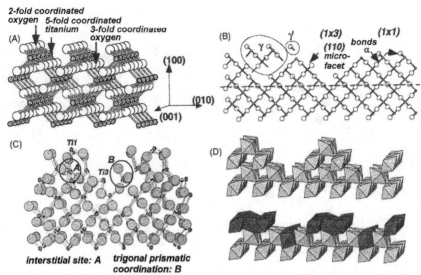

Figure 5-4: (A) Surface structure of the unreconstructed rutile (TiO$_2$) (100)-(1×1) surface, which results when cleavage is along the dashed line shown in (B) with same number of Ti→O as O→Ti bonds broken. Note the surface is strongly corrugated. Large open circles represent the O atoms. (B) Surface structure of the (1×3) microfaceted surface, which results from removal of the volume labeled γ. The same number of Ti→O as O→Ti bonds are broken, again resulting in a stoichiometric surface. (C) A more complex model for the microfaceted (1×3) surface: γ′ O atoms not removed, Ti1 = three-fold coordinated Ti in facet ridges, Ti3 = Ti in an O bridge site, and sites A and B as indicated. (D) The octahedral model of the (1×3) microfaceted surface (top), and an alternate model (bottom) for this (100) surface showing extensive restructuring of the surface where the Ti atoms reside in edge- and corner-sharing octahedral units. *From Diebold (2003)*[5], *copyright © 2003, reprinted with permission from Elsevier. Figure* **(D)** *from Landree et al. (1998)*[6], *copyright © 1998, reprinted with permission from Elsevier.*

In theory, none of the 63 possible rutile (111) surfaces, of which one is shown in Figure 5-5A, are stable. However, experiments have shown that a stable (111) surface does exist.[7] Since the surfaces are prepared in an oxygen-rich environment, the addition of an O atom to the surface is possible. The proposed O adatom is shown in Figure 5-5B, and is the most likely structure for achieving a stable (111) surface on rutile. This autocompensated surface does not allow approximately bond-length-conserving motions of the surface atoms. Accordingly, the autocompensated structure consists of small relaxations of the surface atoms.

Corundum (α-Al$_2$O$_3$):

The surface termination details of the corundum surfaces are based predominantly on theoretical studies because none of the surfaces of corundum is a single cleavage surface. Figure 5-6A illustrates the corundum (0001) surface with various options shown for the location of two unique cleavage planes (labeled A and B), where the cleavage between the Al atoms (labeled A) is estimated to be the most stable by 8 eV per unit cell over the other

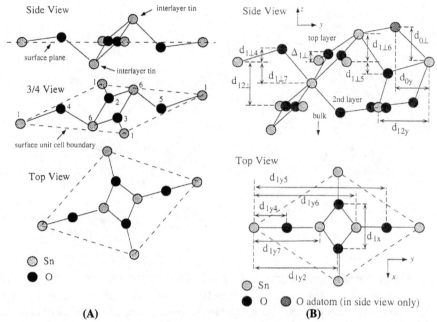

Figure 5-5: Structure of cassiterite (SnO$_2$) (111) surface. Rutile (TiO$_2$) is isomorphic with cassiterite. (A) An ideal (truncated bulk) unrelaxed (111) composite atomic layer of surface shown from three angles and with a bridging cation (no. 6) present. The primitive unit cell of an ideal surface termination is denoted by dashed lines. (B) A relaxed, autocompensated (111) surface following the addition of an oxygen adatom. *From Godin and LaFemina (1994)[7], copyright © 1994, reprinted with permission from Elsevier.*

(0001) termination.[1] This cleavage plane produces two charge neutral and autocompensated surfaces, following the transferring of electrons from the Al atoms to the O atoms (Principle 1). Both Al and O are six-fold coordinated in the bulk, but three-fold coordinated at the surface.

Similarly, the more stable corundum (1$\bar{1}$02) surface occurs with cleavage between the O atoms (8 eV more stable, labeled A in Figure 5-6B). The two resulting (1$\bar{1}$02) surfaces are also charge neutral and autocompensated (Principle 1), with the surface Al being five-fold coordinated and the surface O two-fold coordinated.

The corundum (1$\bar{1}$02) surface topology does not offer any bond-length-conserving motions of the surface atoms for possible relaxations or reconstructions. Conversely, the corundum (0001) surface topology, which has both Al and O atoms in three-fold coordination, does allow for bond-length-conserving motions (Principle 4). Illustrated in Figure 5-7, the Al atoms move down toward the mineral into a nearly trigonal planar conformation (sp^2 hybrid), and the O atoms move away from the mineral into a distorted trigonal pyramidal conformation (p^3 hybrid) (Principle 2). The surface Al atoms relax to new positions 0.04 to 0.07 nm away from their truncated bulk positions, while the surface O atoms are displaced approximately 0.02 nm.[1] This relaxation is better optimized for the Al atoms relative to the O atoms.

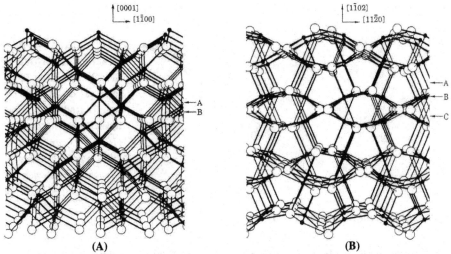

Figure 5-6: Side view of the surface structure of sapphire (the undoped crystal of corundum, α-Al$_2$O$_3$): (A) (0001) surface, (B) ($1\bar{1}02$) surface. Solid filled circles = Al, open circles = O. The arrows labeled A, B and C indicate the various (0001) and ($1\bar{1}02$) cleaving planes. *Reprinted figure with permission from Guo et al. (1992)[8], Phys. Rev. B 45:13647–13656. Copyright © 1992 by the American Physical Society.*

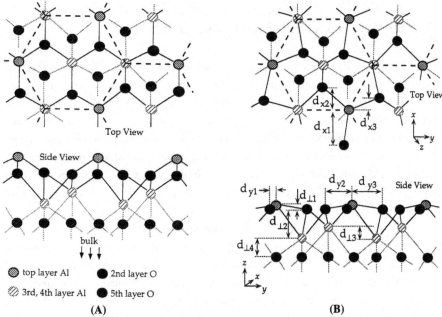

Figure 5-7: (A) Unrelaxed and (B) relaxed surface structure of corundum (α-Al$_2$O$_3$) (0001). The boundaries of the surface unit cell are shown by dashed lines in the top view. Subsurface atoms and bonds are shaded. In the relaxed structure, the surface shows a large, bond-length-conserving displacement of the surface atoms from bulk positions, which allows the surface Al atoms to rehybridize to a nearly perfect *sp^2* configuration. *Reprinted figure with permission from Godin and LaFemina (1994)[9], Phys. Rev. B 49:7691–7696. Copyright © 1994 by the American Physical Society.*

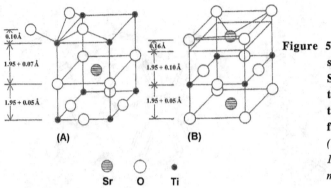

Figure 5-8: Relaxed surface structure of perovskite SrTiO₃ for the (A) TiO₂-terminated and (B) SrO-terminated (001) surfaces. *From Hikita et al. (1993)[10], copyright © 1993, reprinted with permission from Elsevier.*

Perovskite group minerals (XYO₃):

Illustrated in Figure 5-8, the topology of the (100) perovskite surface does not permit motions that will approximately conserve bond lengths. The symmetry of the surface further restricts the motion to be perpendicular to the surface plane. A small surface rumpling is generally observed in these minerals, where the cations move toward the center of the mineral and the anions move out from the surface.

[5.2]   Physical Properties of Real Mineral Surfaces

There is a big difference between "clean" and real (or "dirty") surfaces. Mineral surfaces are typically hydroxylated or hydrated when they are exposed to humid air or come in contact with water. The structure of water near mineral surfaces was discussed briefly in Section 2.3. The presence of water also affects the physicochemical nature of the mineral surface itself. The surface of an oxide or hydroxide mineral will often change when exposed to water primarily because of the chemisorption of water. The surface oxygens plus the chemisorbed H_2O molecules rearrange to form surface–OH groups.[11] Some terms used in reference to these surface–OH groups include *aluminol* groups for AlOH, *ferrol* groups for FeOH, and *silanol* groups for SiOH.

Figure 5-9 illustrates the dry and hydrated surface of γ-Al_2O_3. These hydrated surface–OH sites are highly reactive toward other ions and molecules that may also be present in the liquid phase. This physicochemical change of the dry surface to a hydrated surface is common to all the oxide, hydroxide, and oxyhydroxide minerals. The kinetics of this reaction is fairly rapid, and it is an important first step in the weathering of all minerals.

Since soil environments have extensive exposure to water and the relative humidity in soils is often very high, there is much research interest on the impact of hydrated mineral surfaces on the fate of soil nutrients and contaminants. Not surprisingly, many soil chemists work with soils (or soil constituents) in the form of pastes, slurries, or solids suspended in liquids.

There is a general belief that the surface reactivity of minerals varies with the exposed surface in question. For example, the surface–OH sites on the basal plane may differ in density (that is, sites per square nanometer) and reactivity relative to the particle's edge surface–OH groups. One difficulty in understanding hydrated mineral surfaces arises from the fact that they are not always stable in water and rapidly transform to new phases. For example, the surface of γ-Al_2O_3 will transform to bayerite within 30 days when stored in a water suspension at room temperature.[12] Furthermore, some surface cations and anions may undergo reduction and oxidation reactions while the bulk of the mineral remains unchanged. Adsorption of ions, carbon dioxide, and other molecules will also change the character of the mineral surface.

Brown (1990)[13] noted that atomic adsorption on clean metal oxides generally takes place preferentially at sites of high symmetry and high coordination to metal atoms, and that adsorption tends to cause expansion of underlying metal–metal bonds in cases where these bonds are already shortened on the clean surface. Unfortunately, cleaning a real surface in an ultra-high-vacuum system for surface studies is not a trivial task for it may lead to other complications. High-temperature annealing and ion bombardment may lead to reconstruction of the surface and/or a change in its composition. Cleavage may result in the release of dissolved gas. Dry crushing may result in the aggregation of particles.

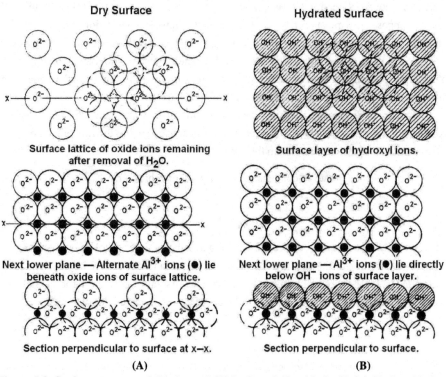

Figure 5-9: Surface structure of (A) dry and (B) hydrated Al oxide (γ-Al_2O_3). Each oxide ion occupies an area of about 0.08 nm^2 on the (100) plane of this mineral. *Reprinted from Peri (1965)[14], copyright © 1965, with permission from the American Chemical Society.*

[5.3]   BET Surface Area Determinations

The amount of a given solid surface that is in contact with the soil solution clearly impacts the magnitude of the various effects of the solid on the soil matrix. Hence, a quantitative measure of the surface area present is imperative to an understanding of the various dynamic reactions present in the soil at any given time. There are essentially two popular methods of determining surface areas: one based on the adsorption of a gas (the BET method), the other on the adsorption of a volatile liquid (e.g., the EGME method and others discussed in subsequent sections).

In 1938, Brunauer, Emmett, and Teller described an analytical technique for determining the surface area of dry solid samples.[15] The method is now known as the BET method, where the letters denote each of their names. Their contribution was so significant that it should be extremely well understood by all students in the various disciplines of environmental science.

The derivation of the BET equation is similar to the kinetic derivation of the Langmuir equation. It can also be derived by statistical mechanics but, due to two necessary assumptions involved in the derivation by statistical mechanics, it is often assumed that the BET equation must apply only when adsorption takes place on a uniform surface and when there is no lateral interaction between the adsorbate molecules. Brunauer (1961)[16] himself strongly opposed this transfer of assumptions from the statistical derivation to the kinetically based derivation because neither the BET theory nor the Langmuir theory, on which it is based, make such assumptions. The BET and Langmuir equations do assume that the heat of adsorption is constant over the entire surface, which is quite different from the misleading assumptions made by the more constrained statistical derivation.

The BET kinetic derivation begins by noting that at equilibrium, the rate of condensation (or adsorption) of a molecule on a surface is equal to the rate of evaporation (or desorption) of the molecule from the surface:

$$\text{rate of condensation } = \text{ rate of evaporation} \qquad \qquad [5\text{-}1],$$

$$a_1 p s_0 \; = \; b_1 s_1 e^{\frac{-E_1}{RT}} \qquad \qquad [5\text{-}2],$$

where a_1 and b_1 = constants, p = pressure, RT = gas constant times absolute temperature, E_1 = heat of adsorption of the first layer, and s = surface area that is covered by the adsorbed molecules (s_0 for the bare surface layer area, s_1 for the first layer area, and s_i for the ith layer area). Extending Equation 5-2 to other ith layers of adsorbed molecules, we obtain:

$$a_i p s_{i-1} \; = \; b_i s_i e^{\frac{-E_i}{RT}} \qquad \qquad [5\text{-}3].$$

The total volume adsorbed (V) and other volume calculations needed are:

$$V \; = \; V_0 \sum_0^{\infty} i s_i \qquad \qquad [5\text{-}4],$$

$$V_m = V_0 \sum_0^\infty s_i \qquad [5\text{-}5],$$

and

$$\frac{V}{V_m} = \frac{\displaystyle\sum_0^\infty i s_i}{\displaystyle\sum_0^\infty s_i} \qquad [5\text{-}6].$$

where V_0 = volume of gas adsorbed if a complete monolayer surface coverage occurs on a square centimeter of the adsorbent surface, and V_m = volume of gas adsorbed if a complete monolayer surface coverage occurs on the adsorbent surface. The only value that can be measured easily in Equation [5-6] is the total volume adsorbed, V, while the summation of the s_i values needs to be expressed differently. Accordingly, we continue as follows. The BET equation assumes that the heat of adsorption of all the layers is the same except for the very first layer. That is,

$$E_2 = E_3 = \dots E_i = E_L \qquad [5\text{-}7],$$

and

$$\frac{b_2}{a_2} = \frac{b_3}{a_3} = \dots \frac{b_i}{a_i} = g \qquad [5\text{-}8],$$

where E_L = heat of liquefaction, and the evaporation–condensation properties of the molecules in all the adsorbed layers above the first are the same as the liquid state. Equation [5-2] is now rearranged to

$$s_1 = s_0 \left[\frac{a_1}{b_1} pe^{\frac{E_1}{RT}} \right] = s_0 y \qquad [5\text{-}9],$$

where

$$y = \frac{a_1}{b_1} pe^{\frac{E_1}{RT}} \qquad [5\text{-}10].$$

Combining Equations [5-3], [5-7], and [5-8]

$$s_i = s_{i-1} \left[\frac{a_i}{b_i} pe^{\frac{E_i}{RT}} \right] = s_{i-1} \left[\frac{p}{g} e^{\frac{E_L}{RT}} \right] = s_{i-1} x \qquad [5\text{-}11],$$

where

$$x = \frac{p}{g} e^{\frac{E_L}{RT}} \qquad [5\text{-}12],$$

and, by extension,

$$s_i = s_{i-1} x = s_{i-2} x^2 = \dots = s_1 x^{i-1} \qquad [5\text{-}13].$$

For example, $s_3 = x s_2 = x^2 s_1$. Equation [5-13] is further combined with Equation [5-9] to yield

$$s_i = s_1 x^{i-1} = s_0 y x^{i-1} = c s_0 x^i \qquad [5\text{-}14],$$

where,

$$c = \frac{y}{x} = \frac{a_1 g}{b_1} e^{\frac{E_1 - E_L}{RT}} \qquad [5\text{-}15].$$

Using Equation [5-14], the numerator of Equation [5-6] becomes

$$\sum_0^\infty is_i = 0 + \sum_1^\infty is_i = cs_0 \sum_1^\infty ix^i \qquad [5\text{-}16],$$

and the denominator becomes

$$\sum_0^\infty s_i = s_0 + \sum_1^\infty s_i = s_0 + cs_0 \sum_1^\infty x^i = s_0\left[1 + c\sum_1^\infty x^i\right] \qquad [5\text{-}17].$$

The summations in Equations [5-16] and [5-17] are replaced by their equivalent geometric progressions for $x < 1$. That is, Equation [5-18] below is a standard arithmetic–geometric series, and Equation [5-19] below is a standard geometric series:

$$\sum_1^\infty ix^i = \frac{x}{(1-x)^2} \qquad [5\text{-}18],$$

$$\sum_1^\infty x^i = \left(\sum_0^\infty x^i\right) - 1 = \frac{1}{1-x} - 1 = \frac{x}{1-x} \qquad [5\text{-}19].$$

Substituting and rearranging these series into Equations [5-16] and [5-17], Equation [5-6] can be expressed as:

$$\frac{V}{V_m} = \frac{cs_0\dfrac{x}{(1-x)^2}}{s_0\left[1 + c\dfrac{x}{1-x}\right]} = \frac{cx}{(1-x)[1 + (c-1)x]} \qquad [5\text{-}20].$$

To express x in terms of pressure, note that $V = \infty$ when an infinite number of layers are allowed, which occurs when $p = p_o$ (where p_o = saturation pressure) and $x = 1$ (note that the summation of Equation [5-18] equals infinity when $x = 1$). From Equation [5-12], x is resolved as follows:

$$\frac{x}{1} = \frac{\dfrac{p}{g}e^{\frac{E_L}{RT}}}{\dfrac{p_o}{g}e^{\frac{E_L}{RT}}} = \frac{p}{p_o} \qquad [5\text{-}21].$$

Substituting the partial pressure (p/p_o) from Equation [5-21] into [5-20] yields the well-known BET isotherm equation:

$$\frac{V}{V_m} = \frac{c\dfrac{p}{p_o}}{\left(1 - \dfrac{p}{p_o}\right)\left[1 + (c-1)\dfrac{p}{p_o}\right]} \qquad [5\text{-}22],$$

which is often expressed as

$$\frac{p/p_o}{V[1 - (p/p_o)]} = \frac{1}{cV_m} + \frac{[c-1](p/p_o)}{cV_m} \qquad [5\text{-}23].$$

A plot of $(p/p_o)/(V[1 - (p/p_o)])$ versus (p/p_o) yields a highly linear regression analysis with an intercept $= 1/(cV_m)$ and a slope $= (c - 1)/(cV_m)$. The partial pressure (p/p_o)

range is typically 0.05 to 0.3. Also, slope + intercept = $1/V_m$. Now for the final steps, the surface area of the sample is easily determined from the monolayer volume of the condensed gas. The number of moles of gas adsorbed at monolayer surface coverage (X_m) is

$$X_m = \frac{PV_m}{RT} \qquad [5\text{-}24],$$

where $R = 82.054$ cm^3 atm deg^{-1} mol^{-1}, T = absolute temperature of the condensing gas (such as 77 K when condensing N_2 on the surface of the sample), and P = ambient or room pressure (in atmospheres) where the BET measurements are being performed.

Using Table 5-1, the number of moles of gas adsorbed at monolayer surface coverage is finally converted to surface area (SA) of the surface based on the projected area per molecule adsorbed (such as 0.162 nm^2 per molecule of N_2 adsorbed, or 9.756×10^4 m^2 per mole of N_2 adsorbed). For example, SA = $X_m(9.756 \times 10^4)$ m^2 if the condensing gas is N_2 in the analysis.

Clearly, an accurate assessment of the size of the adsorbed molecule is critical to the surface area calculations. However, quantifying the size of the adsorbed molecule is not an easy endeavor. One of several approaches to quantify this value may be used. Assuming a spherical shape, Emmett and Brunauer (1937)[17] proposed that the projected surface area of the molecule (σ) can be deduced from

$$\sigma = F(W)^{2/3} = F\left(\frac{M}{N\rho}\right)^{2/3} \qquad [5\text{-}25],$$

where F = packing factor, W = volume factor, M = molecular weight, N = Avogadro's

Table 5-1: Projected area per molecule for surface area (SA) analysis. Note, −195 °C = temperature of liquid N_2. *BET data from McClellan and Harnsberger (1967).*[18] *Gravimetric data from Heilman et al. (1965).*[19]

Molecules for SA, BET Analysis		Projected surface area (σ), Å^2 molecule^{-1}	
nitrogen	N_2	16.2	at −195 °C
argon	Ar	13.8	at −195 °C
		14.1	at −183 °C
krypton	Kr	20.2	at −196 °C
xenon	Xe	23.2	at −184 °C
water	H_2O	12.5	at 25 °C
methanol	CH_3-OH	21.9	at 20–25 °C
ethanol	C_2H_5-OH	28.3	at 25 °C
butane	n-C_4H_{10}	44.4	at 0 °C
benzene	C_6H_6	43.0	at 20 °C
carbon monoxide	CO	13.1	at −195 °C
		16.0	at −183 °C
carbon dioxide	CO_2	21.8	at −78 °C
oxygen	O_2	13.6	at −195 °C
Molecules for SA, Gravimetric Methods		**Conversion Factor, g m^{-2}**	
EGME [a]	C_2H_5-O-C_2H_4-OH	0.000286	
ethylene glycol	HO-CH_2CH_2-OH	0.00031	

[a] EGME = ethylene glycol monoethyl ether.

number ($= 6.022 \times 10^{23}$), and ρ = density (g cm^{-3}) of the condensing liquid or solid at the temperature of adsorption. Based on the structural geometry of the adsorbed molecules, various packing factors are listed by Livingston (1949)[20], such as $F = 1.091$ for hexagonal close packing. For example, for hexagonal close packing of N_2 at 77 K, $M = 28.01$ g mol^{-1}, $\rho = 0.81$ g cm^{-3}, and $\sigma = 1.091[28.01/(6.022 \times 10^{23} \times 0.81)]^{0.667} = 16.2 \times 10^{-16}$ cm^2 molecule^{-1} = 16.2 Å^2 molecule^{-1}.

Another approach uses Hill's (1948)[21] formula to evaluate the two-dimensional van der Waals' constant b (units are in Å^2 molecule^{-1}) based on the critical temperature (T_c, in K) and pressure (P_c, in atm) constants of the bulk substance:

$$b = 6.354 \left(\frac{T_c}{P_c} \right)^{2/3}$$
[5-26].

For example, for CO_2, $T_c = 30.99$ °C $= 304.14$ K, $P_c = 72.786$ atm, and $b = 6.354 \times (304.14/72.786)^{0.667} = 16.5$ Å^2 molecule^{-1}.

A third and the most reliable approach to determining the size of an adsorbed molecule is to numerically calibrate the size based on the amount adsorbed by a sample with known surface area as measured by a standard molecule (such as N_2). But even this very reliable approach is not free of some problems because the projected area of the standard molecule must be well characterized and constant on all surface types (and the latter condition does not occur). For example, some variation on the projected surface area of the adsorbed N_2 molecule is found in the literature. McClellan and Harnsberger (1967)[18] concluded that the size of an adsorbed molecule is not constant, but instead varies with the adsorbent and temperature of adsorption. For this reason, it would be a good idea to specify the adsorbent used and the presumed projected area of the adsorbed molecule when reporting the results of a surface area analysis. For example, the report may say "the surface area was determined by the BET method using N_2 at -195°C (projected area $= 0.162$ nm^2/molecule) ... ". In practice, however, this high degree of detail on the BET experimental methods used in the analysis is rarely found in the published literature, which is unfortunate.

Emmett and Brunauer (1937)[17] assumed N_2 $\sigma = 16.2$ Å^2 molecule^{-1} based on the density of liquid N_2, or $\sigma = 13.8$ Å^2 molecule^{-1} based on the density of solid N_2. Although Livingston (1949)[20] argued that the best value for N_2 is $\sigma = 15.4$ Å^2 molecule^{-1}, the value of 16.2 appears to be treated as a universal truth in most published literature today. It is also the most common default value used by software sold with commercially available BET equipment. It remains, nevertheless, a questionable value sensitive to the type of surface being studied and the temperature of the experimental set-up. To its credit, however, the variability in its adsorbed size is very, very small compared with those of various other possible condensed gases that could be used.

The BET constant (c), which was defined in Equation [5-15], is a function of the difference between the first monolayer bonding energy (E_1) and the energy of the succeeding adsorption layers (E_L). It is also a function of various condensation and evaporation rate constants. The numerical value of c will affect the shape of the predicted adsorption isotherms, as is illustrated in Figure 5-10. Note that at high c and low (p/p_o) values, the

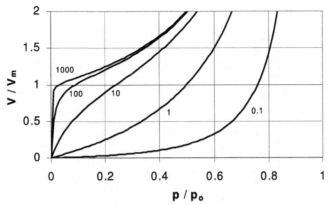

Figure 5-10: Impact of the magnitude of the BET constant (c) on the shape of the adsorption isotherm. The c value for each curve is shown.

adsorption isotherm is similar in shape to the Langmuir isotherm. That is, Equation [5-22] will approximate the Langmuir equation at these conditions.

An adsorbate with a BET c value that is either too high or too low will be subject to considerable error in the BET surface area analysis. When the BET c value is large, the adsorbate is strongly tied to the surface so as to be constrained to specific adsorption sites. The adsorbate cross-sectional area is then strongly dependent on the adsorbent lattice structure, and low surface areas are obtained when these sites are far apart. The BET c value must be low enough to avoid appreciable localization of the adsorbate. Xenon is sometimes nice to use for BET measurements because of its low vapor pressure, which is good for surface area studies, but it may suffer from extensive localization on the surface.

Conversely, a BET c value that is too low results in the adsorbate having a high lateral mobility with no organized structure and appearing more as a two-dimensional gas. Calculations on the size of the molecule, such as those used for Equation [5-25] for hexagonal close packing, would not apply. The BET c value must be high enough to ensure adequate separation between monolayer and multilayer formation. Argon, for example, will sometimes exhibit low c values.

Several adsorbates (such as n-pentane and n-butane) exhibit a significant variation in their cross-sectional areas as a function of the BET c value variation among different adsorbents (Figure 5-11). Clearly, this is very unwelcome when seeking accurate surface area data. Krypton is popular to use with samples that have a very low surface area because of its low saturation vapor pressure ($p_o \approx 2$ Torr). But Kr is tricky to use because 77 K (-195 °C, the temperature of the liquid N_2 cooling bath) is below the triple point of Kr, which is at 116 K, typically resulting in nonlinear plots. There is a wide range of BET c values for Kr, which suggests that there is a considerable variation in the degree of localization. If possible, use N_2 instead.

Nitrogen (N_2) has a reasonable BET c value on nearly all surfaces (ranges from 50 to 300), and it is accepted as a standard with a cross-sectional area set at 0.162 nm^2 at its boiling point of -195.6 °C (77 K). As noted earlier, be aware that this N_2 molecular area

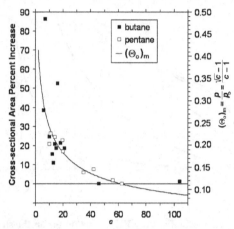

Figure 5-11: Variation of the cross-sectional areas of *n*-butane and *n*-pentane on different adsorbents, each having a different BET *c* constant. The data were calibrated against the N_2 BET surface area results obtained with each adsorbent, where N_2 $\sigma = 16.2$ Å^2 molecule^{-1}. A 0% increase in molecular cross-sectional area corresponds to 39.8 and 53 Å^2 molecule^{-1} for butane and pentane, respectively. The surface area corrections follow the axis on the left, while the curve of $(\Theta_o)_m$ follows the axis values on the right. *Data from Lowell et al. (1982).*[22]

value can vary. For example, a value as high as 0.20 nm^2 has been proposed for adsorption on a graphitized carbon surface, which is close to the value calculated for N_2 molecules rotating freely in a plane parallel to the surface (Gregg and Sing, 1982).[23]

The cross-sectional area of an adsorbed molecule increases hyperbolically as the BET *c* value decreases. For reasons that are not yet clear, the increase follows the $(\Theta_o)_m$ curve drawn in Figure 5-11, where $(\Theta_o)_m$ is the partial pressure when $V = V_m$. Starting with Equation [5-22], it can be shown that if $V/V_m = 1$, then $(\Theta_o)_m = p/p_o$ where

$$(\Theta_o)_m = \frac{p}{p_o} = \frac{\sqrt{c}-1}{c-1} \qquad [5\text{-}27].$$

[5.4]   Gravimetric Surface Area Determinations

The surface area of some dried mineral samples are often much lower than what the mineral would have in a moist or saturated environment. This is particularly true of 2:1 clay minerals, which tend to expand when wet. As a result of this expansion, the aqueous ions in the solution surrounding the mineral will have a larger reactive surface area than the one estimated by measurements on the dried mineral sample. Measuring the total surface area (external plus internal surface area) of an expanding mineral is accomplished with solvents that are able to penetrate the interlayer and expand the clay mineral. These solvents have a high dielectric constant in order to expand the interlayer. If the solvent cannot dissolve or hydrate the interlayer cations, then the interlayer will remain collapsed and held in place by the binding interlayer cations. The gases such as N_2 used in BET surface area calculations do not penetrate and measure the interlayer of clays.

The most common solvent used for measuring the total surface area of minerals is ethylene glycol monoethyl ether (EGME, $C_2H_5OCH_2CH_2OH$, also known as 2-ethoxy-ethanol). Another commonly used solvent for specific surface area determinations is ethylene glycol ($HOCH_2$–CH_2OH). The term "specific" means that the surface area is reported on a per-weight basis. The basic principle is similar to that used by the BET theory. Namely, the heat of adsorption of all the layers is the same except for the very first layer. Hence, the energy of adsorption of EGME molecules by the bare mineral surface is much higher than the energy of adsorption of EGME onto the second and subsequent layers, which are assumed to be equivalent to the heat of liquefaction (see Equations [5-7] and [5-8]). This solvent and other similar solvents work well when the surface monolayer coverage is complete and held strongly, such that the area contributed by bare surfaces or multilayered regions are minimal by comparison to the area covered by a single layer of EGME. Note that this gravimetric method is often referred to as the EGME method when EGME is the solvent used.

The EGME surface area procedure is as follows. A known weight of dried solid sample is mixed with EGME and made into a slurry. Several replicates are prepared and the samples are placed in a vacuum desiccator. A container of $CaCl_2$ with EGME solvate is added alongside the sample containers to control the vapor pressure inside the desiccator. The loss of weight of the solid samples is monitored over time. Heilman et al. (1965)[19] recommended that a few weight measurements be done daily. Note that the vacuum is reestablished between sample weighings.

The weight of the solid sample will decrease rapidly, followed by an abrupt stop in weight loss that occurs when the weight of EGME on each solid sample in the desiccator is equivalent to a monolayer coverage of the mineral's surface. Record this weight and convert the weight of EGME to specific surface area using the 0.000286 g m^{-2} conversion factor (Table 5-1). This monolayer EGME condition will remain on each sample at least until all the solid samples have reached this monolayer condition. In making the initial slurry, the amount of EGME added to each sample will vary, causing some samples to reach monolayer conditions sooner than others. An illustration of this is shown in Figure 5-12.

It is possible, particularly with samples that do not have expanding interlayers, that a rapid EGME weight loss on all the samples will begin anew immediately *after* all the solid samples have reached their particular monolayer condition. Accordingly, for minerals such as oxides, which do not have expanding interlayers, experience shows that it is best to monitor the weight measurements methodically and at consistent, regular intervals. The intent here is to catch the weight of the monolayer condition prior to the second phase of evaporation, which is the evaporation of the monolayer film. This second phase is rapid if there is no other sample in the desiccator that is keeping its atmosphere easily saturated with the EGME solvent (that is, if there is no other sample present that still has multilayer coverage of EGME on its surface). This evaporation of the monolayer EGME is observed on Measuring Cycle 34 in Figure 5-12.

A unique difference in the EGME method described above and the BET method described in Section 5.3 is the identity of the phase that is being measured. In the BET method, one measures the change in volume of the gas that is condensing on the solid sample

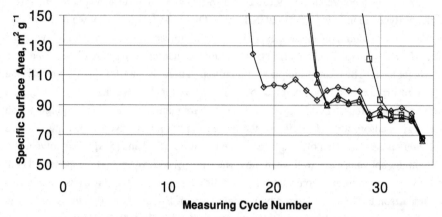

Figure 5-12: EGME surface area (SA) analysis of an acid/base washed Al oxide (δ-Al$_2$O$_3$). Each measuring cycle was 5 min of applied vacuum, followed by 15 min standing and then weighing the samples. Results of four samples shown. Notice how all samples converge to a common final surface area, each reaching it at different times because the amount of EGME present with the starting slurries was different in each sample. However, at Cycle No. 34 they all begin to lose their monolayer coverage as the evaporation of EGME continues. Average SA = 83.1 ± 2.1 m^2/g of washed samples. Average of unwashed samples (data not shown) was 94.9 ± 1.3 m^2/g versus 100 ± 15 m^2/g using N$_2$ BET as specified by the manufacturer for this product line (Aluminum Oxide C by Degussa Corp., Teterboro, NJ). *Data from Schulthess (1986).*[24]

(or, instead, the change in volume caused by evaporating back into the gas flow stream all of the gas that had condensed in a previous step). In the EGME method, one measures instead the change in weight of the liquid that remains on the solid sample after each evaporation increment. In the absence of modern automated equipment, the gravimetric procedure for determining the total surface area is significantly simpler than the volumetric procedure used for determining the BET external surface area. However, with the aid of various modern but expensive instruments, BET surface area analyses are very easy to perform.

[5.5]   Other Techniques for Surface Area Determinations

Visual Estimations:

The physical dimensions of a particle can be used to estimate the specific surface area of a solid sample. The particle's physical dimensions can be estimated based on an electron microscopy image. We then convert surface area per particle to surface area per gram using

$$S_o = \frac{S_p}{10^4 \, \rho \, V_p}$$

[5-28],

where ρ = density of the particle (g/cm^3, which is measured independently or taken from published data), V_p = particle volume according to the microscopy image (cm^3 particle^{-1}), S_p = surface area according to the microscopy image (cm^2 particle^{-1}), 10^4 cm^2 = 1 m^2, and S_o = crystallographic specific surface area (m^2 g^{-1}).

This approach for surface area determinations is not useful with whole soil samples due to the large variety and range of particle shapes and dimensions that may exist in any natural sample. Furthermore, the area of the interlayer region of expanding 2:1 clays is very difficult to estimate crystallographically. There are also numerous small dips and hills on a particle's surface that may go unnoticed or incorrectly estimated. Plus, the size and shape (including the microtopography) of the particle may change in a humid or aqueous environment compared to that existing under the electron microscope's vacuum conditions.

Ion Adsorption:

Soil particles will adsorb aqueous ions, and the amount adsorbed is a function of the sample's specific surface area. This is not a commonly used method to determine surface area. The "projected area" measured is dependent on the crystal's topography and the variable spacing of surface sites between minerals. Another concern with this approach is the difficulty in confirming that the maximum amount adsorbed is in fact representative of a complete monolayer coverage of the surface. In aqueous solutions, there are numerous secondary reactions that could interfere or skew the results (such as pH-dependent surface characteristics and competitive adsorption effects of other ions present in the mixture). Nevertheless, some attempts to estimate surface areas by ion-adsorption have been made. An example with phyllosilicates is the use of N-cetylpyridinium bromide (CPB, also called 1-hexadecylpyridinium bromide, $C_5H_5N^+$–$C_{15}H_{30}$–CH_3 Br$^-$), which has a positively charged N and a negatively charged Br.[25] The CPB molecule adsorbs both on the external and internal surfaces of clays, and the resultant adsorption isotherm follows the Langmuir equation. The specific surface area is calculated based on 0.27 nm^2 molecule^{-1} and on the adsorption maximum extrapolated from the Langmuir equation.

Ion Exclusion:

Another approach to determine surface area is the ion-exclusion method, also referred to as the negative adsorption method, which is based on the basic premise that negatively charged solid particles will repel negatively charged aqueous anions by coulomb forces. Conversely, positively charged solid particles will repel positively charged aqueous cations. For simplicity, the method is described here in terms of anion exclusion. The repulsion results in a region near the surface (referred to as the exclusion volume, V_{ex}) that is low in the anion's aqueous concentration, while the bulk aqueous phase is enriched in the anion's concentration. Measurement of the increase in concentration in the bulk liquid phase is then correlated to the particles surface area. This method is highly controversial and very unreliable due to the numerous assumptions that must be made about the exclusion effects.

For example, the aqueous conjugate cations that are also "hovering" near the negatively charged surface may modify the exclusion effect in unpredictable ways. Aqueous ion concentrations also change as a result of the change in hydration volume of the adsorbed ions. And much overlooked, the particle being studied probably contains numerous preexisting anions on its surface (which we will call here autochthonous ions), which can easily increase the aqueous ion concentrations and result in an apparent ion exclusion effect. It is very difficult to obtain ultrapure, clean surfaces for this type of study. For all these reasons and many more, this highly questionable method is really only useful for academic discussions and not for real life applications. Do not worry much about this method in spite of the fact that it is often presented as a viable method for surface area analysis.

[5.6]   Pore Sizes & Pore Volumes

[5.6.1] Adsorption Isotherms & Hysteresis Loops

The presence of pores affects the adsorption isotherm of gases and aqueous ions and, depending on their size, may cause characteristic hysteresis loops. According to the International Union of Pure and Applied Chemistry (IUPAC) classification, there are three kinds of pores based on their average width: *micropores* (<2 nm), *mesopores* (2 to 50 nm), and *macropores* (>50 nm). The micropores are sometimes further subdivided into *ultramicropores* (very small micropores) and *supermicropores* (upper range of the micropore sizes). The physical characteristics of pores are typically elucidated based on the interpretation of the adsorption isotherm hysteresis.

As Figures 5-13 and 5-14 illustrate, there are several types of adsorption isotherms. The I–V isotherm classification was proposed by Brunauer et al. (1940).[26] The Type VI isotherm, or the stepped isotherm, is relatively rare. The letter classification for adsorption isotherms (L, S, C, and H) was proposed by Giles et al. (1960)[27] for solid–aqueous phase interactions, while the Roman numeral classification is used more for interaction of gaseous condensation on solids (namely, BET isotherms). As in most classifications, some isotherms are difficult to assign to one group or another. That is, there are numerous other isotherms that are borderline cases and difficult to fit into this classification system.

The Type I isotherm is similar to the L- or H-type isotherm. The L- and H-type isotherms reflect a high affinity between the adsorbate and adsorbent. The H-type isotherm is an extreme case of the L-type isotherm, which is sometimes also referred to as a Langmuir-type isotherm. Type I isotherms occur when the micropores in a solid powder are very small (at most, a few adsorbate molecular diameters) and the adsorption surface is found within these micropores rather than on the exposed external surfaces. Hence, the number of adsorption layers that can be present on these solids is limited, and the Type I isotherm reaches a maximum adsorption value at high relative pressures.

Type II isotherms occur when the solid powder does not have micropores as small as those encountered in Type I isotherms. Monolayer coverage usually occurs near the inflection point followed by multilayer coverage at higher relative pressure, reaching an

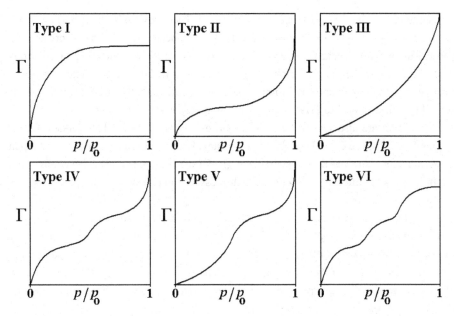

Figure 5-13: Adsorption isotherm classification for gaseous condensation on solid powders, where
Γ = amount adsorbed.

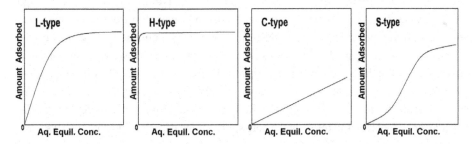

Figure 5-14: Adsorption isotherm classification for solid–aqueous phase interactions. Axes are
amount adsorbed by the solid phase versus the equilibrium concentration remaining in
the liquid phase.

infinite number of layers at saturation.

In Type III isotherms, the heat of adsorption (E_1) is less than the adsorbate heat of
liquefaction (E_L). Type IV isotherms occur when the solid powder has pores in the radius
range of approximately 1.5 to 100.0 nm. Adsorption increases sharply as the pores are being
filled. Type V isotherms are due to small adsorbate–adsorbent interaction potentials (similar
to Type III), but also have pores in the same range as the Type IV isotherms. Type VI
isotherms may be suggestive of a combination of conditions believed present in the Type I
and Type IV isotherms.

For solid–aqueous phase interactions, the S-type isotherm is similar to the Type III
isotherm because it suggests that the adsorbate–adsorbate interaction is stronger than the
adsorbate–adsorbent interaction. The possibility of clustering of adsorbate molecules on the

surface is likely when these isotherms are observed. The C-type (also known as constant-partitioning) isotherm is usually observed only at low concentrations of the adsorbate. The partitioning is rarely linear at higher concentrations. Nevertheless, the C-type isotherm is used often to describe the partitioning of chemicals in soils because the concentration of many compounds (particularly nonpolar organic compounds that have a low aqueous solubility constant) tend to exist in soil solutions at very low concentrations.

There are also several kinds of hysteresis loops, which are letter coded A to E (Figure 5-15).[28] Type C and D rarely occur in nature. Furthermore, a high desorption branch at saturation pressure was described in the original Type B loop, but this is never observed. Accordingly, a modified Type B with the high-pressure end redrawn is also shown, and this is commonly observed.

Since hysteresis exists, it follows that pores of a given radius will fill at a higher relative pressure than they will empty, which is rather difficult to explain. One theory postulated by Zsigmondy (1911)[29] argues that the contact angle during adsorption is larger than the contact angle during desorption. This theory is, in broad terms, still accepted today. Stated differently, the desorption process is not exactly a reversed mirror image of the adsorption process.

Figure 5-16 illustrates one way of explaining why hysteresis occurs. Initially (drawing A), adsorption of the molecule (say, N_2) on the cold particle is randomly distributed. These molecules do not easily move laterally, particularly if they have a high BET c value. Ideally, as more N_2 molecules adsorb, a monolayer coverage is reached (drawing B). If the partial pressure is high, then adsorption continues, particularly on the external planes of the particle where collisions with the N_2 gas molecules are more numerous than on the difficult-to-reach walls of the pores. Unlike the first monolayer, the adsorbed multilayers of N_2 molecules are able to move laterally easily, behaving more and more like a bulk liquid solution. Drawing C illustrates lateral mobility and a preference of the multilayer N_2 molecules to reside in the pore spaces of the particle. Note that $\cos(\theta) \approx 1$ for

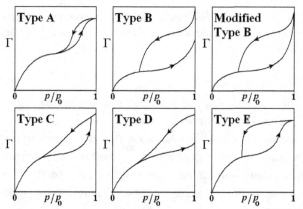

Figure 5-15: The five types of hystereses as originally classified by de Boer (1958).[28] Γ = amount adsorbed during the adsorption step, or amount remaining during the desorption step. Type B has been modified and redrawn according to Gregg and Sing (1982).[23] The three common types of hystereses are A, modified-B, and E.

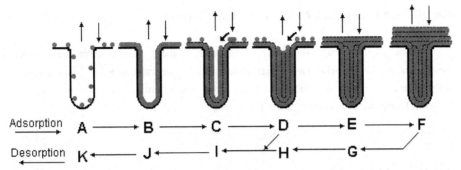

Figure 5-16: (A to F) Condensation (adsorption) of $N_2(g)$ on a porous mineral surface at 77 K. (F to K) Evaporation (desorption) of $N_2(\ell)$ from the porous mineral surface at warmer temperatures. **(A)** Adsorption begins, mostly on the external planes of the mineral where collisions with the surface are most likely to occur. Slow lateral movement of adsorbed N_2 is possible. **(B)** Monolayer coverage of the surface is complete. **(C)** Condensation of $N_2(g)$ continues, again mostly on the external planes of the mineral where collisions by the gas molecules with the surface are most likely to occur. Lateral movement of the condensed N_2 occurs easily and N_2 moves toward the pore spaces. Multilayer coverage is present in the pore spaces, while the external planes only have monolayer coverage. **(D)** Pore spaces are now filled. From **(A)** to **(F)**, almost all of the condensation processes are on the external surfaces, with lateral mobility of N_2 filling the pore spaces. Evaporation of N_2 from **(F)** to **(H)** mimics in reverse the condensation process. Connect **(D)** to **(I)**, instead of **(E)**, if the Gurvich Rule applies. Evaporation of N_2 from **(H)** to **(K)** is not the reverse of the condensation process because most of the evaporation is from pore spaces rather than from the external planar sites. Accordingly, hysteresis is observed in the corresponding BET adsorption–desorption isotherm.

the external planar region, but $\cos(\theta) < 1$ in the pore regions as evidenced by the curvature of the pore walls. That is, the multilayer N_2 molecules are more stable when they reside in the curved pore areas than when they reside in the flat external areas. As adsorption continues (Drawings D, E, and F), the pores fill and multilayer adsorption continues up to the nth layer depending on the specifics of the system involved.

When desorption is initiated on the warmed sample, the N_2 molecules return to the gaseous phase via a path that may be similar to the adsorption path (Drawings F to G). Following this, however, the N_2 molecules must evaporate directly from the pore space (Drawing H) because the laterally mobile N_2 molecules prefer to stay on the pore walls rather than move to the external flat planes where they first "landed". Since they are now more difficult to remove than expected, a lower p/p_o condition is needed for this evaporation to occur, and hysteresis is observed. The desorption process finally reaches the monolayer condition in all parts of the particle (Drawing J), followed by very low surface coverage (Drawing K) on its way toward a clean surface. Hysteresis should not be observed if the adsorption and desorption processes in the micropores proceed by the same path. But if hysteresis is observed (which it nearly always is), it follows that the adsorption and desorption paths are not the same.

[5.6.2] Pore Diameters & Volumes using Gases

The larger the pore size in a given mineral, the higher is the relative pressure needed to fill it. The p/p_0 values where hysteresis is observed are generally above 0.3, which is well past monolayer coverage conditions and in the range of deep multilayer formation. This can be shown theoretically beginning with the Kelvin Equation:

$$\ln\frac{p}{p_0} = \frac{-2\gamma V_L}{r_m RT} \qquad [5\text{-}29],$$

where p/p_0 = relative pressure of vapor in equilibrium with a meniscus having a radius of curvature r_m, γ = surface tension, V_L = molar volume, R = gas constant (8.314×10^7 ergs K^{-1} mol^{-1}), and T = absolute temperature. For N_2 at $T = 77$ K, $\gamma = 8.85$ ergs cm^{-2}, and $V_L = 34.6$ cm^3 mol^{-1}. The surface tension of the adsorbed film is assumed to be identical with that of the liquid. Using Figure 5-17 to illustrate the geometric relationship between the curvature of a concave meniscus (r_m) to the radius of the pore (r) , Equation [5-29] is rewritten as

$$\ln\frac{p}{p_0} = \frac{-2\gamma V_L}{rRT}\cos\theta \qquad [5\text{-}30],$$

where θ = contact angle formed by the liquid on the pore walls. We have seen earlier in Section 2.13.2 how this Kelvin Equation modifies Henry's Law constant in soils.

The value of θ, which can range from 0 to 180°, is difficult to determine for a liquid droplet resting on a plate, and virtually impossible to measure directly for a liquid present inside the small pore minerals. For simplicity, therefore, let $\theta = 0$ and $\cos(\theta) = 1$. Although these assumptions on the values of θ and γ are needed for practical reasons, one should not assume that the estimated r values derived from the Kelvin Equation are error free. Nevertheless, note that as p/p_0 increases, r also increases.

The pore radius and pore volume calculations are best done when working with Type VI adsorption isotherms with desorption hysteresis (Figure 5-18). The total volume of the pores is easily deduced from the amount of gas adsorbed near saturation ($p/p_0 \to 1$) when the isotherm follows a finite angle toward saturation (F–G, but not F–J in Figure 5-18). This F–G region of the isotherm represents the filling of all the pores with liquid adsorbate. Proof of this is based on the generalization that the same maximum adsorption value is obtained regardless of the liquid used when expressed as a volume of liquid based on its normal liquid density (see examples given in Table 5-2). This generalization is known as the *Gurvich Rule*, named after Gurvich (1915)[30] who first noted it. When this rule applies, maximum adsorption can be represented by drawing D in Figure 5-16 (surface conditions illustrated by drawings E, F and G do not form), and desorption begins with drawing D.

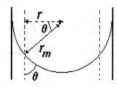

Figure 5-17: Relation of the curvature of the meniscus (r_m) to the radius of the pore (r) for use in Equations [5-29] and [5-30] ($r = r_m \cos\theta$, where θ = contact angle).

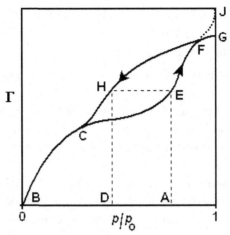

Figure 5-18: Typical Type IV and VI adsorption and desorption isotherms showing hysteresis.

Table 5-2: Uptake of various liquids at saturation by two porous materials. Uptake of the liquid is expressed as a volume of the liquid (V_s), where $V_s = V_L$ at saturation ($p/p_o = 1$). Notice that the standard deviations are small (< 5 %). *Data from Broad and Foster (1946)[31] for ferric oxide gel, and from McKee (1959)[32] for silica gel.*

Adsorbate		Ferric oxide gel [a] V_s, cm³ g⁻¹	Silica gel [b] V_s, cm³ g⁻¹
Benzene	C_6H_6	0.281	0.440
Carbon tetrachloride	CCl_4	0.270	0.421
Chloroform [a]	$CHCl_3$	0.282	—
Cyclohexane	C_6H_{12}	0.295	0.421
Deuterium oxide	D_2O	0.306	—
Diisopropyl ether	$((CH_3)_2CH)_2O$	0.290	—
2,3-Dimethylbutane	$CH_3CH(CH_3)CH(CH_3)CH_3$	—	0.429
Dioxane	⌬O [c]	0.302	—
Ethanol	CH_3CH_2OH	0.300	—
Ethylcyclohexane	$(C_6H_{11})CH_2CH_3$	—	0.426
Ethyl iodide	CH_3CH_2I	0.295	—
n-Heptane	$CH_3(CH_2)_5CH_3$	—	0.431
n-Hexane [a]	$CH_3(CH_2)_4CH_3$	0.308	0.431
Methylcyclohexane	$(C_6H_{11})CH_3$	—	0.425
2-Methylpentane	$(CH_3)_2(CH_2)_3CH_3$	—	0.431
Morpholine	⌬NH	0.282	—
Nitroethane	$CH_3CH_2NO_2$	—	0.434
Nitromethane	CH_3NO_2	—	0.449
n-Octane	$CH_3(CH_2)_6CH_3$	0.278	0.434
Toluene	C_6H_5-CH_3	0.272	—
Triethylamine	N-$(CH_2CH_3)_3$	0.300	—
2,2,3-Trimethylbutane	$CH_3C(CH_3)_2CH(CH_3)CH_3$	—	0.420
2,2,4-Trimethylpentane	$CH_3C(CH_3)_2CH_2CH(CH_3)CH_3$	—	0.439
2,3,4-Trimethylpentane	CH_3-$(CH(CH_3))_3$-CH_3	—	0.425
Water	H_2O	0.302	—
	Average V_s	0.291 ± 0.012	0.430 ± 0.008

[a] Adsorption isotherms are at 25 °C, but 20 °C for chloroform and n-hexane.

[b] Adsorption isotherms are at the boiling point of the gases.

[c] 1,4-dioxane is drawn here, but authors were not specific to which dioxane this V_s value applies.

From Equation [5-30], if $p/p_o = 0.99$ while using N_2 at 77 K, then

$$r = \frac{-2(8.85)(34.6)}{(8.314 \times 10^7)(77)(\ln 0.99)} = 952 \times 10^{-8} \text{ cm} = 95.2 \text{ nm} \qquad [5\text{-}31].$$

That is, the total pore volume obtained from an N_2 isotherm at $p/p_o = 0.99$ includes the volume of all the pores up to 95-nm radius. Note that hysteresis loops are typically closed at the low end of the loop, usually at relative pressures of 0.3 (using N_2 at 77 K). At this value, the Kelvin equation would estimate a pore radius of 0.8 nm. This value is actually too small to establish the validity of the Kelvin equation because only one or two molecular diameters could fit in such a small pore opening.

The pore size distributions are measured using the desorption isotherm because it thermodynamically corresponds to the more stable adsorbate condition. An exception is made for the "bottleneck" pores exhibited by Type E hysteresis because adsorbent in the wide portion of the pore is unable to evaporate until the narrow neck empties. The molar free energy change for condensation of vapor (ΔG_{ads}) is given by

$$\Delta G_{ads} = RT (\ln p_{ads} - \ln p_o) \qquad [5\text{-}32],$$

while the molar free energy change for evaporation of the liquid (ΔG_{des}) is given by

$$\Delta G_{des} = RT (\ln p_{des} - \ln p_o) \qquad [5\text{-}33].$$

Since hysteresis results in $p_{des} < p_{ads}$, it follows that $\Delta G_{des} < \Delta G_{ads}$, which means that the desorption value at any given relative pressure corresponds to the more stable adsorbate condition.

In contrast to BET surface area determinations where the partial pressure range analyzed is below 0.3, the pore volume and radius distribution determinations use the partial pressure range that is above 0.3 (that is, the values where hysteresis is observed). If the loop is closed or if there is no hysteresis, then the calculations are terminated. The calculations run from the saturated conditions downward until the loop closes. In simple terms, the adsorbed volume change per incremental change in partial pressure is used to estimate the average Kelvin pore radius (r_K) for the averaged partial pressure segment change (see Equation [5-31] for an example calculation). The pore volume is also based on the adsorbed volume change. The estimated thickness of the adsorbed film (t) is added to r_K to obtain the actual pore radius (r_p). Each layer of N_2 is estimated to be 0.354 nm thick, which is slightly less then its estimated diameter based on a cross-sectional area of 0.162 nm^2. This is because the N_2 in each layer sits in the depression present between the molecules in the adjacent layers. The thickness for N_2 (in nm units) can be estimated by the Halsey (1948)[33] equation as

$$t = 0.354 \left[\frac{5}{\ln (p_o/p)} \right]^{1/3} \qquad [5\text{-}34].$$

[5.6.3] Pore Diameters & Volumes using Hg

Another common method for determining the pore structure of particles is mercury (Hg) intrusion porosimetry (MIP). Liquid mercury has a high contact angle with clean surfaces, ranging from 112 to 142°, with 130° used widely when this specific information is lacking. Note that the receding angles are often 30° less than the advancing ones. This high contact angle means that Hg does not wet most materials and does not penetrate pores by capillary action, but rather must be forced to penetrate into the pores. The resistance by Hg on entering the pore is:

$$\text{Resistance} = -\pi D \gamma \cos(\theta) \qquad [5\text{-}35],$$

where D = pore diameter, γ = surface tension, and θ = contact angle, as illustrated in Figure 5-19. Note that surface tension (γ) is force (or resistance) per unit length (or circumference of the pore, πD). As noted earlier in Section 2.6, a high surface tension means that the molecular forces on the Hg surface film tend to contract its volume to a form with the least surface area.

The force due to the applied pressure (P) is expressed as

$$\text{Applied Force} = \frac{\pi D^2 P}{4} \qquad [5\text{-}36].$$

Note that the pressure is force per unit area. Hence, the applied force is pressure times the area, where area = $\pi r^2 = \pi(D/2)^2 = \pi D^2/4$. At equilibrium, these are equal, and, for a given pressure, D is obtained by combining and rearranging Equations [5-35] and [5-36]:

$$D = \frac{-4\gamma \cos \theta}{P} \qquad [5\text{-}37],$$

which is known as the *Washburn Equation*, after Washburn (1921).[34] For Hg, γ = 485 dyne cm^{-1} when P is measured in MPa and D is expressed in nm. Mercury will intrude into pores 360 μm in diameter when a force of 0.0034 MPa (0.5 psia) is applied. Pressures as high as 414 MPa (60,000 psia) are applied by various instruments (such as those made by Micromeritics) to detect pore sizes as small as 0.003 μm (or 3 nm). It is important that the samples be thoroughly evacuated prior to the addition of Hg because compression of the entrapped air will adversely affect the results. (Note: psia is pound force per square inch absolute, psig is pound force per square inch gauge, and psia = psig + 14.696 when ambient pressure is 1 atm.)

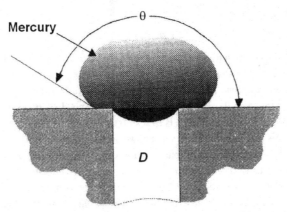

Figure 5-19: Mercury in contact with a porous solid. *From Webb and Orr (1997)[35], with permission.*

The volume of the pore (V) is measured directly from the volume of Hg intrusion using a sample holder known as a penetrometer (Figure 5-20). Assuming the pore is cylindrical, then the wall area is estimated as $A = 4V/D$. Note that the pores measured using a porosimeter will include the void spaces between the particles. Hence, the large-diameter peaks on a Hg-intrusion plot (Figure 5-21) are probably particle-to-particle void spaces, rather than the sought-after intraparticle pore spaces. For these large-diameter peaks, some caution must be exercised in the interpretation of the data based on all the information available on the particles.

The pore diameters (D) can be measured based on Hg intrusion as pressure is applied or Hg extrusion as pressure is released. Almost always, hysteresis loops will be observed when both are done. The reason for this is that some of the Hg is permanently trapped in the sample when pressure is released from the bottle neck pore structures (Figure 5-22). The Hg is also trapped in the void spaces between the particles. When the Hg is trapped in the pores and void spaces, various characteristic hysteresis loops may be observed (Figure 5-23).

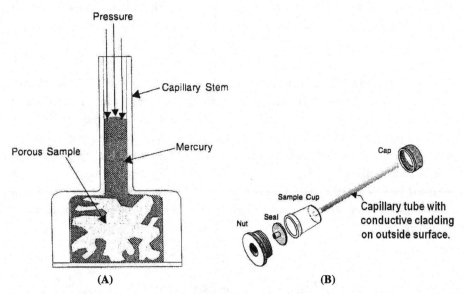

(A)   **(B)**

Figure 5-20: (A) Porous sample in penetrometer after Hg has been forced into pores. The sample pore diameters are exaggerated for clarity. (B) Details of penetrometer and closure components. All parts are made of glass or electrically nonconducting components, except for the metallic seal and the thin layer of metal cladding around the outside portion of the stem. The sample is first evacuated and then filled with Hg, beginning with the low pressure readings. As the pressure increases, the Hg is forced down the tube and into the sample pore spaces. An electrical potential is applied to the seal and the cladding, causing the stem to function as a coaxial capacitor whose capacitance is directly related to the amount of Hg remaining in the stem. Its accuracy is <0.1 µL. For safety in the event of a failure, the high pressures are generated hydraulically. *From Webb and Orr (1997)*[35], *with permission.*

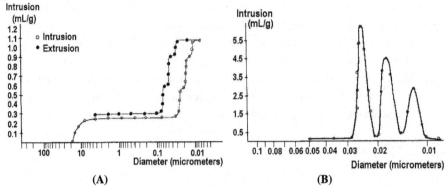

Figure 5-21: (A) Example of Hg intrusion and extrusion plots. Notice the hysteresis, which results from the fact that not all of the Hg returns to the stem when pressure is incrementally released. (B) Log differential plot of the Hg intrusion data shown in (A). *From Webb and Orr (1997)[35], with permission.*

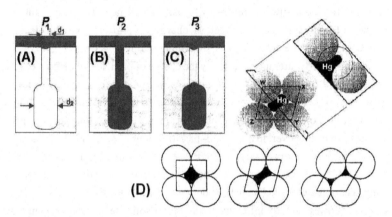

Figure 5-22:(A,B,C) Mercury intrusion into and extrusion out of bottleneck pore structures. (A to B) The total intrusion volume is a function of the pore diameter d_1. (B to C) Extrusion leaves Hg trapped in the void space even when $P_3 < P_1$. Large pore spaces are observed between particles (resulting in "interparticle" pore spaces) and very small pore spaces are observed in the channels and crystal faults of individual particles (resulting in "intraparticle" pore spaces). (D) Cross-sections of void spaces between spherical particles, with the upper right drawing showing the Hg intruding into the void at the "breakthrough" pressure. *From Webb and Orr (1997)[35], with permission.*

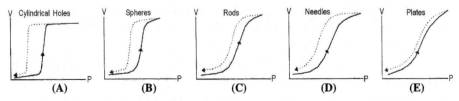

Figure 5-23: Characteristic hysteresis loops for cylindrical holes and for pores formed within aggregates. *From Webb and Orr (1997)[35], with permission.*

Mercury intrusion porosimetry measures the pore diameter and the amount of Hg that goes into its corresponding pore space, which, as we just saw, can be a larger cavity on the other side of the pore opening. The calculations discussed earlier for determining the pore wall areas assumed that the pores are cylindrical in shape. Clearly, however, the pore channels in soil particles come in all kinds of shapes, and alternate estimation techniques may be warranted. More comprehensive reviews on the subjects of surface area and pore size analyses are found in Gregg and Sing (1982)[23], Lowell and Shields (1991)[36], Webb and Orr (1997)[35], and others, where additional theories and computational methods are discussed.

[5.7]   CEC, AEC, & Surface Charge

An important behavior observed with nearly all minerals is their ability to exchange cations and anions. Thompson first observed the retention of Ca^{2+} and the removal of NH_4^+ from a soil column in 1850, which was reported in the classic papers written by Way in 1850 and 1852.[37,38] This common observation, which is now known as cation exchange, is one of several reasons for the assumed presence of positive and negative charge on particles. Another reason for the assumed presence of charge on particles is their electrokinetic behavior under an applied electric field or shearing motion.

Numerous early attempts to quantify the charge of particles involved measurements of *cation-exchange capacity* (CEC) and *anion-exchange capacity* (AEC) based on ion-exchange studies, as well as proton retention estimations based on acid–base titration studies. Acid–base titration experiments resulted in rather unusual results and, accordingly, fell in disfavor. The truth is, however, that the acid–base titrations of suspended soil samples were done incorrectly.[39] Not surprisingly, a prominent soil scientist by the name of Schofield (1949)[40] challenged the "old" titration methods for surface charge determinations and suggested using a cation–anion exchange method with acid NH_4Cl. Schofield's work, and that of others who followed, greatly influenced the field of soil science. Ion exchange methods to determine surface charge (CEC versus AEC) became acceptable, while potentiometric titrations were generally reserved for point of zero charge (PZC) analyses only.

The CEC and AEC of soils is measured on the fine-earth fraction, which includes the organic matter, sand, silt, and clay. There are several methods used to measure CEC, and at times there is much disagreement among research scientists as to how to perform and interpret the results. Two common methods are the *summation method* and the *direct displacement method*. The summation method exchanges the cations present on the soil with a concentrated salt solution (e.g., 0.1 M $BaCl_2$), followed by an analysis of the extracted cations in the supernatant (e.g., Ca^{2+}, Mg^{2+}, K^+, Al^{3+}), where CEC = Σ(aqueous cation concentrations). The direct displacement method washes the soil with an index cation (e.g., three times with 0.002 M $BaCl_2$) followed by exchanging the index cation with another ion (e.g., 0.005 M $MgSO_4$). An analysis of the index cation concentration extracted yields an estimate of the CEC, while an analysis of the index anion concentration extracted yields an estimate of the AEC. Another commonly used method measures the exchangeable cations extracted with ammonium acetate (NH_4OAc) buffered at pH 7. The results are usually expressed in centimoles or millimoles of charge per kilogram of exchanger ($cmol_c$ kg^{-1} or $mmol_c$ kg^{-1}).

When the CEC is measured at native pH values (or, more specifically, when the acidity is controlled by the salt extractable acidity), as is the case in the first two methods described above, the result is called the *effective cation-exchange capacity* (ECEC). Often, however, the term CEC is used when the term ECEC is intended. The pH of the extraction procedure is critical and it is best to clearly state the pH used for extraction to avoid misunderstandings.

The *total exchange capacity* of a soil sample is a very elusive quantity. Much of the problem is due to the presence of soluble salts and carbonates that are not exchangeable ions, yet their dissolution does skew the measurement of exchangeable cations and anions. Another analytical problem is the role of protons and hydroxyls on the exchange sites. Figure 5-24 illustrates the dependence of CEC and AEC on the pH of the extracting medium. *The reason for the pH dependence of the CEC and AEC values is that the surface sites occupied by exchangeable H^+ and OH^- ions are not measured and are not included as part of the extractable cations and anions of the mineral.* Some CEC methods do attempt to measure extractable acidity, but such attempts are flawed as is evidenced by the pH dependence of the results obtained. In actuality, the number of reactive surface sites (or exchange sites) in a given mineral is a constant number at all pH values when all the exchange species are properly identified and quantified. Exceptions are made for the extremely high or low pH values, where the mineral surface sites may be significantly decreased by dissolution processes.

Traditionally, the surfaces of soil particles are assumed to have sites that are either negatively charged (at high pH values) or positively charged (at low pH values), or both (at intermediate values). A neutral condition may be one where no charge exists or, more likely, where an equal number of positive and negative sites exist. For weakly adsorbing ions, the positively charged sites adsorb (or exchange) anions, while the negatively charged sites adsorb (or exchange) cations. Strongly adsorbing ions may adsorb regardless of the surface charge. In Figure 5-24, the pH value at the point of intersection between the positive charge line represented by the AEC results and the negative charge line represented by the CEC results is referred to as the *point of zero net charge* (PZNC). That is, CEC = AEC when pH = PZNC.

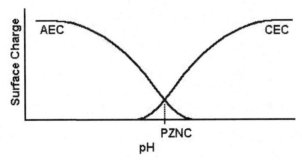

Figure 5-24: Cation-exchange capacity (CEC) and anion-exchange capacity (AEC) versus pH. The surface charge is based here on the measured exchange of cations and anions with the exchanger. The sign of the charge is ignored here. The pH of the intercept of these two measurements is the point of zero net charge (PZNC).

The data collected from cation adsorption or cation exchange studies will often follow the pattern illustrated in Figure 5-25, which illustrates the two types of charges found in soils: (1) *permanent charge*, and (2) *variable charge*. Permanent charge is surface charge that is pH independent. Smectites and other 2:1 clays typically exhibit permanent charge, such as a constant CEC value over a broad pH range. Permanent charge, which is due predominantly to isomorphic substitutions in the mineral, is not present in all minerals. Variable charge is present in most, if not all, minerals, particularly oxides, as well as in soil organic matter. Variable charge is also referred to as pH-dependent charge. Organic matter and various 1:1 clays, such as kaolinite, exhibit variable charge. An increase in CEC values as the pH increases is an expression of variable charge behavior.

The variable charge is due to surface sites (S) that release or gain protons as the pH of the medium varies, and this is typically represented by

$$SOH + H^+ \rightleftharpoons SOH_2^+ \qquad\qquad\qquad [5\text{-}38],$$

$$SOH \rightleftharpoons SO^- + H^+ \qquad\qquad\qquad [5\text{-}39].$$

These surface sites are found in the planar regions of minerals and on the edges of minerals. The pH-dependent charge of organic matter is due to the ionization of various organic components, particularly organic acids, phenolic groups, and amines. The organic matter is usually negatively charged and the counter cations are nearby in the solution. Although these counter cations are no longer considered to form a part of the organic molecule, it is not possible to separate the two components from each other unless an electrical field is applied. For example, if you centrifuge the anion out of solution using an ultracentrifuge, the cation is also removed with it. This is also true with minerals. If the mineral settles to the bottom of a beaker, it does so with the aqueous counter ions near it.

The permanent charge is typically assumed to be due to charge imbalances that are buried deep in the mineral and cannot be satisfied by aqueous protons (otherwise it would be pH dependent). With clay minerals, the charge resides in the octahedral or tetrahedral layers due to isomorphic substitution. The counter ions reside in the interlayer of the clay or are found very close to the clay in the surrounding liquid. Since the actual identity of the

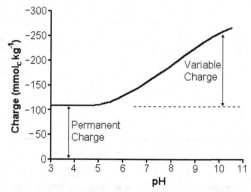

Figure 5-25: Permanent charge (pH-independent) and variable charge (pH-dependent) components of soil minerals. The surface charge of minerals is typically expressed in mmoles of charge per kilogram of exchanger.

counter ions present does not affect the composition or the structure of the solid phase, we do not generally refer to these counter ions as being a part of the mineralogical composition of the clay. They are, nevertheless, inseparable from the clay. There are exceptions made for some clay minerals, such as micas, where the counter ions are considered to be a part of the mineral because these counter ions do affect the structure of the solid phase.

The concept of surface charge needs to be reexamined. First of all, it is absurd to assume that aqueous protons (H_3O^+), which are the most active cations of all cations in the liquid phase, are not capable of neutralizing a charge such as that associated with permanent-charge minerals. All minerals, therefore, should exhibit variable charge or pH-dependent charge. If no pH dependence was observed, then the pH range of the analysis was probably too narrow. In Figure 5-25, for example, the "permanent charge" component of the sample would be recognized as a pH-dependent component if the pH analysis were taken to lower values, where its value would probably drop to zero. The sample shown in Figure 5-25 is merely poorly buffered in the pH range of 5 to 3, or lower.

Smectites are said to display "permanent" charge. Since they are not considered stable at low pH values (pH < 4), the pH-dependent charge behavior at very low pH is viewed as an artifact of this instability. Presumably, the permanent-charge components have escaped or have dissolved away at these low pH values, while the unreactive Si sites remain behind. Note that the isoelectric point (IEP) of Si oxides and 2:1 clay minerals (which have a large fraction of their total exposed surface consisting of Si sites) is around pH 2, which means that the reactivity of protons with Si surface sites is much weaker than with other types of sites, such as aluminol or ferrol surface sites. The reactivity of protons with ferrol or aluminol groups is orders of magnitude stronger than that of most cations. The reactivity of protons with SiO_2 or silanol groups is comparable in strength to that of most cations.

The 1:1 clay minerals should not display a "permanent" charge because they lack isomorphic substitution in the octahedral and tetrahedral layers. The variable-charge behavior observed by these minerals is presumed to originate from the broken edges of the clay particles, or from the presence of amorphous oxide impurities coating the surfaces of the clay particles. Accordingly, the 1:1 clay minerals are noted for their very low CEC values.

Clays with no isomorphic substitution ($X = 0$) will have low CEC values. Conversely, clays that have isomorphic substitution will have high CEC values. Table 5-3 lists the typical CEC values for various clays. In the 2:1 clay series, notice that the low isomorphic substitution of pyrophyllite and talc ($X = 0$) result in low CEC values, while the high isomorphic substitution of montmorillonite and vermiculite ($X = 0.2$ to 0.9) result in high CEC values. The micas are collapsed clays, and the trapped K^+ cations neutralize the isomorphic substitution, resulting in low CEC values. The focus here is that the amount of CEC of clays is directly related to the amount of isomorphic substitution of the clays, and the data in Table 5-3 seem to support this concept.

The X values of clays are in *per half-cell* units, which has, by extension on their impact on ion exchange, a *per surface area* effect. Although the CEC values are the result of the mineral's surface reactivity, they are expressed based on the mineral's mass instead. In other words, the CEC values in Table 5-3 are based on a misleading unit scheme.

Table 5-3: Typical cation-exchange capacity (CEC), surface area, and surface charge density values for various soil minerals. Layer charge (X) is in equivalents of charge per unit half-cell. *The CEC and surface area data are from Goldberg et al. (2000)[41], Talibudeen (1981)[42], Skopp (2000)[43], Sparks (2003)[44], and White and Zelazny (1986).[45]*

Solid	X	CEC, $\mu mol_c\, g^{-1}$	Surface Area, $m^2\, g^{-1}$	Charge Density, $\mu mol\, m^{-2}$
Allophane	0	500–1000	500–700	1–1.4
Al & Fe oxides	0	20–40	70–250	0.16–0.28
1:1 Clays:				
Kaolinite	0	20–60	10–20	2–3
2:1 & 2:1+1 Clays:				
Pyrophyllite	0	< 10	65–80	< 0.15
Talc	0	< 10	65–80	< 0.15
Montmorillonite	0.2–0.6	800–1200	280–500	2.4–2.8
Illite	0.6–0.9	100–400	65–195	1.5–2.1
di-Vermiculite	0.6–0.9	100–1500	50–800	1.8–2
tri-Vermiculite	0.6–0.9	1000–2000	600–800	1.7–2.5
Biotite	1	<10	40–100	< 0.25
Muscovite	1	<10	60–100	< 0.17
Chlorite	variable	100–400	25–150	2.6–4

Dividing the CEC values by the surface area values yields the *surface charge density* values, listed in the last column in Table 5-3. This time, notice that the charge-per-surface-area values for the 2:1 and 2:1+1 clays are roughly the same as the values for kaolinite, allophane, and metal oxides. Talc and pyrophyllite do yield low charge density values, but there is no noticeable trend between montmorillonite and vermiculite or chlorite. Talc, pyrophyllite, micas, and Al and Fe oxides all yield low charge density values relative to the others listed.

Let's look now at kaolinite. In theory, zero isomorphic substitution should have zero CEC, with anything over zero being attributed to oxide impurities coating the surface. But in charge density units, we see that the values for kaolinite are comparable to the 2:1 clays and much higher than the values anticipated for the Al and Fe oxides.

As a result of this, the traditional concepts involving isomorphic substitution, CEC, and "permanent" charge can be misleading. Isomorphic substitution does cause a high concentration of counter cations to reside nearby, which can be exchanged to yield measurable CEC values. But the amount of isomorphic substitution does not correlate well with the measured CEC value when it is expressed in charge density units. In the absence of this isomorphic substitution (or even concurrently with it), other mechanisms of retaining exchangeable cations near the mineral surfaces must exist if CEC values are observed. An example would be exchange of surface protons for aqueous cations.

The CEC values are high for smectites, mostly because of the particle size. That is, there is more surface area per gram of the smectite sample measured. Similarly, the CEC values are low for kaolinite, mostly because the particles are large and the surface area is low per gram of kaolinite sample measured.

An alternate and less misleading approach is to view the *net* particle charge as always equal to zero. That is, the surfaces of clays and other minerals are net neutral particles. To do this, you must view the aqueous counterions as being an integral part of the clay, albeit they are not a part of the taxonomic identity of the clay. Electroneutrality must be maintained in aqueous solutions (see Section 2.18.1). An important component of the bulk structure of minerals (discussed in Chapter 4) is that electroneutrality must be maintained. As was noted earlier in Section 5.1, dry stable surfaces are autocompensated and charge neutral. Similarly, all solid–liquid interface reactions must maintain electroneutrality. The CEC of minerals should be viewed as an *exchange* reaction on a net charge neutral surface. Since the particle never has any net surface charge, all surface interface (solid–liquid) reactions are reactions on neutral surfaces and not reactions that neutralize a particle's charge imbalance. The clays that have isomorphic substitution are already charge neutral as a result of the various conjugate cations present on their surfaces when the clays were first formed or substituted later, but always in strict stoichiometric proportions. All particles are created charge neutral in nature, with counter ions hovering nearby as needed in the melt or in the solution. All natural surface components will be net charge neutral regardless of any treatment or pretreatment that they may have undergone. As a result of this, it is impossible to collect a mineral paste and have it behave as an anode (negative pole) in an electrical circuit, with the separated liquid portion acting as the cathode (positive pole).

References Cited

[1] Gibson, A.S., and J.P. LaFemina. 1996. Structure of mineral surfaces. p. 1–62. *In* P.V. Brady (ed.) Physics and chemistry of mineral surfaces. CRC Press, Boca Raton, FL.

[2] Duke, C.B. 1988. Atomic geometry and electronic structure of tetrahedrally coordinated compound semiconductor interfaces. p. 69–118 *In* D.A. King and D.P. Woodruff (ed.) The chemical physics of solid surfaces and heterogeneous catalysis. Volume 5: Surface properties of electronic materials. Elsevier Sci. Publ., Amsterdam, The Netherlands.

[3] Duke, C.B., and Y.R. Wang. 1988. Surface structure and bonding of the cleavage faces of tetrahedrally coordinated II–VI compounds. J. Vac. Sci. Technol. B 6:1440–1443.

[4] Godin, T.J., and J.P. LaFemina. 1993. Surface atomic and electronic structure of cassiterite SnO_2 (110). Phys. Rev. B (Condensed Matter) 47:6518–6523.

[5] Diabold, U. 2003. The surface science of titanium dioxide. Surface Sci. Rep. 48:53–229.

[6] Landree, E., L.D. Marks, P. Zschack, and C.J. Gilmore. 1998. Structure of the TiO_{2-x}(100)-1×3 surface by direct methods. Surface Sci. 408:300–309.

[7] Godin, T.J., and J.P. LaFemina. 1994. Atomic structure of the cassiterite SnO_2 (111) surface. Surface Sci. 301:364–370.

[8] Guo, J., D.E. Ellis, and D.J. Lam. 1992. Electronic structure and energetics of sapphire (0001) and ($0\bar{1}02$) surfaces. Phys. Rev. B (Condensed Matter) 45:13647–13656.

[9] Godin, T.J., and J.P. LaFemina. 1994. Atomic and electronic structure of the corundum (α-alumina) (0001) surface. Phys. Rev. B (Condensed Matter) 49:7691–7696.

[10] Hikita, T., T. Hanada, M. Kudo, and M. Kawai. 1993. Structure and electronic state of the TiO_2 and SrO terminated $SrTiO_3$ (100) surfaces. Surface Sci. 287/288:377–381.

[11] Goldberg, S., J.A. Davis, and J.D. Hem. 1996. The surface chemistry of aluminum oxides and hydroxides. p. 271–331. *In* G. Sposito (ed.) The environmental chemistry of aluminum, 2nd ed. CRC Press, Boca Raton, FL.

[12] Wijnja, H., and C.P. Schulthess. 1998. ATR-FTIR and DRIFT spectroscopy of carbonate species at the aged γ-Al_2O_3/water interface. Spectrochim. Acta, Part A, 55:861–872.

[13] Brown, G.E., Jr. 1990. Spectroscopic studies of chemisorption reaction mechanisms at oxide–water interfaces. Rev. Mineral. 23:309–363.

[14] Peri, J.B. 1965. A model for the surface of γ-alumina. J. Phys. Chem. 69:220–230.

[15] Brunauer, S., P.H. Emmett, and E. Teller. 1938. Adsorption of gases in multimolecular layers. J. Am. Chem. Soc. 60:309–319.

[16] Brunauer, S. 1961. Solid surfaces and the solid–gas interface. p. 5–17. *In* R.F. Gould (ed.) Advances in chemistry, Series no. 33: Solid surfaces and the gas–solid interface. Am. Chem. Soc., Washington, DC.

[17] Emmett, P.H., and S. Brunauer. 1937. The use of low temperature van der Waals adsorption isotherms in determining the surface area of iron synthetic ammonia catalysts. J. Am. Chem. Soc. 59:1553–1564.

[18] McClellan, A.L., and H.F. Harnsberger. 1967. Cross-sectional areas of molecules adsorbed on solid surfaces. J. Colloid Interface Sci. 23:577–599.

[19] Heilman, M.D., D.L. Carter, and C.L. Gonzalez. 1965. The ethylene glycol monoethyl ether (EGME) technique for determining soil-surface area. Soil Sci. 100:409–413.

[20] Livingston, H.K. 1949. The cross-sectional areas of molecules adsorbed on solid surfaces. J. Colloid Sci. 4:447–458.

[21] Hill, T.L. 1948. Statistical mechanics of multimolecular adsorption. IV. The statistical analog of the BET constant $a_1 b_2 / b_1 a_2$. Hindered rotation of a symmetrical diatomic molecule near a surface. J. Chem. Physics 16:181–189.

[22] Lowell, S., J. Shields, G. Charalambous, and J. Manzione. 1982. Adsorbate cross-sectional areas as a function of the BET C constant. J. Colloid Interface Sci. 86:191–195.

[23] Gregg, S.J., and K.S.W. Sing. 1982. Adsorption, surface area and porosity. 2nd ed. Academic Press, New York, NY.

[24] Schulthess, C.P. 1986. Unpublished data: Laboratory notes from the University of Delaware, Newark, DE.

[25] Greenland, D.J., and J.P. Quirk. 1964. Determination of the total specific surface areas of soils by adsorption of cetyl pyridinium bromide. J. Soil Sci. 15:178–191.

[26] Brunauer, S., L.S. Deming, W.E. Deming, and E. Teller. 1940. On a theory of the van der Waals adsorption of gases. J. Am. Chem. Soc. 62:1723–1732.

[27] Giles, C.H., T.H. MacEwan, S.N. Nakhwa, and D. Smith. 1960. Studies in adsorption. Part XI. A system of classification of solution adsorption isotherms, and its use in diagnosis of adsorption mechanisms and in measurement of specific surface areas of solids. J. Chem. Soc. 3973–3993.

[28] de Boer, J.H. 1958. The shapes of capillaries. p. 68–94 *In* D.H. Everett and F.S. Stone (ed.) The structure and properties of porous materials. Butterworths Scientific Publ., London, UK.

[29] Zsigmondy, R. 1911. Über die Struktur des Gels der Kieselsäure. Theorie der Entwässerung. Z. Anorg. Chemie 71:356–377.

[30] Gurvich. L.G. 1915. On the physical-chemical force of attraction (*in Russian*). Zhurnal Russkago Fiziko-Khimicheskago Obshchestva, Chast Khimicheskaia 47:805–827.

[31] Broad, D.W., and A.G. Foster. 1946. The sorption of vapours by ferric oxide gel: Part II. J. Chem. Soc. 446–450.

[32] McKee, D.W. 1959. The sorption of hydrocarbon vapors by silica gel. J. Phys. Chem. 63: 256–1259.

[33] Halsey, G. 1948. Physical adsorption on non-uniform surfaces. J. Chemical Physics 16:931–937.

[34] Washburn, E.W. 1921. Note on a method of determining the distribution of pore sizes in a porous material. Proc. Natl. Acad. Sci. 7:115–116.

[35] Webb, P.A., and C. Orr. 1997. Analytical methods in fine particle technology. Micromeritics Instrument Corp., Norcross, GA.

[36] Lowell, S., and J.E. Shields. 1991. Powder surface area and porosity. 3rd ed. Chapman & Hall, New York, NY.

[37] Way, J.T. 1850. On the power of soils to adsorb manure. J. Roy. Agric. Soc. Engl. 11:313–379.

[38] Way, J.T. 1852. On the power of soils to adsorb manure. J. Roy. Agric. Soc. Engl. 13:123–143.

[39] Schulthess, C.P., and D.L. Sparks. 1986. Backtitration technique for proton isotherm modeling of oxide surfaces. Soil Sci. Soc. Am. J. 50:1406–1411.

[40] Schofield, R.K. 1949. Effect of pH on electric charges carried by clay particles. J. Soil Sci. 1:1–8.

[41] Goldberg, S., I. Lebron, and D.L. Suarez. 2000. Soil colloidal behavior. p. B195–B240. In M.E. Sumner (ed.) Handbook of soil science. CRC Press, Boca Raton, FL.

[42] Talibudeen, O. 1981. Cation exchange in soils. p. 115–177. In D.J. Greenland and M.H.B. Hayes (ed.) The chemistry of soil processes. John Wiley & Sons, New York, NY.

[43] Skopp, J.M. 2000. Physical properties of primary particles. p. A3–A17. In M.E. Sumner (ed.) Handbook of soil science. CRC Press, Boca Raton, FL.

[44] Sparks, D.L. 2003. Environmental soil chemistry. 2nd ed. Academic Press, San Diego, CA.

[45] White, G.N., and L.W. Zelazny. 1986. Charge properties of soil colloids. p. 39–81. In D.L. Sparks (ed) Soil physical chemistry. CRC Press, Boca Raton, FL.

[46] Keii, T., T. Takagi, and S. Kanetaka. 1961. A new plotting of the BET method. Anal. Chem. 33:1965.

Questions

1. Figure 5-5B shows the rutile (111) surface with an oxygen adatom. Why was this adatom needed to stabilize the dry surface? What reasons were given to justify the availability of the adatom in the environment? In real environments, do you expect the rutile surface to also be easily autocompensated?

2. With the EGME method, the amount of EGME in the container with $CaCl_2$ often increases over time. Why does this occur?

3. Show how Equation [5-22] reduces to the Langmuir equation at high c and low p/p_o values:

$$\frac{V}{V_m} = \frac{c \frac{p}{p_o}}{1 + c \frac{p}{p_o}}$$ [5-40].

4. Keii et al. (1961)[46] commented that there can be much error in the estimation of the intercept of the plot of Equation [5-23], $1/(cV_m)$, because of the numerically small answer caused by the large c values (most c values are about 100). Based on

$$\frac{1}{V\left(1 - \frac{p}{p_o}\right)} = \frac{1}{V_m} + \frac{1 - \frac{p}{p_o}}{c V_m \frac{p}{p_o}}$$ [5-41],

they proposed that a plot of $1/[V(1 - (p/p_o))]$ versus $(1 - (p/p_o))/(p/p_o)$ would yield more reliable results. Showing each step carefully, show how Equation [5-41] was derived. Comment on alternate ways to estimate the V_m value in Equation [5-22] using linear and nonlinear regression methods.

5.   Keii et al. (1961)[46] comment that for p/p_o relative pressures commonly used for BET assays, the second term in Equation [5-41] is very small compared with the first term, and can therefore be ignored. They consequently proposed a single point method for solving the BET equation based on

$$V_m = V\left(1 - \frac{p}{p_o}\right) \qquad\qquad [5-42].$$

Using $c = 0.1, 1, 50, 100, 500,$ and 1000, plot the actual amount of error generated by Equation [5-42] as a function of p/p_o values ranging from 0 to 1. Scale the y axis appropriately for best legibility of the corresponding numerical values and, if helpful, split your plot into a second exploded plot to optimize its legibility.

6.   As noted earlier, assuming $\theta = 0$ will lead to some error in the Kelvin Equation (Equation [5-30]). What is the error if the wetting angle is in fact as large as 20°? Is this error a major concern compared to a lack of knowledge regarding the exact pore shape?

7.   Why are minerals porous?

8.   Is the CEC value of a soil sample or mineral a reliable estimator of its ability to adsorb other cations, such as Zn^{2+} and Cu^{2+}?

      Answer: The CEC values are pH dependent, as are the adsorption of other cations. However, their dependence on pH is not identical. In measuring the CEC, one generally is measuring the exchange of NH_4^+ with various cations, such as Ca^{2+} and Mg^{2+}. The exchange of H^+ ions is also involved, but is probably not measured explicitly, hence the dependence of pH on CEC. Heavy metal cations like Zn^{2+} and Cu^{2+} are much more aggressive cation exchangers than NH_4^+ or the alkaline earth elements. Accordingly, the ability of Zn^{2+} and Cu^{2+} to compete and exchange with ions like H^+ will also be more aggressive. The CEC is not a measure of the *total* adsorption capacity of the exchanger and it is not a reliable estimator of the amount of adsorption of other cations that you will be observing.

Author Index

Numbers in brackets [ ] are chapter and reference numbers; these are followed by the page numbers.

Chemical Formula Index

Chemical Names

For the corresponding pages, refer to the Chemical Formula Index on pages 268–274.

acetate ion	$C_2H_3O_2^-$, CH_3COO^-	arabinose	$C_5H_{10}O_5$
acetic acid	$C_2H_4O_2$, CH_3COOH	arginine	$C_6H_{15}N_4O_2^+$
acetone	C_3H_6O, $(CH_3)_2C{=}O$	argon	Ar
acetonitrile	C_2H_3N, CH_3CN	arsenate ion	AsO_4^{3-}
acetylacetone	$C_5H_8O_2$, $CH_2(COCH_3)_2$	arsenic	As
actinium	Ac	arsenic acid	H_3AsO_4
Al dihydroxide ion	$Al(OH)_2^+$	arsenic(V) ion	As^{5+}
Al hydroxide species	$Al(OH)_x$	arsenide ion	As^{3-}
Al monohydroxide ion	$AlOH^{2+}$	arsine	AsH_3
Al polymer ion	$Al_{13}O_4(OH)_{24}(H_2O)_{12}^{7+}$	asparagine	$C_4H_8N_2O_3$
Al tetrahydroxide ion	$Al(OH)_4^-$	aspartic acid	$C_4H_7NO_4$
alanine	$C_3H_7NO_2$	astatine	At
alcoholic groups	R-OH	atrazine	$C_8H_{14}N_5Cl$
aluminol surface group	$=$AlOH		
aluminum	Al	barium	Ba
aluminum hydroxide	$Al(OH)_3$	barium chloride	$BaCl_2$
aluminum ion	Al^{3+}	barium ion	Ba^{2+}
aluminum tetrachloride ion	$AlCl_4^-$	barium zirconate	$BaZrO_3$
americium	Am	benzene	C_6H_6
aminoacetate ion	$C_2H_4NO_2^-$, $NH_2CH_2COO^-$	benzenedicarboxylate ion	$C_8H_4O_4^{2-}$, $C_6H_4(COO)_2^{2-}$
amino acids	$R{-}CH(NH_3^+)(COO^-)$	benzoate ion	$C_7H_5O_2^-$, $C_6H_5COO^-$
4-aminopyridine	$C_5H_6N_2$	berkelium	Bk
ammonia	NH_3	beryllium	Be
ammonium acetate	$C_2H_7O_2N$, CH_3COONH_4	beryllium chloride	$BeCl_2$
ammonium chloride	NH_4Cl	beryllium ion	Be^{2+}
ammonium iodide	NH_4I	bicarbonate ion	HCO_3^-
ammonium ion	NH_4^+	bichromate ion	$HCrO_4^-$
ammonium nitrate	NH_4NO_3	bismuth	Bi
ammonium oxalate	$C_2H_8N_2O_4$, $(COONH_4)_2$	bismuth ion	Bi^{3+}
ammonium sulfate	$(NH_4)_2SO_4$	bisulfate ion	HSO_4^-
amylopectin	$(C_6H_{10}O_5)_n$	bisulfite ion	HSO_3^-
amylose	$(C_6H_{10}O_5)_n$	borane	BH_3
antimony	Sb	borate ion	BO_4^{5-} (for BO_3^{3-} see orthoborate ion)
antimony ion	Sb^{5+}	boric acid	(see orthoboric acid)
apiose	$C_5H_{10}O_5$	boron	B

boron ion B^{3+}
boron tetrahydroxide ion $B(OH)_4^-$
bromate ion BrO_3^-
bromide ion Br^-
bromine Br
bromine pentafluoride BrF_5
bromochlorodifluoromethane $CBrClF_2$
bromomethane CH_3Br
butanate ion $C_4H_7O_2^-$, $CH_3(CH_2)_2COO^-$
n-butane C_4H_{10}, $CH_3(CH_2)_2CH_3$
3-butenoate ion $C_4H_5O_2^-$, $CH_2CHCH_2COO^-$

cadmium Cd
cadmium chloride ion $CdCl^+$
cadmium iodide CdI_2
cadmium ion Cd^{2+}
cadmium selenate $CdSeO_4$
calcium chloride $CaCl_2$
calcium Ca
calcium carbonate $CaCO_3$
calcium fluoride CaF_2
calcium hypochlorite $Ca(ClO)_2$
calcium iodide CaI_2
calcium ion Ca^{2+}
calcium sulfate $CaSO_4$
californium Cf
carbaryl $C_{12}H_{11}NO_2$
carbon C
carbon dioxide CO_2
carbon disulfide CS_2
carbon monoxide CO
carbon tetrachloride CCl_4
carbon tetrafluoride CF_4
carbonate ion CO_3^{2-}
carbonic acid H_2CO_3
carboxylic groups $R\text{-}COOH$
cellulose $(C_6H_{10}O_5)_n$
cellulose xanthate $C(OR)S_2Na$
cerium Ce
cerium(III) ion Ce^{3+}
cerium(IV) ion Ce^{4+}
cesium Cs
cesium chloride $CsCl$
cesium iodide CsI
cesium ion Cs^+
cesium fluoride CsF
N-cetylpyridinium bromide (CPB)
$C_{21}H_{38}NBr$, $(C_5H_5N^+\text{-}C_{15}H_{30}\text{-}CH_3)Br^-$
chlorate ion ClO_3^-
chloride ion Cl^-
chlorine Cl, Cl_2
chlorine trifluoride ClF_3
chlorite ion ClO_2^-
chloroacetate ion $C_2H_2ClO_2^-$, CH_2ClCOO^-
chlorobenzoate ion $C_7H_4ClO_2^-$, $C_6H_4ClCOO^-$
chloroethane C_2H_5Cl, CH_3CH_2Cl

chloroform $CHCl_3$
chloromethane CH_3Cl
chloroplatinate ion $PtCl_6^{2-}$
chromate ion CrO_4^{2-}
chromium Cr
chromium ion Cr^{3+}
citrate^{3-} ion
$C_6H_5O_7^{3-}$, $^-OOCCH_2C(OH)(COO^-)CH_2COO^-$
H(citrate)$^{2-}$ ion
$C_6H_6O_7^{2-}$, $^-OOCCH_2C(OH)(COOH)CH_2COO^-$
H$_2$(citrate)$^-$ ion
$C_6H_7O_7^-$, $HOOCCH_2C(OH)(COOH)CH_2COO^-$
clonitrilide (see niclosamide)
clorox $NaClO$
cobalt ion Co^{2+}
condensed tannin polymer $(C_{15}H_{12}O_6)_n$
(Congo Red)$^{2-}$ ion $C_{32}H_{22}N_6O_6S_2^{2-}$
copper Cu
copper ion Cu^{2+}
CPB (see N-cetylpyridinium bromide)
cupferron $C_6H_9N_3O$
cyanate ion NCO^-
cyanide ion CN^-
cyclohexane C_6H_{12}
cyclopentanone C_5H_8O, $(CH_2)_4C=O$
cyclopropane C_3H_8
cysteine $C_3H_7NO_2S$
cystine $C_6H_{12}N_2O_4S_2$

2,4-D (see 2,4-dichlorophenoxyacetic acid)
DDT (see dichlorodiphenyltrichloroethane)
deoxy-heptulosonic acid $C_7H_{12}O_7$
deoxy-octulosonic acid $C_8H_{14}O_8$
deuterium oxide (heavy water) D_2O
dialuminum dihydroxide ion $Al_2(OH)_2^{4+}$
diamminesilver(I) ion $Ag(NH_3)_2^+$
dibromotyrosine $C_9H_9Br_2NO_3$
dichloroacetate ion $C_2HCl_2O_2^-$, $CHCl_2COO^-$
dichlorodiphenyltrichloroethane (DDT) $C_{14}H_9Cl_5$
2,4-dichlorophenoxyacetic acid (2,4-D) $C_8H_6Cl_2O_3$
dichromate ion $Cr_2O_7^{2-}$
dicloran $C_6H_4Cl_2N_2O_2$
dicofol $C_{14}H_9Cl_5O$
didecyldimethylammonium chloride $C_{22}H_{48}N^+Cl^-$
dienochlor $C_{10}Cl_{10}$
diethyl ammonium ion $C_4H_{12}N^+$, $(C_2H_5)_2NH_2^+$
dihydoxoiron(II) $Fe(OH)_2$
dihydrogen arsenate ion $H_2AsO_4^-$
dihydrogen phosphate ion $H_2PO_4^-$
dihydrogen sulfide H_2S
dihydroxodioxosilicon(IV) ion $SiO_2(OH)_2^{2-}$
dihydroxoiron(III) ion $Fe(OH)_2^+$
dihyroxybutanedioate ion $C_4H_4O_6^{2-}$, $(CHOHCOO)_2^{2-}$
diiodotyrosine $C_9H_9I_2NO_3$
diisopropyl ether $C_6H_{14}O$, $((CH_3)_2CH)_2O$
dimethylammonium ion $C_2H_8N^+$, $(CH_3)_2NH_2^+$

2,3-dimethylbutane	C_6H_{14} , $(CH_3)_2CHCH(CH_3)_2$
dimethyl formamide	C_3H_7NO, $(HCO)N(CH_3)_2$
dimethyl sulfoxide (DMSO)	C_2H_6OS, $(CH_3)_2S{=}O$
dioxane	$C_4H_8O_2$, $(CH_2CH_2O)_2$
diphenylacetate ion	$C_{13}H_{11}O_2^-$, $(C_6H_5)_2CHOO^-$
dipotassium copper(I) chloride	$K_2CuCl_3(s)$
dipropyl ammonium ion	$C_6H_{16}N^+$, $(C_3H_7)_2NH_2^+$
dithionate ion (hyposulfate ion)	$S_2O_6^{2-}$
dithionite ion	$S_2O_4^{2-}$
divinylbenzene (DVB)	$C_{10}H_{10}$, $(C_6H_4)(CH{=}CH_2)_2$
DMSO	(see dimethyl sulfoxide)
DVB	(see divinylbenzene)
dysprosium	Dy
dysprosium ion	Dy^{3+}
EDTA	(see ethylenediaminetetraacetic acid)
EGME	(see ethylene glycol monoethyl ether)
einsteinium	Es
ellagitannin	$C_{34}H_{24}O_{22}$
entacyanothiosulfatocobalt(III) ion	
	$[Co(S_2O_3)(CN)_5]^{4-}$
erbium	Er
erbium ion	Er^{3+}
ethane	C_2H_6 , CH_3CH_3
ethanol	C_2H_6O, C_2H_5OH
2-ethoxyethanol	$C_4H_{10}O_2$, $C_2H_5OC_2H_4OH$
ethylammonium ion	$C_2H_8N^+$, $C_2H_5NH_3^+$
ethylcyclohexane	C_8H_{16} , $(C_6H_{11})CH_2CH_3$
ethylenediaminetetraacetic acid (EDTA)	
	$C_{10}H_{16}N_2O_8$, $(CH_2N)_2(CCH_2OOH)_4$
ethylene glycol	$C_2H_6O_2$, $C_2H_4(OH)_2$
ethylene glycol monoethyl ether (EGME)	
	$C_4H_{10}O_2$, $C_2H_5OC_2H_4OH$
ethyl ether	$C_4H_{10}O$, $CH_3CH_2OCH_2CH_3$
europium	Eu
europium ion	Eu^{3+}
fermium	Fm
ferrol surface group	$={FeOH}$
flavanoid	$C_{15}H_{13}O$
flavonoid	$C_{15}H_{10}O_2$
fluoride ion	F^-
fluorine	F
formaldehyde	CH_2O, $H_2C{=}O$
formate ion	CHO_2^-, $HCOO^-$
formic acid	CH_2O_2 , $HCOOH$
francium	Fr
fructose	$C_6H_{12}O_6$
fucose	$C_6H_{12}O_5$
fulvic acid	$\frac{1}{6}(C_{65}H_{40}O_{58})$, $C_{27}H_{26}O_{18}$
gadolinium ion	Gd^{3+}
galacturonic acid, galacturonose	$C_6H_{10}O_7$
gallium	Ga
gallium ion	Ga^{3+}
germanium	Ge

germanium ion	Ge^{4+}
glucose	$C_6H_{12}O_6$
glucuronic acid	$C_6H_{10}O_7$
glutamic acid	$C_5H_9NO_4$
glutamine	$C_5H_{10}N_2O_3$
glycerol	$C_3H_8O_3$, $HOCH_2CHOHCH_2OH$
glycine	$C_2H_5NO_2$, $NH_3^+CH_2COO^-$
glycine, protonated	$C_2H_6NO_2^+$, $NH_3^+CH_2COOH$
glycogen	$(C_6H_{10}O_5)_n$
gold	Au
gold ion	Au^{3+}
graphite carbon	C
hafnium	Hf
hafnium ion	Hf^{4+}
helium	He
hemicellulose	$(C_6H_{10}O_5)_n$
n-heptane	C_7H_{16} , $CH_3(CH_2)_5CH_3$
heptanedioate ion	$[C_7H_{10}O_4]^{2-}$, $[OOC(CH_2)_5COO]^{2-}$
hexaamminechromium(III) ion	$[Cr(NH_3)_6]^{3+}$
hexaamminecobalt(III) ion	$[Co(NH_3)_6]^{3+}$
hexacyanoferrate(II) ion	$[Fe(CN)_6]^{4-}$
hexacyanoferrate(III) ion	$[Fe(CN)_6]^{3-}$
n-hexane	C_6H_{14}
hexanedioate ion	$C_6H_8O_2^{2-}$, $(CH_2CH_2COO)_2^{2-}$
histidine	$C_6H_9N_3O_2$
HPAN	(see hydrolyzed polyacrylonitrile)
holmium	Ho
holmium ion	Ho^{3+}
humic acid	$C_{44}H_{31}O_{20}$, $C_{58}H_{41}O_{27}N$, $C_{64}H_{80}O_{25}N_4$
hydriodic acid	HI
hydrochloric acid	HCl
hydrochlorous acid	HClO
hydrofluoric acid	HF
hydrogen	H, H_2
hydrogen arsenate ion	$HAsO_4^{2-}$
hydrogen bromide	HBr
hydrogen ion (proton)	H^+
hydrogen peroxide	H_2O_2
hydrogen phosphate ion	HPO_4^{2-}
hydrogen sulfide ion	HS^-
hydrogen selenate ion	$HSeO_4^-$
hydrolyzed polyacrylonitrile (HPAN)	
	$(C_3H_3O_2^-)_m(C_3H_3N)_n$, $(CH_2CHCOO^-)_m(CH_2CHCN)_n$
hydronium ion	H_3O^+
hydroxide ion	OH^-
hydroxoiron(II) ion	$FeOH^{2+}$
hydroxoiron(II) ion	$FeOH^+$
hydroxybenzoate ion	$C_7H_5O_3^-$, $C_6H_4OHCOO^-$
hydroxylysine	$C_6H_{15}N_2O_3^+$
hydroxyproline	$C_5H_8NO_3$
8-hydroxyquinoline	C_9H_7NO
indium	In
indium hydroxyoxide	InOOH
indium ion	In^{3+}

PAM	(see polyacrylamide)
paraquat	$C_{12}H_{14}N_2^{2+}$
pectin acid/polymer	$(C_6H_7O_6)_n(H,CH_3)_m$
pentaamminechlorocobalt(III) ion	$[Co(NH_3)_5Cl]^{2+}$
pentacyanonitrosylferrate(III) ion	$[Fe(CN)_5NO]^{2-}$
pentaammineaquocobalt(III) ion	$[Co(NH_3)_5H_2O]^{3+}$
n-pentane	C_5H_{12} , $CH_3(CH_2)_3CH_3$
pentanedioate ion	$C_5H_6O_2^{2-}$, $CH_2(CH_2COO)_2^{2-}$
perchlorate ion	ClO_4^-
periodate ion	IO_4^-
permanganate ion	MnO_4^-
peroxodisulfate ion	$S_2O_8^{2-}$
perrhenate ion	ReO_4^-
persulfide ion	S^-
phenol	C_6H_6O , C_6H_5–OH
phenolic groups	R-OH
phenolphthalein	$C_{20}H_{14}O_4$
phenoxide ion	$C_6H_5O^-$, C_6H_5–O^-
phenylacetate ion	$C_8H_7O_2^-$, $C_6H_5CH_2COO^-$
phenylalanine	$C_9H_{11}NO_2$
phenylmercuric acetate (PMA)	$C_8H_8HgO_2$
phosphate ion	PO_4^{3-}
phosphoric acid	H_3PO_4
phosphorus	P
phosphorus(V) chloride	PCl_5
phosphorus ion	P^{5+}
platinum	Pt
plutonium	Pu
plutonium ion	Pu^{3+}
PMA	(see phenylmercuric acetate)
polonium	Po
polyacrylamide (PAM)	
	$(C_3H_5NO)_n$, $[CH_2CH(CONH_2)]_n$
poly(methyl methacrylate)	$(C_5H_5O_3)_n$
poly(phenol formaldehyde)	$(C_9H_8OR)_n$
poly(styrene-divinylbenzene)	$(C_{18}H_{18})_n$
poly(vinyl acetate) (PVAc)	
	$(C_4H_6O_2)_n$, $[CH_2CH(OOCCH_3)]_n$
poly(vinyl acetate-maleic anhydride) (VAMA)	
	$(C_4H_6O_2)_m(C_4H_2O_4)_n$
poly(vinyl alchohol) (PVA)	
	$(C_2H_4O)_n$, $[CH_2CH(OH)]_n$
polyvinylpyrrolidone (PVP)	$(C_6H_9NO)_n$
potassium	K
potassium bromide	KBr
potassium chloride	KCl
potassium dichromate	$K_2Cr_2O_7$
potassium ferricyanide	$K_3Fe(CN)_6$
potassium fluoride	KF
potassium hydrogen phthalate (KHP)	
	$C_8H_5O_4^-K^+$, HOOC–(C_6H_4)–COO$^-K^+$
potassium iodide	KI
potassium ion	K^+
potassium nitrate	KNO_3
potassium tantalum(V) oxide	$KTaO_3$
potassium thiocyanate	KCNS

praseodymium	Pr
praseodymium ion	Pr^{3+}
proline	$C_5H_9NO_2$
promethium	Pm
promethium ion	Pm^{3+}
propane	C_3H_8 , $CH_3CH_2CH_3$
propyl ammonium ion	$C_3H_{10}N^+$, $(C_3H_7)NH_3^+$
protactinium	Pa
PVA	(see poly(vinyl alchohol))
PVAc	(see poly(vinyl acetate))
PVP	(see polyvinylpyrrolidone)
pyrogallol	$C_6H_6O_3$, C_6H_3–$(OH)_3$
quinone	$C_6H_4O_2$, $O{=}(C_6H_4){=}O$
radium	Ra
radium ion	Ra^{2+}
radon	Rn
rhamnose	$C_6H_{12}O_5$
rhenium	Re
rhenium ion	Re^{4+}
rhodium	Rh
rubidium	Rb
rubidium ion	Rb^+
ruthenium	Ru
salicylic acid	$C_7H_6O_2$, o-$C_6H_4(OH)(COOH)$
samarium	Sm
samarium ion	Sm^{3+}
scandium	Sc
scandium ion	Sc^{3+}
selenate ion	SeO_4^{2-}
selenide ion	Se^{2-}
selenium	Se
selenium(VI) ion	Se^{6+}
serine	$C_3H_7NO_3$
silanol surface group	$={}SiOH$
silicate ion	SiO_4^{4-}
silicon	Si
silicon dioxide	SiO_2
silicon hydroxide	$Si(OH)_4$
silicon ion	Si^{4+}
silver	Ag
silver ion	Ag^+
silver fluoride	AgF
silver iodide	AgI
silver phosphate	$(AgPO_3)_x(s)$
sodium	Na
sodium bicarbonate	$NaHCO_3$
sodium carbonate	$Na_2CO_3(s)$
sodium chloride	NaCl
sodium chloride dihydrate	$NaCl{\cdot}2H_2O(cr)$
sodium dihydrogen arsenate	NaH_2AsO_4
sodium dihydrogen phosphate	NaH_2PO_4
sodium hydroxide	NaOH
sodium hypochlorite	NaOCl

sodium ion	Na^+	thorium ion	Th^{4+}
sodium niobate	$NaNbO_3$	threonine	$C_4H_9NO_3$
sodium nitrate	$NaNO_3$	thulium	Tm
sodium pyrophosphate	$Na_4P_2O_7$	thulium ion	Tm^{3+}
sodium sulfate	Na_2SO_4	thyroxine	$C_{15}H_{11}I_4NO_4$
sodium sulfate decahydrate	$Na_2SO_4 \cdot 10H_2O$	tin (stannium)	Sn
stannium	(see tin)	tin dioxide (stannic oxide)	SnO_2
starch	$(C_6H_{10}O_5)_n$	tin ion	Sn^{2+}, Sn^{4+}
streptomycin	$C_{21}H_{39}N_7O_{12}$	titanium	Ti
strontium	Sr	titanium dioxide	TiO_2
strontium ion	Sr^{2+}	titanium ion	Ti^{3+}, Ti^{4+}
strontium metatitanate	$SrTiO_3$	toluene	$C_7H_8 , C_6H_5CH_3$
styrene	$C_8H_8 , C_6H_5-CH=CH_2$	s-triazines	$C_3H_2N_3R_3$
succinate ion	$C_4H_4O_4^{2-}, (CH_2COO)_2^{2-}$	triborate ion	$B_3O_3(OH)_5^{2-}$
sucrose	$C_{12}H_{22}O_{11}$	trichloroacetate ion	$C_2Cl_3O_2^-, CCl_3COO^-$
sulfarsenide ion	AsS^{2-}	1,1,1-trichloroethane	$C_2H_3Cl_3 , Cl_3CCH_3$
sulfate ion	SO_4^{2-}	2,4,5-trichlorophenoxyacetic acid (2,4,5-T)	
sulfide ion	S^{2-}		$C_8H_5Cl_3O_3$
sulfite ion	SO_3^{2-}	triethylamine	$C_6H_{15}N,$ N-$(CH_2CH_3)_3$
sulfur	S	triethyl ammonium ion	$C_6H_{16}N^+, (C_2H_5)_3NH^+$
sulfur dioxide	SO_2	trihydroxoiron(III)	$Fe(OH)_3$
sulfur hexafluoride	SF_6	trihydroxooxosilicon(IV) ion	$SiO(OH)_3^-$
sulfur(VI) ion	S^{6+}	trimethyl ammonium ion	$C_3H_{10}N^+, (CH_3)_3NH^+$
sulfur tetrafluoride	SF_4	2,2,3-trimethylbutane	
sulfuric acid	H_2SO_4		$C_7H_{16} , CH_3C(CH_3)_2CH(CH_3)CH_3$
		2,2,4-trimethylpentane	
2,4,5-T (see 2,4,5-trichlorophenoxyacetic acid)			$C_8H_{18} , CH_3C(CH_3)_2CH_2CH(CH_3)CH_3$
tantalum	Ta	2,3,4-trimethylpentane	
tantalum ion	Ta^{5+}		$C_8H_{18} , CH_3-(CH(CH_3))_3-CH_3$
technetium	Tc	tripropyl ammonium ion	$C_9H_{22}N^+, (C_3H_7)_3NH^+$
tellurium	Te	tris(ethylenediamine)cobalt(III) ion	
telluride ion	Te^{2-}		$[Co(NH_2(CH_2)_2NH_2)_3]^{3+}$
terbium	Tb	tryptophane	$C_{11}H_{12}N_2O_2$
terbium ion	Tb^{3+}	tungstate(VI) ion	WO_4^{2-}
tetraamminedinitrocobalt(III) ion		tungsten	W
	$[Co(NH_3)_4(NO_2)_2]^+$	tungsten ion	W^{4+}, W^{6+}
tetraammineplatinum(II) ion	$Pt(NH_3)_4^{2+}$	tyrosine	$C_9H_{11}NO_3$
tetraborate ion	$B_4O_5(OH)_4^{2-}$		
tetrachloroethylene	$C_2Cl_4 , Cl_2C=CCl_2$	uranium	U
tetracyanodisulfitocobalt(III) ion		uranium ion	U^{4+}, U^{6+}
	$[Co(SO_3)_2(CN)_4]^{5-}$		
tetracyanonickelate(II) ion	$Ni(CN)_4^{2-}$	valine	$C_5H_{11}NO_2$
tetracyanozinc(II) ion	$Zn(CN)_4^{2-}$	VAMA (see poly(vinyl acetate-maleic anhydride))	
tetraethyl ammonium ion	$C_8H_{20}N, (C_2H_5)_4N^+$	vanadium	V
tetrahedral aluminum	AlO_4^{5+}	vanadium ion	$V^{2+}, V^{3+}, V^{4+}, V^{5+}$
tetrahedral phosphate chain	$(NaPO_3)_x$	vinylpyrrolidone	C_6H_9NO
tetrahydroxoiron(III) ion	$Fe(OH)_4^-$		
tetramethyl ammonium ion		warfarin	$C_{19}H_{16}O_4$
	$C_4H_{12}N^+, (CH_3)_4N^+, [N(CH_3)_4]^+$	water, ice	H_2O
tetrapropyl ammonium ion	$C_{12}H_{28}N^+, (C_3H_7)_4N^+$	wolfram	(see tungsten)
thallium	Tl		
thallium ion	Tl^+, Tl^{3+}	xenon	Xe
thiocyanate ion	NCS^-, SCN^-	xenon difluoride	XeF_2
thiosulfate ion (hyposulfite ion)	$S_2O_3^{2-}$	xenon tetrafluoride	XeF_4
thorium	Th	xylose	$C_5H_{10}O_5$

ytterbium	Yb	zinc	Zn
ytterbium ion	Yb^{3+}	zinc ion	Zn^{2+}, Zn^{4+}
yttrium	Y	zirconium	Zr
yttrium ion	Y^{3+}	zirconium ion	Zr^{4+}
yttrium orthoaluminate	$YAlO_3$		

Mineral Index

Subject Index

Printed in the United States
By Bookmasters